Natural Neuroscience

Natural Neuroscience

Toward a Systems Neuroscience of Natural Behaviors

Nachum Ulanovsky

The MIT Press
Cambridge, Massachusetts
London, England

The MIT Press
Massachusetts Institute of Technology
77 Massachusetts Avenue, Cambridge, MA 02139
mitpress.mit.edu

The MIT Press would like to thank the anonymous peer reviewers who provided comments on drafts of this book. The generous work of academic experts is essential for establishing the authority and quality of our publications. We acknowledge with gratitude the contributions of these otherwise uncredited readers.

This book was set in Times New Roman by Westchester Publishing Services. Printed and bound in the United States of America.

Library of Congress Cataloging-in-Publication Data

Names: Ulanovsky, Nachum, 1973- author.
Title: Natural neuroscience : toward a systems neuroscience of natural behaviors / Nachum Ulanovsky.
Description: Cambridge, Massachusetts : The MIT Press, [2025] | Includes bibliographical references and index.
Identifiers: LCCN 2024020197 (print) | LCCN 2024020198 (ebook) | ISBN 9780262044998 (paperback) | ISBN 9780262361712 (pdf) | ISBN 9780262361729 (epub)
Subjects: MESH: Neurosciences—methods | Behavior, Animal—physiology | Brain—physiology | Perception—physiology
Classification: LCC QP376 (print) | LCC QP376 (ebook) | NLM WL 100 | DDC 612.8/2—dc23/eng/20240909
LC record available at https://lccn.loc.gov/2024020197
LC ebook record available at https://lccn.loc.gov/2024020198

10 9 8 7 6 5 4 3 2 1

EU product safety and compliance information contact is: mitp-eu-gpsr@mit.edu

To Hagit—For Everything

Contents

Preface

Neuroscience is one of the fastest-growing fields in modern science, because it directly addresses the inherent desire of humans to "know thyself": to understand behavior, perception, and cognition. To reach these goals, brain research traditionally focused on reductionist approaches, such as presenting simple sensory stimuli to anesthetized animals, or studying neural activity in animals performing rather simple behaviors. This reductionism had two sources. First, major technological limitations have historically precluded brain studies under real-world conditions. However, rapid technological advances in recent years have now finally allowed scientists to pursue more naturalistic experiments. Second, simplicity and controllability in scientific research have profound conceptual appeal and deep philosophical roots, which can be traced to René Descartes and Francis Bacon in the seventeenth century. Indeed, simplified and highly controlled experiments have yielded centuries of great achievements in understanding the fundamental laws of nature, most notably in the exact sciences and molecular biology. However, as will be argued in this book, neuroscience is different from many other fields of science. In physics, it is clear that to study the properties of the hydrogen atom, one must investigate it in a highly controlled laboratory environment, with little concern to the context of the experiment; but this is not the case when studying the neural bases of behavior and cognition. The hallmarks of behavior are that it is highly nonlinear, context dependent, history dependent, complex, and multidimensional (i.e., many aspects of behavior are happening simultaneously). Therefore, elements of behavior may become aberrant if they are studied in overly controlled, oversimplified conditions. Thus, the focus of brain research on simplicity and control came at a high cost. Most of mainstream neuroscience has largely neglected what real brains evolved for—namely, guiding behavior in real-world, complex natural environments. This created a fundamental problem for the field because conclusions drawn from simplified laboratory experiments do not necessarily transfer to more naturalistic scenarios. This book will survey dozens of specific examples of experimental studies showing that brain activity changes dramatically under natural versus artificial conditions, in the neurobiology of navigation (chapter 4), sensory neuroscience (chapters 5 and 6), and many other fields. Therefore, there is a clear need to focus on the study of the neural bases of natural behaviors.

Granted, thousands of simplified, controlled laboratory experiments have provided an enormous body of information that is highly useful as a foundation for understanding the brain. What I will claim in this book is that having *only* simplified controlled experiments is not enough. The emphasis on controlled experiments was an important first step, but now we should dramatically expand the number of experiments performed under more complex, naturalistic conditions. The core problem that will be discussed here is that behavior is rich and complex. Therefore, highly simplified laboratory experiments, as much as they allow for tight control, may also impoverish too much the very thing that we are trying to understand: behavior. In particular, the mechanistic emphasis of neuroscience, whereby mechanisms of behavior are being studied in ever-greater detail—at the molecular, cellular, and network levels—may have focused on oversimplified, sometimes even aberrant behaviors. We may have been studying the neural mechanisms of the wrong behaviors.

As noted here, this overcontrolled and oversimplified research tradition was in part due to technological limitations on neuronal studies of truly natural behaviors. Importantly, however, recent years have seen rapid technological advances that are now allowing us to pursue more naturalistic experiments. These advances are surveyed in chapter 3 and include methods for *recording neural activity* in freely behaving, freely moving animals (e.g., wired and wireless electrophysiology and imaging), methods for *manipulating neural activity* in freely moving animals (e.g., wired and wireless optogenetics), and methods for *quantifying the details of behavior.* Thus, there are now both a conceptual realization of the importance of studying the brain closer to the animal's natural behaviors, and the technological tools needed to do it—and together, these conceptual and technological developments form the basis for the approach outlined in this book.

The book will focus on reviewing a growing corpus of data that demonstrate that the brain acts very differently once tested under more complex, more natural behaviors and stimuli. There are dozens of such examples, and the majority of this book will be devoted to surveying them. For example, in sensory neuroscience, many researchers adopted the "active sensing" approach, stressing the key importance of natural movements for the sensory information available to the animal (chapter 5), while other researchers highlighted the importance of natural stimulus statistics and demonstrated that the responses of sensory neurons to traditional simple stimuli cannot predict well their responses to natural stimuli (chapter 6). And in behavioral neuroscience, there is a growing realization of the importance of introducing complex natural aspects of behavior (chapters 4 and 8).

But even before considering the empirical data, there are three conceptual reasons for adopting such a naturalistic approach. First, natural behavior is very rich and complex, and therefore we should worry that by limiting the animal to performing a highly simplified, single behavior in an impoverished laboratory setup, we may be distorting the very thing that we are trying to understand: the neural bases of behavior. One problem is the poorness of typical laboratory setups, which lack many natural cues and contextual elements that may be crucial for real-life behavior. Another problem is the very small size of laboratory setups, which is worrying because animal behavior is often fundamentally different in small versus large environments (see chapters 2 and 8). Further, a too-simplified laboratory experiment alters not just the animal's behavior, because of the impoverished conditions, but also the related neural activity; and it may structurally alter the involved brain circuits due to overtraining and the associated synaptic plasticity. All of these cast doubt on the relevance of such experiments.

Second, the brain is a highly interconnected network, and therefore if one tries to control all aspects of the task except the one under study, then this likely affects many parts of the network. This is because the control actions by the experimenter are directly affecting multiple brain regions and may reverberate through the entire brain—including the subnetwork that the experimenter is studying. In other words, when studying complex networks, it is impossible to clamp off most of the nodes of the network and claim that this allows one to study the remaining nodes in isolation—there is no such thing as "in isolation" in a network like the brain.

Third, biological neural networks are highly nonlinear, and they also exhibit strong history dependence (Movshon & Lennie 1979; Sejnowski et al. 1988; Ulanovsky et al. 2003; Hamm & Yuste 2016; Lu & Vicario 2017; Cohen et al. 2020). The nonlinearity and history dependence imply that the *context* in which you measure the system may strongly interact with your measurements and can influence them; therefore, one cannot easily generalize from one context to another. Hence, the most reasonable choice would be to use a natural context and allow the animal to perform close-to-normal, naturalistic behaviors (i.e., allow the brain to solve natural real-world problems that are relevant for the species). This will put the network in a normal, unimpeded regime. I will call this approach "natural neuroscience."

This may remind some readers of the neuroethological approach, as neuroethology also focuses on studying the neural basis of natural behaviors. However, as will be discussed in chapter 2, neuroethology is in fact quite different from natural neuroscience. Briefly, the focus of many neuroethologists has classically been on nonstandard "champion species," which excel at certain behaviors that are relatively simple, such as the escape behavior of cockroaches or the hunting behavior of barn owls (Lorenz 1982; Hoyle 1984). By contrast, the scope of natural neuroscience is much broader: natural neuroscience does not focus necessarily on nonstandard species, nor on simple behaviors. Instead, it aims at all fields and every field of neuroscience. Natural neuroscience posits that when studying any brain region in any animal—be it the visual cortex of mice, the hippocampus of rats, or the auditory system of bats—it is crucial to consider the natural conditions under which this brain region needs to function, and what natural, real-world problems it needs to solve. This book will survey numerous recent examples of this approach, drawing from research conducted by many laboratories on multiple species—focusing primarily on the standard mammalian model species in neuroscience, such as mice, rats, and primates (but also treating nonstandard models, such as the animal that I use for my own research: the bat).

This book is aimed primarily at the audience of active neuroscientists (principal investigators, postdocs, and graduate students), as well as advanced undergraduates interested in neuroscience. But the book is also intended for researchers from diverse fields outside of neuroscience who are interested in behavior, including behavioral ecology, ethology, zoology, cognitive science, and psychology. The overarching aim of the book is to convince the reader that it is crucial to study the brain under evermore naturalistic conditions using natural stimuli and focusing on natural behaviors.

The structure of the book is as follows. It begins with an introduction and a survey of the general approach (chapter 1), and is followed by a comparison to related approaches such as neuroethology and comparative neuroscience (chapter 2), and then a discussion of the technological advances that have enabled the natural neuroscience approach (chapter 3). The next four chapters (chapters 4–7) discuss examples of how a detailed consideration of the natural world and the natural behaviors of animals has fundamentally changed our understanding of the relevant brain functions. These examples include spatial cognition and the brain's navigation system (chapter 4), active sensing (chapter 5), and natural stimulus statistics and sensory ecology (chapter 6). I then discuss some topics in human neuroscience, highlighting the importance of also focusing on natural stimuli and natural behaviors in humans, in particular in the study of human perception and human memory (chapter 7). Next, in chapter 8, I will attempt to provide a synthesis of neuroscience, psychology, ethology, and ecology, taking into account the complex interactions between the animal and its environment and the social interactions between animals—and will finish with a future outlook for natural neuroscience. The epilogue will then discuss in a less formal manner—using a question-and-answer format—some of the key outstanding issues related to natural neuroscience. I also provide a glossary, which explains some of the nonobvious terms that I used in the text but for which I did not have space to explain in detail (these terms are marked in bold in the main text).

* * *

Many people improved the book substantially through their comments. I am extremely grateful to Liora Las, my long-standing scientific partner, and to Avner Wallach, both of which have read all the chapters and gave very thorough comments. My warmest thanks to my former and current lab members, who have read all or parts of the book and gave detailed criticism; I was privileged to work with such wonderful people, and I thank them dearly: Dori Derdikman, Tamir Eliav, Arseny Finkelstein, Maya Geva-Sagiv, Gily Ginosar, Maximilian Kirschhock, Shir Maimon, David Omer, Shaked Palgi, Saikat Ray, Ayelet Sarel, Yuval Waserman, Itay Yona, and Yossi Yovel. I also am deeply indebted to many colleagues, who have read and commented on parts of the book, or with whom I had valuable discussions: Ehud Ahissar, Uri Alon, Amos Arieli, Murat Aytekin, Felix Baier, Carlos Brody, Yoram Burak, Matteo Carandini, Ron Dekel, Yadin Dudai, Micah Edelson, Ahmed El Hady, George Ellis, Ehud Fonio, Jonathan Fritz, Ilan Golani, Liron Gruber, Uri Hasson, Adil Khan, Tali Kimchi, Rafi Malach, Cindy Moss, Eli Nelken, Dubi Sagi, Hugo Spiers, Alessandro Treves, Misha Tsodyks, Shimon Ullman, Eric Warrant, Gerald Westheimer, Ofer Yizhar, Noga Zilkha, and Yaniv Ziv. Finally, I would like to thank the anonymous reviewers for the MIT Press, who gave very helpful comments.

This book was written all over the globe, and I would like to thank André Longtin and Leonard Maler for hosting me at the University of Ottawa, Joseph Raimondo for hosting me at the University of Cape Town, and Netanel Mor and Avraham Shragai for hosting me in their "monasteries" in Israel. (Well, they weren't real monasteries, but they gave me lots of peace and quiet and a safe haven to escape to for writing the book.) Most of all, I would like to thank my wife, Hagit, and my children—Shachaf, Ben, and Alon—for putting up with this crazy project for so many years and for giving me the space to go through with it. You are my real haven!

Finally, many thanks to Genia Brodsky for the figure graphics and to my editors at the MIT Press: Robert Prior, Anne-Marie Bono, Jitendra Kumar, and Roger Wood. This was quite a journey—and I thank you for it!

Nachum Ulanovsky
Rehovot, Israel
Summer 2024

1

Introduction and Overall Approach

Why does the brain captivate our mind? There can be many answers: the brain is a fascinating biological system; it poses challenging problems in physics, being a complex dynamical system; and it is the hardware from which cognition emerges and behavior is generated. But above all, our brain is who we are—and therefore, it is the key to understanding our perceptions, behaviors, memories, emotions, and thoughts. We would like to understand how the brain underlies our daily **natural behaviors** and how it shapes and is being shaped by our everyday life experiences. However, brain research typically does not address directly the complexities of natural behavior—instead, the classical approach for studying brain functions is to break them into small subproblems and sub-subproblems and study each of them separately in highly controlled laboratory settings. This approach is termed *reductionism,* and I will describe it in more detail shortly. It has been used with great success in the exact sciences, such as physics. In neuroscience, too, the reductionist approach has led to amazing discoveries and exponentially growing knowledge about brain mechanisms. However, in what follows, I will argue that relying *exclusively* on the reductionist approach also poses fundamental problems, and I will discuss a quite different, complementary approach—namely, to study the brain closer to the natural behavior of animals and humans. (Note: Terms marked in bold are explained in the glossary.)

The reductionism of classical neuroscience had two sources. First, major technological limitations have historically precluded brain studies under real-world situations. However, recent years have seen rapid technological advances that are now allowing to pursue more naturalistic experiments: These advances are surveyed in chapter 3 and include methods for *recording neural activity* in freely behaving, freely moving animals (e.g., wired and wireless electrophysiology and imaging); methods for *manipulating neural activity* in freely moving animals (e.g., wired and wireless optogenetics); and methods for *quantifying the details of behavior.* A second source of reductionism in neuroscience is that simplicity and controllability have profound conceptual appeal and philosophical roots. However, there has been a growing conceptual appreciation that simplicity and control also carry a major cost. The basic argument is that natural behavior is very rich and complex: therefore, we should be worried that by limiting the animal to performing a highly simplified single behavior in an impoverished laboratory setup—as is done classically in behavioral neuroscience—we may be distorting the very thing that we are trying to understand; namely, the neural basis of behavior. In other words, a too-simplified laboratory experiment may be altering both the behavior and the brain, thus making the results less relevant.

This book will focus on reviewing a large corpus of data that demonstrate that the brain acts very differently when it is tested using more complex, more natural behaviors and stimuli. And indeed, a growing number of neuroscientists are choosing to conduct evermore naturalistic experiments—pursuing brain research that considers the natural behaviors of the animal and its ecology. The overarching aim of the book is to convince readers to join this growing movement.

1.1 The Problem: Reductionism and Control versus Natural Behavior—from Descartes and Bacon to Natural Neuroscience

To illustrate the gap between natural behaviors in the wild and behavior as studied in the laboratory, I will start with an example: visual scene analysis in jumping spiders. This book is *not* going to focus on spiders, bats, or other nonstandard model species—in fact, it will focus mostly on the standard mammalian models (rodents, monkeys, and humans)—but nevertheless, it is useful to start with an example from a more humble creature to illustrate the magnitude of the problem. Jumping spiders are highly visual, with a pair of large frontal eyes, which they use to stalk their insect prey and then rapidly attack it from a close distance. They were shown to be able to plan a complex, three-dimensional (3D) route to their prey (figure 1.1a). For example, the jumping spider may climb down from its perch and then climb up another branch and make several correct choices at branch bifurcations—even on curved branches that meander in 3D, one *behind* the other—and it does all of this in the face of visual occlusions by branches and against the background of complex visual clutter that includes moving leaves that can easily obscure the little fly that the spider is hunting (Tarsitano & Jackson 1994). Remarkably, while the spider is on the move, it seems to know at all times the 3D direction to its target—even when the target is occluded!—as evident when it stops and turns its body to face the 3D direction of the target (Hill 1979). Such feats of 3D spatial perception, spatial memory, and sophisticated 3D route planning are performed daily by the little jumping spiders using their tiny brains. In other cases, the spider would spot a potential mate and have to identify that it belongs to the same species; and if it is a female that approaches a male, she will inspect the courtship dance of the male carefully, and will integrate the visual input with the vibrations created by his dance, to decide whether it is a suitable mate (Elias et al. 2003). To achieve all of this, the spider must solve a set of difficult problems: it needs to construct an internal representation of the 3D layout of the environment, separate foreground from background and clutter, identify a target, decide whether it wants to attack the target or mate with it, plan and execute a complex route in 3D, and integrate over time, while on the move, the visual information about the 3D scene and the target location, all while actively scanning the complex visual scene with its eyes. Very few of these astonishing behaviors are being studied in their full complexity within neuroscience laboratories, whether in spiders or in mammals.

This is but one example of the kinds of real-world scene analysis problems that animals must solve daily (Lewicki et al. 2014). But how do we typically study scene analysis in the laboratory? In most cases, two-dimensional (2D) simple drawings are used, where subjects need to do segregation and grouping in the spirit of the Gestalt school of psychology

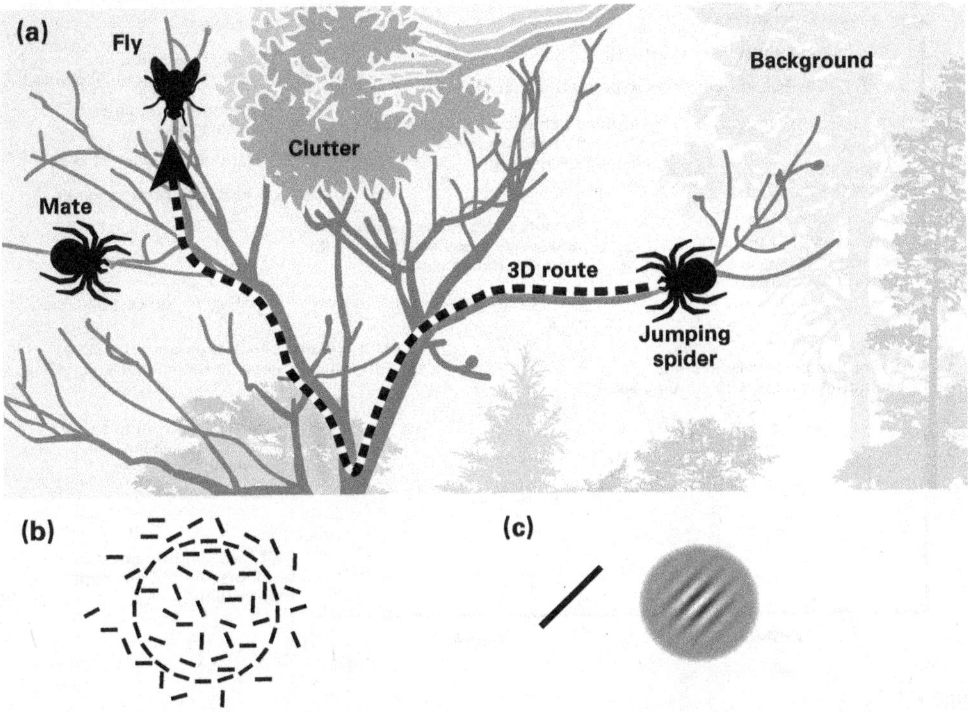

Figure 1.1
The problem addressed in this book: the gap between real-world natural behavior and behavior and perception as studied in the laboratory. (a) The challenges of real-world visual scene analysis facing a jumping spider. (b) Example of visual scene analysis as studied in the laboratory. To identify the circle in this image, the viewer needs to group together the short lines forming the foreground circle and segregate them from the short lines in the background (distractors). (c) Examples of classical stimuli that are commonly used in visual neuroscience research are short lines (left) and Gabor patches (right).

(figure 1.1b). Such laboratory stimuli are studied both psychophysically and at the level of neuronal processing (Gilad et al. 2017). Notably, these are relatively complex stimuli compared to the classical simple stimuli that are commonly used in visual neuroscience, such as isolated simple lines, dots, or Gabor patches (figure 1.1c). Just inspecting figure 1.1 clearly illustrates the gap: figures 1.1b and 1.1c are not truly complex, are not naturalistic, and are not even close to the real-world complexities and problems facing the spider, as illustrated in figure 1.1a. And this gap is even worse in mammals, with their highly evolved brains and more sophisticated behaviors. This book is aimed to start closing this gap: to demonstrate that it is crucial to study the brains of animals and humans as they perform evermore naturalistic behaviors.

* * *

Research in behavioral neuroscience can be classified qualitatively along two axes (figure 1.2): how controlled the behavior is (uncontrolled → controlled axis) versus how natural the behavior is (artificial → natural axis). Historically, the large majority of experiments in neuroscience until the 1990s (and to some degree, well into the 2000s) were conducted either in vitro (in slices, cultures, or isolated ganglia) or in vivo under anesthesia or in artificial, head-fixed conditions. All these experiments occupied the upper-left quadrant

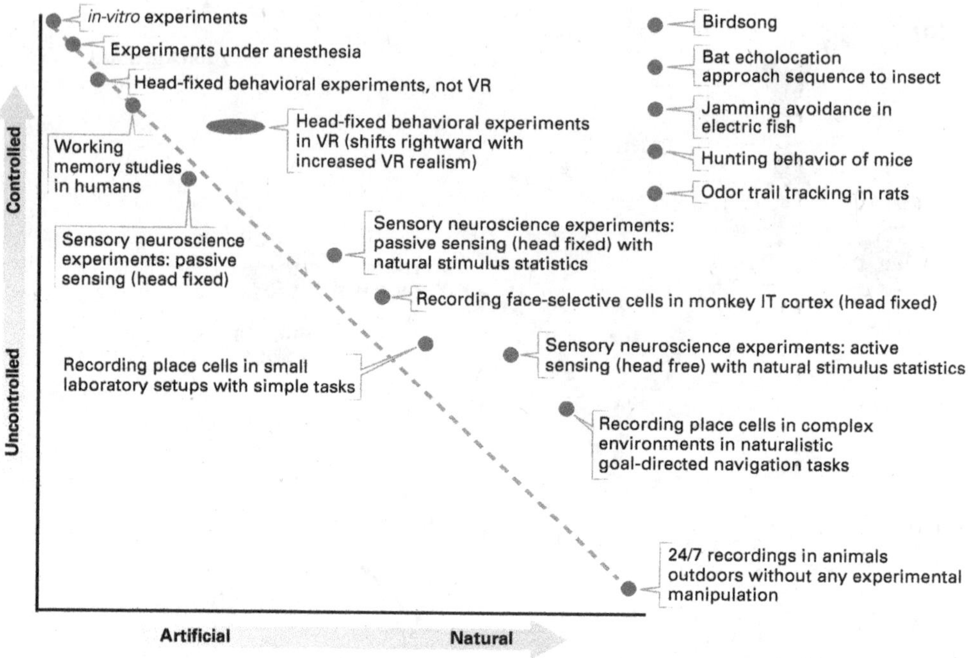

Figure 1.2
Schematic illustration of different experimental styles in systems neuroscience and behavioral neuroscience, displayed as a scatterplot with two axes: the uncontrolled–controlled axis and the artificial–natural axis. Note that, strictly speaking, behavior cannot be controlled directly because it is produced by an animal that has free will; therefore, experimenters *indirectly* control the animal's behavior by extensive training and/or by controlling the experimental setup and experimental procedure, in order to restrict the animal's possible behaviors to a narrow subgroup. In this book, we will refer to this as "controlling the behavior." Note also that "controlled" here could mean "controlled by the experimenter to produce highly reproducible behavior," as described previously, or it could mean "self-controlled by the animal, producing highly reproducible behavior of its own volition" (e.g., birdsong). VR, virtual reality; fMRI, functional magnetic resonance imaging; IT, inferior temporal.

of "controlled but artificial" in this conceptual 2D diagram (figure 1.2; see dots at the top left). Research in behavioral neuroscience also followed this controlled approach: each study focused on one particular behavior of interest, and the experiments usually were done in highly controlled and restricted conditions and often were only loosely related to natural behavior. There were notable exceptions, of course, such as the seminal recordings of hippocampal place cells, which were pioneered by John O'Keefe in the 1970s (see chapter 4), where neurons were recorded in freely moving rats that explored the environment on their own volition—a highly natural behavior for rats—but these experiments were an exception to the rule. The rule was: "The more controlled, the better."

We might pause here and ask: Where did this rule come from? One reason was the technical limitations in the early days of behavioral neuroscience, which required head-fixing the animal in many cases. Another reason was the wish to standardize the experiments, so they can be repeated over time and across labs. But there was also a deeper reason. This rule has been a fundamental axiom of scientific research that can be traced to René Descartes, Francis Bacon, and Galileo Galilei, the trio who laid the conceptual foundations of modern science.

Descartes is well known in mathematics for the invention of analytical geometry, and in philosophy for his extreme dualism: the suggestion that mind and body are strictly separate entities, a notion that most neuroscientists today would reject. But another influential contribution of Descartes was to formulate how the material (physical) world should be studied (Descartes 1644). He posited that one must study objects and physical phenomena that can be *quantified* in terms of size, shape, and motion—the emphasis here is on "quantified"—and indeed, he insisted that all phenomena should be precisely quantified and described by the language of mathematics. But equally important, he emphasized that all objects and physical phenomena can be broken into a small set of units that are common across all realms of the natural world. Descartes was thus a *reductionist.* In fact, he is often credited for introducing the modern approach of *reductionism*, wherein we try to understand nature using the smallest possible set of physical laws and basic entities, all of which should be described mathematically. Reductionism is one of the cornerstones of modern science; it has led to extraordinary successes—and throughout all branches of science, to this day, scientists are seeking to explain natural phenomena via evermore basic mechanisms and through combinations of simple factors. However, although Descartes never gave a "recipe" for how to conduct scientific research, his reductionism had another profound influence: it gave rise to the notion that experiments should be conducted under controlled conditions.[1] The reasoning for this is as follows: if the phenomenon at hand is to be explained through a small set of basic factors that are influencing it, then the best approach would be to study these factors in isolation, by varying just one factor while fixing (or eliminating) all the other factors. As will be argued later, this reasoning is often wrong.

Francis Bacon, Descartes's contemporary, emphasized that to understand the laws of nature, the scholastic method of the Middle Ages should be abandoned: one should stop using introspection and contemplation and *talking* about natural phenomena, and instead we should *act.* In other words, we must study these phenomena empirically, and collect worthwhile facts by performing systematic observations and experiments (Bacon 1620; Rees 2004). He suggested that to elucidate the causes of a physical phenomenon, our observations and experiments must be performed very methodically to *isolate* the causes—and then follow inductive reasoning to understand the governing physical laws. In today's terms, we would say that Bacon stressed the need for *controlled experiments.* Bacon's method—based on careful isolation of causes through systematic observations and controlled experiments, followed by inductive reasoning—has been called the "Baconian method." It formed the cornerstone of what we call today the "scientific method" (Rees 2004).

The third person to have ignited the scientific revolution in the early seventeenth century was Galileo Galilei, who was less of a philosopher than Descartes or Bacon but instead was a highly successful experimental scientist. Independent of Descartes, Galileo also advocated the crucial combination of detailed observation, careful experimentation, and a mathematical formulation of the results. Unlike Descartes, Galileo practiced what he preached as an active experimental scientist, contributing so many fundamental discoveries in physics and astronomy that he is often credited as the father of modern science (Hawking 2009). He is famous for going against the ruling dogma of his time, and his example has motivated generations of scientists to swim against the current of predominant convention. Galileo's success in using controlled experiments to study physics—and the admiration that he received from the scientific world throughout the years for doing so—arguably also have promoted the use of highly controlled experiments in modern science.

All active scientists, myself included, endorse the crucial importance of careful empirical observation and systematic experimentation, which are essential to collect a broad knowledge base of *facts,* which is then used to refute existing theories and generate new ideas. However, we must stop and ponder the accompanying notion, that of *controlled* experiments, where the aim is to focus on just one particular cause of the phenomenon at hand while fixing (or eliminating) all other causes. Is it indeed true that breaking phenomena into their parts, as advocated by Descartes, always leads to better understanding? Is it always true that fixing/eliminating all factors except one—and repeating this process one by one for all the contributing factors—is the best approach for understanding the phenomenon?

In some branches of physics, this reductionist approach is clearly appropriate. For example, if our hypothesis is that the pressure of a gas in a container is influenced separately by the temperature of the gas and by the volume of the container, then we should conduct experiments where we fix the volume and vary the temperature and other experiments where we fix the temperature and vary the volume. By measuring the pressure in both sets of experiments, we can elucidate the separate contribution of each of these well-controlled factors: temperature and volume. However, the key question is: Are there counterexamples? Specifically, can we find natural phenomena whose causes are so complex, coupled, and interwoven that they cannot be separated—that is, where *the whole cannot be broken into parts*? Even in physics, the answer is clearly "yes," as exemplified by condensed-matter physics and statistical physics—where the fundamental physical properties of microscopic particles *cannot* explain many macroscopic phenomena, such as superconductivity and superfluidity, the properties of glasses and of ordinary crystals, and a whole host of phase transitions (Anderson 1972; Parisi et al. 2020).

All of these are examples of **emergent phenomena**, where the whole cannot be broken into parts. As eloquently put by the eminent physicist Philip Anderson: "It seems inevitable to [believe] an obvious corollary of reductionism: that if everything obeys the same fundamental laws, then the only scientists who are studying anything really fundamental are those who are working on those laws [namely, particle physicists and astrophysicists]. The main fallacy of this kind of thinking is that the reductionist hypothesis does not by any means imply a "constructionist" one: The ability to reduce everything to simple fundamental laws does not imply the ability to start from those laws and reconstruct the universe" (Anderson 1972). The failure of the constructionist type of reductionism and the prevalence of emergent phenomena are especially prominent when studying brain and behavior—and, in particular, higher brain functions. One cannot construct complex behavior and cognition by examining the detailed biophysical mechanisms of individual neurons, channels, or molecules. As phrased by Gomez-Marin and Ghazanfar (2019): "The current approach to behavior and its mechanisms could be characterized as the "Frankenstein error," or the failure of the principle that what can be taken apart can be put . . . back together again." It is generally *impossible* to break down a complex neuronal or behavioral phenomenon by trying to isolate and tightly control its constituents one by one, and then meaningfully reconstruct and explain the phenomenon based on the underlying constituents. A corollary of this is that *when studying the neural basis of complex behaviors, one should not necessarily aspire to perform highly controlled experiments as classically done in neuroscience; rather,* *experiments should be made more naturalistic, focusing on the neural basis of natural behaviors.* Of course, some amount of control is always needed, in any experiment, because all experiments, by definition, involve controlling and manipulating something. But what do we wish to control? How much? And what do we lose if we control too much? In what follows, I will argue that we should move away from overcontrolled experiments and allow animals and humans to behave more naturally.

There are at least twelve main reasons why we should move toward more natural experiments (see table 1.1). In brief: (1) There are plenty of experimental data showing that the brain cares about natural behaviors—that is, that neurons often respond very differently if tested under highly controlled versus more natural conditions. Further, neuronal responses under natural conditions often cannot be predicted based on their responses under controlled conditions and artificial stimuli—and this failure directly argues against the reductionist logic. This book will focus on surveying dozens of such examples, which imply that highly controlled experiments yield a very partial and potentially distorted picture of brain function in the real world. (2) The brain is a complex network—a highly nonlinear dynamical system that exhibits **collective behaviors** and emergent phenomena; hence, trying to control individual factors, or nodes of the network, strongly affects the entire network. (3) The world itself is a highly complex network of external factors, which often cannot be easily separated (as schematized in figure 1.3). (4) The brain and the world are bidirectionally connected by multiple loops, mediated via a complex set of behaviors (figure 1.3). Controlled experiments often open these loops, which fundamentally alters both brain activity and behavior. (5) Most brain systems contain very prominent top-down anatomical projections, which convey high-level cognitive information about context, emotions, expectations, attention, planning, and more. These top-down inputs are likely to be distorted in highly simplified behavioral experiments. (6) Brain state and behavioral state are known to profoundly influence network activity, but they might become distorted if the experiments are too simplified. (7) Theoretical studies suggest that we need to use complex naturalistic tasks to understand the richness of brain dynamics, neuronal computations, and algorithms. (8) Long training in highly repetitive, controlled situations may create long-term plasticity that changes the very brain circuits that we are trying to study. It is, therefore, better to utilize more natural behaviors that require less training, or no training at all, thus reducing this unwanted plasticity. (9) Some highly controlled artificial experiments are also stressful for the animal, and stress is known to strongly affect the activity and plasticity of brain circuits. Hence, we should seek more natural and less stressful experimental paradigms. (10) Evolution: the brain evolved to execute particular natural behaviors, so these are the behaviors that we should focus on. (11) Relevance: natural behaviors are the only ones that we know for sure are relevant for the animal. Conversely, many laboratory-based artificial stimuli and behavioral assays are not very relevant for the subject animals. (12) There is a key problem of "unknown unknowns," whereby there may be fundamental aspects of brain function and behavior that we are entirely missing by narrowly focusing on simplified controlled experiments.

These twelve reasons, which are elaborated next, are not necessarily mutually exclusive. In fact, some of them overlap. Further, note that reason 1—the evidence for which forms the bulk of this book—is the key reason why the brain should be studied under natural conditions: it's an *empirical* reason. The remaining reasons are in essence the

Table 1.1
Twelve major reasons why the brain should be studied under natural conditions

1. Extensive data show that neurons respond very differently during natural behaviors and natural stimuli compared to artificial ones. Further, neuronal responses under natural conditions often cannot be predicted from their responses under controlled conditions and artificial stimuli, which directly argues against the reductionist logic.
2. The brain is a complex network, a nonlinear dynamical system that exhibits emergent collective behaviors—and hence cannot be reduced to its parts.
3. The world contains complex networks of external factors.
4. The world and the brain are bidirectionally connected in multiple loops, mediated by complex behaviors.
5. Top-down anatomical projections carry signals that are distorted in impoverished setups.
6. Effects of brain state and behavioral state.
7. Theoretical studies suggest the need to use complex naturalistic tasks to understand the richness of brain computations and algorithms.
8. Reducing plasticity due to overtraining.
9. Reducing stress.
10. Evolution: to understand the brain, we must let animals perform the natural behaviors for which their brains have evolved.
11. Relevance: we need to study behaviors that are relevant for the animal.
12. Unknown unknowns: by preventing rich natural behaviors, we miss key aspects of brain function—and we may not know what we are missing.

explanations for reason 1: the explanations for *why* the brain acts so differently under natural conditions and natural behaviors as compared to artificial laboratory conditions.

Let us now discuss each of these reasons in more detail:

1. Extensive experimental data show that neuronal activity is often very different when tested under highly controlled versus more natural conditions. This is the most important reason, and surveying this experimental evidence forms the core of this book. Thus, the push toward a **natural neuroscience** approach is not just a theoretical argument—but it is backed up by plenty of experimental evidence, as will be discussed throughout the subsequent chapters. For example, neurons in the barrel cortex of freely moving rodents respond very differently to active touch than to passive touch of the whiskers—that is, whether the animal or the experimenter initiates the touch (see details in chapter 5). Moreover, when rodents are allowed to dig burrows in natural soil, their somatosensory barrel cortex undergoes massive functional reorganization, which is unseen in standard laboratory conditions (chapter 5). Natural eye and body movements dramatically affect neuronal activity in the visual cortex (chapter 5). Neurons in both the visual and auditory systems respond very differently to natural visual or auditory stimuli than to artificial ones (chapter 6)—and, importantly, the responses of sensory neurons to simple stimuli usually provide poor prediction for the responses of the same neurons to natural stimuli (chapter 6). In fact, sensory neurons were shown to be exquisitely tuned for representing natural stimuli—be it natural sounds, images, or odors (chapter 6). In the brain's navigation system, hippocampal place cells were discovered by allowing rats to exhibit natural exploratory behaviors (see more on place cells in chapters 2 and 4). Moreover, recent years revealed novel unexpected spatial representations in the hippocampus when using more naturalistic situations, such as navigating to goals, where a vectorial representation of direction and distance-to-goal was discovered; during long-distance navigation, where place cells exhibited multiple place fields (multiple activity zones) with different sizes, comprising a fundamentally different neural code for space than in small environments; or during social interactions with other animals, where many hippocampal neurons represented the position of the other animal (chapter 4). In humans, too, natural behaviors matter. The brain mechanisms involved in the retrieval of natural autobiographical memories are very different from those involved in retrieving the memories of artificial lists of words or syllables (chapter 7); human sensory perception under naturalistic conditions is quite different than under artificial laboratory-based restrained conditions (chapter 7); and human navigation in large, complex, and naturalistic settings is very different from simplified environments (chapter 4). These and many other examples are discussed in detail throughout the book, and form its backbone. The extensive experimental evidence thus provides a strong empirical argument that the brain cares about natural stimuli and natural behaviors, and therefore a larger fraction of neuroscience research should be dedicated to studying brain activity closer to the natural behavior of animals and humans.

2. The brain is a complex network comprised of a web of highly interconnected subnetworks, each of which is very complicated. When it comes to complex networks with high interconnectedness, which act as nonlinear dynamical systems, one cannot easily isolate individual factors. Therefore, if one tries to control all aspects of the behavioral task except for the one aspect under study, this necessarily affects many parts of the brain's network, including likely the subnetwork that the experimenter is studying. In other words, when investigating complex networks, it is impossible to "clamp" most of the nodes of the network and claim that this allows for studying the remaining nodes in isolation; there is no such thing as "in isolation" in a network.

This critique can be illustrated by considering experiments utilizing optogenetics (chapter 3), which enables the use of light to activate or turn off (i.e., inactivate) a specific class of neurons—which are identified by a specific molecular marker—and to observe the effect of this perturbation on the animal's behavior (Aravanis et al. 2007; Cohen et al. 2012; Deisseroth 2015). Thus, if one turns off neuronal class X_1 and sees that behavior Y_1 is abolished, then the typical conclusion would be that "X_1 is *necessary* for Y_1." Conversely, by activating neuronal class X_1 and eliciting behavior Y_1, the conclusion is often that "X_1 is *sufficient* for Y_1." However, there is a fundamental misconception here:[2] neuronal class X_1 is also interconnected with neuronal classes X_2, X_3, X_4, etc. (see figure 1.3), and hence the activation or inactivation of X_1 may also perturb neuronal populations X_2, X_3, or X_4—which may be *equally important* for behavior Y_1. In other words, classical controlled experiments that aim to determine necessity and sufficiency—the hallmarks of classical mechanistic biology—are not easily applicable to **systems neuroscience**: Such experiments cannot be used to elucidate causality in a highly interconnected network like the brain. For further discussion of these problems, see Jazayeri and Afraz (2017) and note 2 (see endnotes); and see the following for alternative approaches to understanding causality in networks, as outlined in the work of Judea Pearl (2009).

3. The world itself is a complex network of factors. These factors often cannot be isolated from each other (figure 1.3; see the world network). Think, for example, of three factors in the world: conspecifics (other individuals of the same species), predators, and food availability

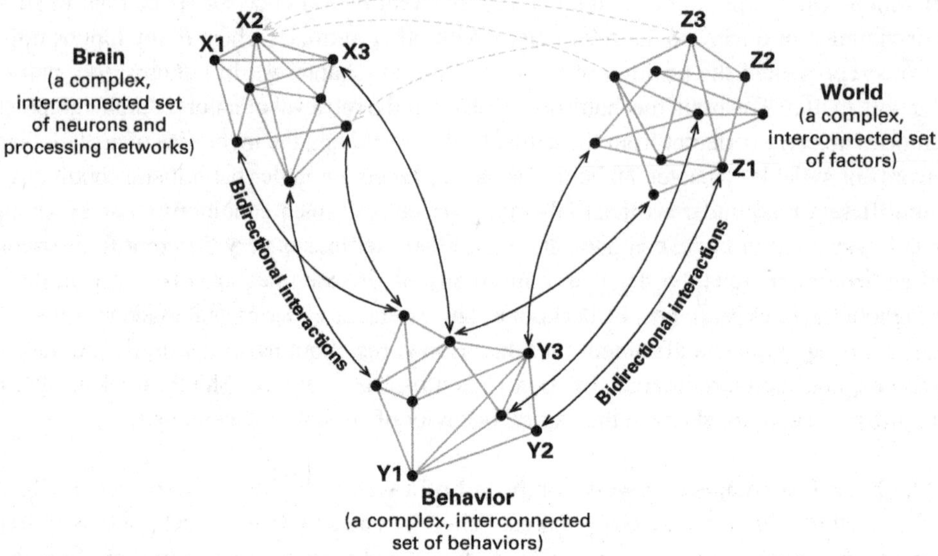

Figure 1.3
Schematic illustration of the complex interactions between the brain, behavior, and the world. The brain is a complex network (or network of networks) of neurons. The world itself forms a complex network of interconnected factors. There are also strong bidirectional interactions between the brain and the world, mediated primarily via complex behaviors: the brain can affect the world only via behaviors (motor actions), and conversely, almost all the sensory information arriving from the world to the brain relies on complex behaviors (i.e., active-sensing behaviors: see chapter 5). The world also creates strong constraints on behavior: for example, the spatial distribution of food affects foraging behaviors, encounters with conspecifics affect social behaviors, and the topography of the world affects navigation behaviors. The nodes X_1, X_2, and X_3 in the brain network denote different subnetworks of the brain, or different cell classes; the nodes Y_1, Y_2, and Y_3 in the behavior network denote different behaviors, such as social interactions, foraging, avoiding predators, and active-sensing behaviors (e.g., whisking or sniffing); and the nodes Z_1, Z_2, and Z_3 in the world network denote different factors in the world, such as conspecifics, food availability, predators, and topography.

(denoted in figure 1.3 as Z_1, Z_2, and Z_3). These three factors are strongly interconnected: for example, in those spatial locations in the environment where there is large food availability, there will often be many predators, and also many social interactions with conspecifics. Thus, high food availability and high social interactions are strongly associated with high predation risk. In classical neuroscience, such factors and behaviors are studied separately, in isolation (and usually by different labs): social neuroscientists study the neurobiology of social interactions without any food or predators available; neuroscientists interested in escape responses and in predator avoidance study the neural bases of escape responses from predators; and neuroscientists interested in foraging or in decision-making study the neural mechanisms of foraging for food without any social interactions or predators. But in real life, all these world factors are nonlinearly interacting with each other: when multiple animals forage for food, this often creates strong competition for food, as well as food-related social interactions with conspecifics and also increased predation risk (because predators gather in areas of high concentration of prey in a predator-prey interaction loop). Similar arguments have been raised in the domain of human cognitive neuroscience by Uri Hasson and colleagues, who also highlighted the prevalence of nonlinear complex interactions between multiple factors in human cognition (Nastase et al. 2020). More generally, many factors in the world nonlinearly interact with each other in multiple loops, forming complex nonlinear networks of interactions.

These types of real-world nonlinear interactions are very likely to be reflected in animals' brain circuits—but our typical isolated experiments tend to ignore them. Because of these nonlinearities and loops, the attempt of classical neuroscience to isolate such world factors one at a time, in so-called controlled laboratory experiments, is futile. It is impossible to isolate one node in a complex nonlinear network. This means that the reductionist agenda of deconstructing the world into individual factors and studying them one by one—with the hope of later reconstructing (assembling) them back together—most likely would fail in the case of brain research. This is why it has proved so difficult to generalize from simplified and tightly controlled laboratory experiments in neuroscience to real-life complex behaviors, as demonstrated time and again by the many dozens of empirical examples that will be reviewed in this book. Notably, the research field of behavioral ecology *does* consider these types of complex interactions—and in chapter 8, I will propose a synthesis between neuroscience and behavioral ecology, which offers new future experiments for natural neuroscience.

Another example relates to real-life memories. Humans remember facts and events that are composed of a multitude of interconnected factors, which together make a rich multidimensional memory. However, the classical approach to study human memory was to present random lists of words or syllables as stimuli and then test the subject's ability to remember them—thus stripping away any emotional, social, or contextual factors, or any personal relevance of the memory (chapter 7). But does it make sense to try to eliminate all these factors and study just one narrow facet of a complex memory, displaced out of context? Such impoverished memories are perhaps akin to what an amnesic might remember from an event: a shallow, one-dimensional (1D) impoverished shadow of a normal person's memory. Should we be studying the neurobiology of highly impoverished memories? We shall come back to these issues in chapter 7.

4. The brain and the world are bidirectionally connected via multiple loops, mediated by a complex set of behaviors (see the arrows in figure 1.3), which, again, cannot be isolated and controlled individually. The closed-loop dynamics between the brain and the world (via the body) has fundamental importance. For example, as will be discussed in detail in chapter 5, on active sensing, we regularly move our eyes to scan the visual scene. Thus, the eye is used to collect visual information into the brain, and then the brain interprets this incoming information and rapidly moves the eye to acquire additional relevant visual information. Therefore, under normal natural conditions, the world (objects in the visual scene) and the brain (visual system) are bidirectionally connected in a loop via the eyes; and the eyes act to channel visual information both into the brain (afferent input) and as motor organs whose position is tightly controlled via motor commands from the brain (efferent input) (Yarbus 1967). This raises the intriguing question: How much of our visual perception depends on our eye movements? However, classical controlled experiments in visual neuroscience have used head-fixed animals that were either anesthetized and paralyzed, and hence could not move their eyes, or awake monkeys that were trained to fixate their eyes on one spot (known as a "fixation point")—thus either completely eliminating eye movements or clamping them. This is justified if we assume that eye movements are irrelevant for visual perception.

However, this assumption may be completely wrong—as some theories of vision suggest—and then it is unjustified to eliminate eye movements when studying vision (see chapter 5). Active-sensing theories go even further, suggesting that *action is an inherent part of perception*, at a very mechanistic level—and therefore motor-sensory interactions should always be studied together (chapter 5). From this perspective, eliminating the normal repertoire of eye-movement behaviors, as typically done in most visual-neuroscience experiments, eliminates the very thing that we ought to be studying if we wish to understand natural visual perception.

Consider another example: the coupling of whisking and sniffing in rats—which will be further discussed in chapter 5. Rodents use their whiskers to actively explore their environment, such that the movement of the whiskers generates somatosensory perception. In turn, the brain actively controls and optimizes whisking movements to acquire the most relevant information. Likewise, rodents actively sniff their environment to acquire olfactory information, and this incoming stream of sensory information is then used to change the sniffing pattern adaptively. However, the classical studies of the rodent olfactory and whisking systems were done by passively presenting odors or passively touching a whisker, without allowing the animal to actively sniff or whisk: a very unnatural condition. For this reason, recent studies of these systems have gradually moved away from these passive-sensing paradigms and toward allowing animals to actively whisk or actively sniff (chapter 5). But there is yet another level of sensory coordination, which is lost in controlled experiments: under natural conditions of exploring an object of interest, *whisking and sniffing are highly coupled*—specifically, they are synchronized and phase locked to each other (Moore et al. 2013). In the terminology used here, we have two behaviors, Y_1 and Y_2, which are highly coupled (see figure 1.3)—and in fact, we know today that these behavioral rhythms are controlled through two tightly coupled oscillators in the brain (Moore et al. 2013; see also chapter 5 of this book). Thus, if one studies the whisking system in isolation and abolishes olfactory sniffing, or vice versa, then this may put both systems in conflict, at least when considering objects close to the animal, for which the animal would normally whisk and sniff together. Experimentally isolating these two systems creates an unnatural and aberrant condition. Why is it justified to do this? Why should we study aberrant situations? It is impossible to understand active and dynamic coupling of perception and action, which are coupled in a recurrent loop, when action is uncoupled from perception. The results can be completely misleading.

Of course, the issues outlined in reasons 2, 3, and 4 are relevant to every kind of nonlinear dynamical system, or **complex system**—such as predator-prey networks in ecology, gene networks in systems biology, and many nonlinear dynamical systems in physics and chemistry (Strogatz 2015). All these systems display strong **collective behaviors**, with **emergent properties** that cannot be elucidated by separating the individual components and trying to control them one by one. The core reason for this limitation is that in a complex network, a given component is influenced by feedback from many other interacting components, and therefore, shutting off ("controlling") any subset of these components may collapse the entire network into an abnormal state. For example, studying an isolated behavior while shutting off all the contextual cues that are naturally associated with this behavior may put the brain networks that generate this behavior into an abnormal state, because major key inputs are lacking. In other words, complex networks are often very *context dependent* (i.e., highly sensitive to diverse external inputs), and therefore, abolishing the normal context for the purpose of experimental control may put the system in a highly aberrant regime. Thus, attempting to impose control on a complex dynamical system may collapse the system's normal collective behaviors, which are the very thing that we are trying to understand—truly, a major measurement problem. Of course, it is nevertheless very customary to perform experiments and collect data under such "abnormal" conditions—but it is crucial to at least acknowledge that such experimental measurements might bear little (or, at best, partial) relevance to the real-world function of the system.

Moreover, complex dynamical systems are characterized by *nonlinear dynamics*, by processes of *self-organization*, and by multilevel *hierarchical structures* with strong *feedback* connections—where phenomena at one level of the hierarchy cannot be easily reduced to the levels below it—implying that the whole is more than the sum of its parts (Anderson 1972; Mazzocchi 2008; Bradbury & Vehrencamp 2014). The problem of complexity versus reductionism is thus a fundamental unresolved challenge in many fields of science—and possibly pertaining to *all* fields of science.[3] However, this issue is arguably of particular importance to neuroscience because of the enormous complexity and interconnectedness of the brain, of the world, and of the behavioral interactions between them.

Finally, the monumental work of Judea Pearl on causal inference, summarized in his book *Causality* (Pearl 2009), showed that in a complex network (graph), one *cannot* understand the causal relations in the system by setting the values of all the nodes to some fixed (controlled) values, while measuring and/or manipulating just one node—which is the typical approach taken in classical controlled experiments. In behavioral neuroscience research, a node could be one brain region that is being manipulated in isolation, if we consider an optogenetics experiment—or it could be a set of several brain regions that are being manipulated in unison by a behavioral manipulation, if we consider a controlled behavioral experiment. In other words, Pearl's work implies that it is not advisable to control the entire system too strongly. Instead, he showed that to understand such networks, the controlled interventions and manipulations should be done by the experimenters in a much more nuanced manner, in a specific set of nodes simultaneously, and the network should be kept in a state that is much closer to its original state. Pearl developed a mathematical formalism to design the correct set of manipulations given the network's structure, which allows for disentangling the interwoven causal paths—and I think that this formalism should be used more broadly in neuroscience to replace the classical single-node (e.g., single-brain-region) experimental manipulations used today. In the case of behavioral neuroscience, this also means that a too-strong behavioral control can in fact undermine the prospects of understanding causality. Thus, a proper mechanistic understanding of causality in a brain-behavior-world network requires the approach of natural neuroscience—or at least a middle way between completely natural experiments and completely controlled, highly reductionist experiments.

5. The brain contains massive top-down anatomical projections. These projections are thought to convey high-level information about context, emotions, expectations, attention, planning, social interactions, and many other cognitive processes (Gilbert & Li 2013). This top-down information may be distorted, or altogether aborted, when conducting simplified experiments in tightly controlled, impoverished laboratory setups. This issue, of course, is related to the previous ones (reasons 2 and 4), but here the emphasis is specifically on the abolition of many top-down *cognitive* inputs in typical neuroscience experiments—inputs that have a paramount importance in everyday behavior. Therefore, if we take Descartes's approach and study the various low-level factors one by one, in isolation, we may never understand the full system as it functions in the natural world because we have neglected the effects of top-down causation[4] (or "downward causation") (Noble 2008; Ellis 2016).

Further, the massive top-down projections may also be important for *learning*. Under this possibility, since a central objective of the brain is to learn the structure of the world, then we are missing a major piece of the brain puzzle if we abolish (or distort) top-down input.

6. The effects of brain state and behavioral state. Many studies have shown that neural activity in multiple brain circuits is strongly influenced by brain state, such as regular slow activity, large irregular activity, or desynchronized brain activity (Kay & Frank 2019); these states can differ among brain regions, such as the hippocampus and the neocortex. These brain states are related to the animal's behavior, such as voluntary behavior, automatic behavior, or immobility. But we can look beyond these gross brain states and gross behavioral modes and consider more finely grained behavioral states, such as "foraging state," "navigation state," "burrowing/nest-building state," "mating state," and "fighting/aggressive state," and each of these behavioral states might have major effects on neural activity in the network. In fact, behavioral-state switches have been shown to exert dramatic functional consequences in invertebrates such as the nematode *Caenorhabditis elegans*, and the neural mechanisms of these switches have been studied in depth (Flavell et al. 2020; Ji et al. 2021). However, much less is known about this in vertebrates. This is something that must be studied in detail.

One interesting clue that supports the possible importance of behavioral states in mammals comes from Michael Graziano and colleagues, who found that the primate (and rodent) motor cortex does not contain one body map (homunculus) as classically thought; instead, it contains many cortical zones, each one eliciting a particular type of ethological movement such as reaching, grasping, object manipulation, climbing, leaping, defense movements, hand-to-mouth movements, chewing, and more (Graziano 2016). If different ethological movements engage different brain circuits—then perhaps, more broadly, different ethological behaviors engage the brain differently. In particular, we cannot just assume that, from the brain's perspective, artificial behavioral states in the lab—such as back-and-forth shuttling on a linear track or repeated lever-pressing for obtaining rewards, and many other lab behaviors—are equivalent to more natural, ethological behavioral states, as occur outdoors. Consequently, the effects of behavioral state on neural activity could be quite different between laboratory behaviors and natural behaviors.

7. Theoretical studies suggest that we need to use complex naturalistic tasks to understand the richness and dimensionality of brain dynamics, its computations, and algorithms. For example, Pitkow and Angelaki (2017) discussed algorithms that the brain might be using to implement probabilistic inference, and they argued that to elucidate such algorithms, experimental neuroscientists should employ naturalistic behavioral tasks of greater complexity than the typical tasks used today in behavioral neuroscience. They also emphasized that the tasks should not be too complex and completely uncontrolled; rather, they should have a medium level of complexity. This call for neuroscientists to use medium-complexity, naturalistic behaviors resonates well with the call expressed here for enhancing behavioral diversity (figure 1.2).

In another study, Gao and Ganguli (2017) proved mathematically that a quantity called "neural task complexity"—which strongly depends on the behavioral task complexity and on the number of stimuli used—provides an upper bound on neural dimensionality: the dimensionality of population activity that can be estimated from the ensemble of neurons recorded during the behavioral task (also see Gao et al. 2015). In other words, *neural dimensionality ≤ neural task complexity.*

This mathematical theorem (Gao et al. 2017) has important implications for the design and analysis of neuroscience experiments. As Gao and Ganguli (2017) pointed out, it suggests that the neural responses and neural dimensionality observed in neuroscience experiments are limited to a large degree by the task complexity, which is often too low. In such cases, just increasing the number of recorded neurons is *not* enough because neuronal data dimensionality is generally independent of the number of recorded neurons (Gao et al. 2017). What instead needs to be done is to record larger numbers of neurons *and* to increase the complexity of behavioral tasks and stimuli.[5] Arguably, the best way to increase behavioral complexity is to focus on *natural behaviors*, which are inherently complex. In other words, the best way to understand the brain is to record large ensembles of neurons in animals performing complex naturalistic behaviors.

The mathematical results of Gao and Ganguli also resonate with experimental studies that have shown that many neurons in higher brain areas are tuned to a large number of behavioral variables—so-called **mixed selectivity** of neurons (Mante et al. 2013; Rigotti et al. 2013; Eichenbaum 2015)—and this complex mixed selectivity can be revealed only by utilizing complex behavioral tasks rather than the overly simple, overly reductionist tasks that are often used today. Likewise, several recent studies showed that in complex social settings, neurons encode many more behavioral and social variables than hitherto thought (Ray et al. 2024; Testard et al. 2024). Together, these theoretical and experimental studies suggest that we should use complex, naturalistic behaviors to understand the brain.[6]

Finally, Uri Hasson and colleagues (Hasson et al. 2020) proposed that while many classical theories in neuroscience and psychology assume very simple models with just a few parameters, in reality, the brain does not work according to simple human-interpretable models and does not learn simple representations. Instead, the brain reflects a process of local computations that learn to interpolate the structure of the real world by optimizing billions of synaptic connections, creating effectively a model of the world with billions of parameters—akin to modern deep neural networks, but on a larger scale. According to this view, there is nothing simple, elegant, or low-dimensional in the manifold that is learned by the brain, and any simplicity or low dimensionality that our experiments reveal in recorded neural activity is due solely to the oversimplified, low-dimensional task structure enforced by experimenters. If we do simple, low-dimensional, and classical controlled experiments, then we shall inevitably find simple, low-dimensional representations in the brain—because the brain learns the low-dimensional idiosyncrasies of the task. But this may have absolutely nothing to do with how the brain operates in the real world. The true high-dimensional and complex operation of the brain could be elucidated only by studying the animal or human as it behaves freely in the natural world, letting the brain operate normally (Hasson et al. 2020). While this notion of the brain as a high-dimensional interpolator of the structure of the world still needs to be tested, the only way to prove or disprove it is to study the brain under naturalistic conditions (i.e., to take the natural neuroscience approach).

8. Reducing unwanted plasticity in the brain due to overtraining. Many (but not all) laboratory-based behavioral assays require extensive training of the animal on performing a behavioral task. Behavioral training alters sensory representations in the mammalian brain, including changes in the intrinsic excitability of neurons, major alterations in **receptive fields** and tuning curves of individual neurons, structural and functional synaptic modifications, and cortical map expansions (Weinberger 1995; Feldman 2009). Besides sensory regions, dramatic training-related changes were documented in the motor cortex, striatum, prefrontal cortex, parietal cortex, the midbrain's dopaminergic system, and many other brain areas (Seger & Miller 2010; Xiong et al. 2015; Constantinidis & Klingberg 2016; Peters et al. 2017). But plasticity can also be very natural. Learning new behaviors and new associations constitutes part of the natural behavioral repertoire of animals and humans, and learning requires brain plasticity. Therefore, if we are interested in the neurobiology of learning, then these plastic changes are of great interest. However, natural learning of real-world behaviors is often very fast, requiring just a few trials (hence dubbed "few-trials learning" or "one-trial learning") and does *not* require thousands of training trials, as is often done in neuroscience experiments. Therefore, it is unclear whether overtrained tasks can yield true insight into real-life, fast learning processes.

Furthermore, if we are interested in the neural bases of behavior per se—with no particular focus on learning—then these dramatic changes in the brain due to extended training protocols are a confound: they may influence unpredictably the neural mechanisms of the behavior that we are trying to understand. Thus, the main problem is the *amount* of training involved and the common practice of using extensive training or overtraining of animal subjects in neuroscience, as massive training leads to massive changes in the brain. How much of what has been discovered so far in behavioral neuroscience is due to overtraining? In most cases, we don't really know, but it is clearly desirable to minimize training as much as possible and focus on tasks that the animal (or human) can learn rapidly (i.e., within a few trials). Or we may forgo training altogether and study the neural bases of natural behaviors.

Another aspect of training is that it usually requires water or food deprivation. Thirst and hunger are known to affect neural activity in multiple brain areas, including sensory-processing areas (Burgess et al. 2018). Food deprivation even reduces coding precision in primary visual cortex (V1) neurons by broadening their orientation **tuning curves** (Padamsey et al. 2022). It is thus quite likely that performing a behavioral-neuroscience experiment in a thirsty or hungry animal may change neural activity and **neural coding** in some of the brain areas that we are interested in and are studying, including both sensory and high-level brain areas. This is clearly undesirable. The solution to this problem is to study natural behaviors, which do not require any water or food deprivation, thus minimizing the effects of thirst and hunger.

9. Reducing stress. Some commonly used behavioral tasks might be stressful for the animal, especially if they use head fixation. Stress has widespread effects on brain and behavior—it affects synaptic plasticity, learning, and memory (Kim & Diamond 2002; Morris 2007), social behaviors (Sandi & Haller 2015), and many other behaviors and cognitive functions. Stress affects multiple neurotransmitters and neuromodulatory systems (Joëls & Baram 2009), and it consequently affects synaptic transmission across multiple brain regions (Popoli et al. 2011). It causes structural changes in individual neurons (mostly in their dendritic arborization) and in entire neural networks (Kim & Diamond 2002; Morris 2007; Sandi & Haller 2015). Moreover, stress can even change the intrinsic cellular biophysical properties of individual neurons, as well as their excitability (Kim & Diamond 2002; Morris 2007). Therefore, if we wish to study the normal brain, it is clearly desirable to reduce stress for animals as much as possible in neuroscience research—and the most natural way to do so is to focus on naturalistic behaviors, which a freely moving animal is willing to perform with little or no training, of its own volition.[7]

10. Evolution. The basic argument for taking an evolutionary perspective when studying the brain is that evolution did not shape the brain to perform the simple binary tasks that are often used for laboratory studies in behavioral neuroscience. Instead, evolution acted on a whole set of natural behaviors—such as predator avoidance, foraging, and a host of social behaviors—some of which are rarely studied in neuroscience and should be studied more.

But there is another reason why it's important to keep an eye on evolution. To illustrate this point, let me use a somewhat contrived example. We can train a person to use their pinky finger to tap twice on the doorway whenever going through it—humans can certainly learn this task very well—and we could record neurons in the brain while subjects perform this "doorway pinky-tapping task." We may then study the mechanisms of this motor behavior at the network, cellular, and molecular levels. But all this doesn't mean that fingers have evolved to tap twice on doorways, of course. We can spend decades studying the detailed mechanisms of this contrived task, down to the smallest mechanistic detail, but it is unclear what this would actually teach us about the brain. Instead, goes the argument, we should study more natural behaviors, which evolution has acted upon. If we are interested in the motor control of fingers, we should study the mechanisms of grasping, throwing, object manipulation, and perhaps also pointing. If we are interested in the neural basis of habits, we should study the mechanisms of species-relevant habits. In other words, to understand the brain, we must challenge the animal to perform the kinds of natural behaviors for which the brain has evolved.

11. Relevance. Relevant stimuli and relevant behaviors are ones that are meaningful and important for an individual animal. These can be (at least in part) different from the natural stimuli and behaviors that are relevant for evolution. For example, the sounds of the approaching animal technician are very relevant for a laboratory mouse or rat, which has learned that these sounds may signal that new food is coming—even though these sounds are unimportant for evolution because the animal's survival and reproduction do not depend on them. When neuroscientists train animals to perform a certain behavioral task, they use stimuli that are inherently neutral to animals, such as the sound of a pure tone or the sight of a lever or a food-port, and through training, they turn them into something relevant for the animal. The advantage of this approach is that these stimuli were originally neutral, or irrelevant to the animal, and in this way, we control in our experiments for stimuli or behaviors that the brain inherently cares about—and thus we can create arbitrary stimulus-reward or behavior-reward associations without having to worry about their internal relevance to the animal. This approach offers the power of neutrality and helps avoid biases. On the other hand, this neutrality—or irrelevance—is also the disadvantage of this approach. If the brain inherently cares about particular stimuli and behaviors without any training, then why shouldn't we study them? Why treat them as "biases" rather than as fascinating objects for investigation, which may reflect evolutionary adaptation or prior learning by the animal, and which should be studied in their own right (in parallel to the classical neutral approach to behavioral neuroscience)? Many stimuli, such as social vocalizations or pheromones, and many behaviors, such as social behaviors, foraging, navigation, or predator avoidance, are both relevant (at least for wild animals) and evolutionarily important. So why not focus on studying the neural basis of such relevant naturalistic behaviors?

12. The problem of unknown unknowns. This is another cardinal reason why highly controlled classical experiments *cannot* be the only way to conduct neuroscience research, and also why we need to promote more natural styles of experimentation. This problem can be illustrated by considering the neurobiology of navigation (which will be discussed in more detail in chapter 4). As already mentioned, hippocampal place cells were originally discovered by abolishing highly controlled setups and instead letting rats exhibit their natural behavior of curiosity-driven spatial exploration (O'Keefe & Nadel 1978; see also chapters 2 and 4 of this book). Thus, O'Keefe was a prophet and a key founder of the natural neuroscience approach, and his naturalistic experiments led to a Copernican revolution in neuroscience: moving away from highly artificial head-fixed setups and overtrained tasks and embracing the study of the neurobiology of natural behaviors. This historical example serves as a powerful reminder why channeling animals to perform oversimplified controlled tasks sometimes may be premature—in this case, the extreme reductionism practiced by hippocampal researchers in the 1960s neglected what turned out to be a crucial feature of the hippocampal system: space. Reductionist simplification seeks to find the truth under a very specific theoretical lamppost, which casts a limited pool of light—but this runs the danger of premature oversimplification, which might miss the truth entirely, because the truth might be under a different lamppost, or it may be lying quietly in the dark. Therefore, it is often advisable to venture into the darkness in search for the unknown. Moreover, if one does choose to place a lamppost and search only around it, then I would argue that one should opt to put the lamppost close to the natural behaviors of the animal. The reason for this is that the space of possible experiments is practically infinite, and we cannot hope to cover all this space and discover all the unknowns. On the other hand, we know that the brain responds very differently to natural stimuli than to artificial ones—for example, the responses of visual and auditory neurons to natural scenes or natural sounds are very poorly predicted by their responses to artificial gratings or pure tones, which argues against the reductionist approach (see chapter 6). Therefore, if we are to focus our attention on particular kinds of experiments, then we'd better concentrate on natural behaviors and natural stimuli: the ones that the brain evolved for, the ones that truly matter in the natural lives of animals. Of course, finding those natural stimuli and natural behaviors is not a trivial pursuit, but we should at least try. Thus, you should always ask yourself: Which aspects of behavior *should* we investigate in neuroscience? What are the interesting questions—both behaviorally and evolutionarily? As will be shown throughout this book, the answer often reveals itself by observing the natural behavior of animals (outdoors or in more naturalistic environments indoors).

It is useful to consider here some recent concrete examples of unknown unknowns, which were revealed by going to more naturalistic experiments. These examples come from the hippocampal system. As previously noted, place cells—hippocampal neurons that represent the animal's location—were originally discovered by allowing rats to perform a very natural behavior for this species: exploration of space (O'Keefe & Dostrovsky 1971; O'Keefe 1976; O'Keefe & Conway 1978; O'Keefe & Nadel 1978; Moser et al. 2017). These early place-cell experiments were performed in small environments, typically 1-meter linear tracks or T-mazes, or 1 × 1-meter empty boxes. In such environments, it was found that each place cell is usually activated in a single location, called the "place field" of the neuron However, while O'Keefe's first step was fundamentally important, it was insufficient because subsequent research did not proceed to become naturalistic enough. Today, it is still customary to study place cells (and other spatial cell types) in rodents running in small, empty environments, much like O'Keefe's early experiments. Yet, real-world navigation differs dramatically from such typical laboratory setups, in several ways:

i. The world is not empty; rather, it contains objects and goals. For many years, almost nothing was known about how neural circuits represent goal location, which is essential for navigating toward the goal. We performed naturalistic experiments in bats to address this major gap. Bats navigated in a large room toward an explicit spatial goal, and these experiments revealed hippocampal neurons that encode egocentric direction and distance to the goal—that is, the vector to the goal (Sarel et al. 2017; see also chapter 4 of this book). This new functional cell class may underlie goal-directed navigation, and it was discovered thanks to the use of a more naturalistic situation, involving actual goal-directed spatial orientation. Similar encoding of egocentric direction to objects was also recently found in the hippocampal formation of rodents (Wang et al. 2018; Ormond & O'Keefe 2022).

ii. The world is not 1 meter in size, and many animals navigate over kilometer-scale distances. Until recently, nothing was known about how the brain supports such long-distance navigation. To address this major gap, we have built a long flight facility for bats (200 meters), allowing wireless recordings of hippocampal neurons in bats flying long distances (Eliav et al. 2021). Single hippocampal neurons were found to exhibit multiple place fields, and a multiscale representation of space—whereby place fields of the *same* neuron differed up to twenty-fold in size. In other words, a single neuron could have a 1-meter place field in one location and a 20-meter place field in another location. This multifield, multiscale code constitutes a surprising neuronal representation of space, which is radically different from the classical place code found in small environments—and it would have never been predicted from studies in small environments (see chapter 4 for further details, and for why these results are so surprising). Theoretical analysis showed that this multiscale representation provides a greater capacity to represent large environments and reduces decoding errors, specifically for large environments. This surprising coding scheme was discovered thanks to pursuing the natural navigational behavior of bats, which naturally navigate long distances in almost straight flights (Tsoar et al. 2011).

As chapter 4 will discuss in detail, these and other studies from multiple labs are starting to bridge the gap between hippocampal laboratory studies and real-world natural navigation. Importantly, this set of findings could *not* have been easily predicted from laboratory experiments on animals running in small boxes—they were discovered only by conducting naturalistic navigation experiments in very large environments or in environments with goals. Chapters 4–7 of this book will describe many dozens of additional examples where unexpected results were found once experimenters ventured to use more naturalistic stimuli, naturalistic environments, or naturalistic behaviors. There are certainly many new unknowns that await being discovered by following the natural neuroscience approach.

* * *

This set of twelve problems was agonizingly noted by many neuroscientists, who started to shift away from classical, highly controlled experiments in the late 1990s and early 2000s, a movement that has accelerated over the last few years (Krakauer et al. 2017; Dennis et al. 2021; Miller et al. 2022; Cisek & Green 2024). This shift comprised a steady growth in research that could be classified in the "natural and controlled" quadrant in figure 1.2 (top-right), or "natural but uncontrolled" quadrant (bottom-right)—namely, the quadrants that capture the natural neuroscience approach.

Now is the time to define "natural neuroscience": *Natural neuroscience is the science of studying the neural mechanisms of ecologically relevant natural behaviors—in freely moving, unrestrained animals.* Of course, there can be many definitions of natural and ecologically relevant behaviors. Here, by *natural behaviors,* I refer to behaviors that the animal (or human) normally exhibits in its natural environment, such as navigation, exploration, foraging for food, escape from predators, and social behaviors. By **ecologically relevant behaviors,** I mean behaviors that the animal performs to survive and reproduce in its ecological niche outdoors—these are the behaviors that are selected by evolution. Like many terms, these can be interpreted in various ways—so there are certainly many shades of natural neuroscience experiments. But in any case, these definitions of natural behaviors and ecologically relevant behaviors, as well as of natural neuroscience, rule out many of the classical experimental approaches in behavioral neuroscience, including head fixation and the use of overtrained tasks. These classical experiments can surely teach us a great deal about brain mechanisms, but much less about brain mechanisms of natural behaviors.

The change toward natural neuroscience has occurred on many fronts simultaneously, and subsequent chapters will discuss in detail many of these changes in specific subdisciplines of neuroscience. This change was driven in part by multiple *technological* developments, which enabled evermore naturalistic experiments; these developments will be reviewed in chapter 3. Of equal importance, this was also a *conceptual* change, based on the realization that studying *natural behaviors* is important if we wish to fully understand the neural basis of behavior. Next, I briefly discuss several examples of such recent developments, which will be further elaborated throughout the book.

In sensory neuroscience, the shift to more natural experiments was expressed in at least two directions. First, there is growing appreciation that the movement of the sensory organ influences perception or may even form an integral part of perception—be it eye movements in primate vision, whisking movements in rodent somatosensation, or sniffing in olfaction. We cannot understand these sensory systems without letting the animal saccade, whisk, or sniff. In other words, we have to let the animals express their natural sensory-acquisition behaviors. This understanding spurred the active sensing approach, whereby primates are allowed to actively move their eyes, rodents are allowed to actively move their whiskers or sniff freely, and so on. As will be reviewed in chapter 5, it revealed that neural activity in sensory brain areas is very different when the animal is studied using an active-sensing approach compared to classical passive sensing. Some of these experiments allowed the animal to actually move and behave rather freely, while other experiments examined head-fixed animals performing behavioral tasks, which allows better control than the free-movement situation, but at the expense of naturalness. Figure 1.2 schematically summarizes some of these approaches.

Second, a growing body of both experimental and theoretical work has demonstrated that it is very difficult to explain the responses of sensory neurons to natural stimuli—such as natural images or natural soundscapes—based on the responses of the same neurons to simple artificial stimuli, such as moving gratings or pure tones. This has spurred a whole field of research focusing on *natural stimuli*, as will be reviewed in chapter 6. Some experiments have even combined both an active-sensing approach and a natural-stimulus approach, investigating neuronal responses in animals that are behaving and actively sensing naturalistic stimuli.

In behavioral neuroscience, an increasing number of experiments are being done in freely moving, freely behaving animals, enabled by the technological developments that are reviewed in chapter 3. Historically, the use of freely moving setups has been particularly prominent in the fields of the neurobiology of navigation (chapter 4) and the birdsong system (chapter 2).[8] Today, however, free-behavior approaches are becoming increasingly employed in many other fields of behavioral neuroscience and are used in many studies on small animals, including rodents, lizards, birds, and bats, and they are also starting to be pursued in primates. Many of these developments are reviewed in chapters 2, 3, and 4.

Another branch of neuroscience in which recordings in freely moving, freely behaving animals are gradually becoming the norm is the field of neuroethology. As chapter 2 will discuss, the credo of neuroethology has long been to focus on stereotyped natural behaviors, which are both natural and internally controlled behaviors (see the top-right quadrant in figure 1.2). Neuroethology has sought to study this primarily in "champion species," which perform these behaviors very robustly and reproducibly. A good example is the study of birdsong. Adult male songbirds—in particular zebra finches—naturally produce a highly stereotyped song, with a crystallized structure and extremely precise timing that the bird tightly controls and reproduces with near-millisecond precision every time it sings. The study of birdsong thus allows the experimenter to enjoy both worlds—namely, to perform experiments that are both highly natural and highly reproducible and controlled. Indeed, the birdsong field has been utilizing these advantages very effectively. Nature abounds with many other examples of highly reproducible, stereotyped behaviors—from the echolocation approach sequence of a bat attacking an insect (chapter 5), through the jamming avoidance response of electric fish (chapter 2), to the hunting behavior of mice or shrews. However, not all these behaviors are as highly reproducible as birdsong (see the top-right quadrant of figure 1.2). Importantly, in most of the classical neuroethological models, very few experiments were performed in freely behaving animals. Neuroethology has been *motivated* by natural behaviors, but most of the actual experiments in neuroethology have been typically performed in highly artificial conditions—in anesthetized or head-fixed animals or in other reduced preparations—again primarily due to technical limitations. However, these limitations have been largely overcome in recent years (chapter 3). Thus, neuroethology is finally starting to pursue its holy grail: to understand the neural basis of simple natural behaviors in unrestrained, freely behaving animals.

This book, however, will *not* focus on nonstandard neuroethological animal models. Instead, it will focus primarily on the standard mammalian models, such as rodents and primates, and will posit that for *every* question in systems neuroscience and in behavioral neuroscience—in *any* species—there are immense benefits to studying this question not just under classical, well-controlled conditions, but also under conditions that are more naturalistic. Indeed, studies of naturalistic, uncontrolled behaviors—in both mice and *Drosophila* flies—have already revealed new behavioral patterns that could not be uncovered by traditional controlled experiments (Coen et al. 2014; Wiltschko et al. 2015; see chapter 3 for details). However, I do *not* wish to make the extreme claim that we should all abolish control and just let animals loose in the lab or in the wild and record their brain activity 24/7 (see the bottom-right point in figure 1.2). Although such extreme, fully natural conditions will surely yield important insights, they also suffer from being completely uncontrolled, which makes it harder to use Francis Bacon's inductive-reasoning method, which relies on repeated observations of the same phenomenon. In completely natural situations, the exact same observation may never repeat itself twice. However, I will claim that there is a large range of possible middle-ground experiments that are "naturalistic enough" (even if not fully natural outdoors), but are also "controlled enough," which could yield deeper insight into the neural basis of natural behavior. This book will review the emerging approaches for conducting such experiments, across multiple subfields of neuroscience, and will propose that, as a field, neuroscience should be *diverse* and that different experiments should together cover the entire broad range of the "phase space" depicted schematically in figure 1.2, including in the bottom-right quadrant and especially in the top-right quadrant (natural and controlled).

There are two distinct approaches for how best to move toward the top-right corner of figure 1.2—that is, to achieve control and reproducibility when studying the neuroscience of natural behaviors. The first approach is to train the animals on operant behavioral tasks, as done classically in behavioral neuroscience, but to make these experiments more naturalistic in some aspect. For example, if you are studying sensory perception, you can train the animal to do a discrimination task, as done classically, but allow it to walk around and move its whiskers freely (if you study somatosensation) or move its eyes freely (if you study vision), to allow the animal to use its full repertoire of active sensing, as discussed in chapter 5. Or if you are studying the neurobiology of navigation and are recording from place cells or grid cells, you can make the environment more complex to study the neural coding of complex spaces, or you can increase its size to 1 kilometer to study the neural coding of natural-scale environments. You might also introduce navigational goals to understand how goals are represented in the brain (chapter 4). The second approach is to abolish training altogether and instead focus on those natural behaviors that the animal naturally repeats many times in a roughly reproducible manner. That is, we should choose the behaviors where the animal is "doing the control for us," of its own volition. An example of this kind has already been mentioned: songbirds such as zebra finches sing in a highly reproducible manner, hundreds of times per day, thus controlling the experiment for us—allowing the researcher to reliably measure the neuron's responses triggered by the song onset (e.g., construct a **spike raster**). This is not to say that we cannot intervene in these experiments to encourage more repetitions: in fact, birdsong researchers often introduce a female in front of the male to encourage him to sing more, thus obtaining more repeats, which yields stronger statistics. But no explicit training is involved or needed. Many other natural behaviors are often repeated reproducibly many times. This is the case for many social behaviors, navigation behaviors, active sensing, hunting, escape, and communication, among many other natural behaviors. Thus, in this scenario, behavioral reproducibility—the foundation that allows solid statistical analysis—would be obtained not by *training* animals to repeat a specific behavior in a task, but rather by the *natural tendency* of animals to repeat certain behaviors over and over. Importantly, the animal may naturally perform *several* distinct behaviors, each of them being repeated over and over—for instance, several distinct social behaviors or several types of songs (albeit not in zebra finches)—and this multiplicity of distinct behaviors then allows *comparisons* of neural responses among them. In other words, instead of a human experimenter controlling the animal by training it to perform repeatedly two distinct artificial behaviors over and over (e.g., lick left versus lick right), we will utilize the natural tendency of animals to repeat *natural* yet distinct behaviors over and over. In this way, the animal will control the experiment for us. Each of these two approaches—training for a behavioral task that is more naturalistic than classical tasks or focusing on untrained yet reproducible natural behaviors—has its advantages. I think that we should simultaneously pursue *both* of these approaches. These and additional approaches will be discussed throughout the book, and in particular in chapter 8, which will outline future directions and challenges.

It is noteworthy that many natural behaviors are much less stereotyped than the examples shown in the top-right corner of figure 1.2. Nevertheless, even for such variable behaviors, there are analysis methods that allow for disentangling the behavioral complexity and elucidating what are the neurons responding to. These analysis methods will be discussed in chapter 8.

Is this an antireductionist book? The answer is no. The thrust of the book is to make the point that we need *diversity*: that is, both reductionist and nonreductionist experiments and everything in between. The reductionist approach is very powerful; it allows one to study in great detail the brain circuits and mechanisms underlying certain well-defined and simplified behaviors. We can learn many details about the behavior, the neuronal activity, the circuits involved, and their causal relations. We can elucidate networks, synapses, molecules, and genes. But so long as we keep using behavioral paradigms that are reproducible yet highly simplified and artificial, we will never know if all this mechanistic wealth of detail is at all relevant to real-world behaviors in the real-life natural conditions that the brain evolved to deal with. There are some examples demonstrating that what we learned through the reductionist approach bears little relevance to real-world situations, while in other examples, it has great relevance. This book will discuss both kinds of examples and highlight the urgent need for a major effort by the neuroscience community to explore, in each subfield of brain research, whether the brain behaves similarly under naturalistic conditions as under the standardized reductionist experimental assays.

A full disclosure to the reader is due here about my personal biases and their influence on this book. As stated already, I fully believe that there should be diverse approaches for studying the brain, such that neuroscience research across the globe should collectively tile the entire phase-plane of figure 1.2—and it should tile it uniformly. There is a *continuum* of possible research approaches between highly controlled but unnatural and highly natural but uncontrolled—and there is much to be said in favor of many of the intermediate points, and certainly much to be said in favor of control in general. This is true not only for neuroscience as a field, but for the study of any specific brain system in particular: I believe that the key to unravelling the functions of any brain region is to move back and forth between reductionist, controlled experiments and more natural experiments that focus on nonreduced behaviors. For example, in the neurobiology of navigation, we should study the hippocampal formation in animals performing challenging navigation tasks in complex environments and compare the results to animals tested in simplified setups. We should also use reduced in vitro preparations to better understand circuit mechanisms, generate predictions from the mechanisms to the behavior, and test these predictions in the animal's natural behavior—and vice versa, we should generate predictions from the intact system's behavior back to the underlying mechanisms. A similar diverse, multilevel approach can be envisioned for any brain system.

In other words, I claim that a diversity of approaches is paramount for any and all brain systems and for any kind of animal behavior. However, today there is a massive overrepresentation of approaches that focus on "controlled but artificial" experiments (see the top-left region of figure 1.2), although this state of affairs is starting to change at an accelerating rate. Therefore, I will *not* dwell on defending these classical approaches, important as they are. Instead, my personal bias in this book will be to highlight what is rarely done and to promote what is missing—namely, to try to convince the reader that it is crucially important to conduct naturalistic experiments that cover the right side of figure 1.2 (both top-right and bottom-right). The presentation in this book, therefore, will be tilted toward natural neuroscience—in the hope of swinging the pendulum in the direction of natural behaviors and eventually reaching a new steady state, whereby the field of neuroscience will exhibit a more uniform and diverse coverage of the phase plane in figure 1.2. This diversification is schematized as a 1D cartoon in figure 1.4. The current state of affairs is that the vast majority of neuroscience experiments are conducted in highly controlled and highly artificial

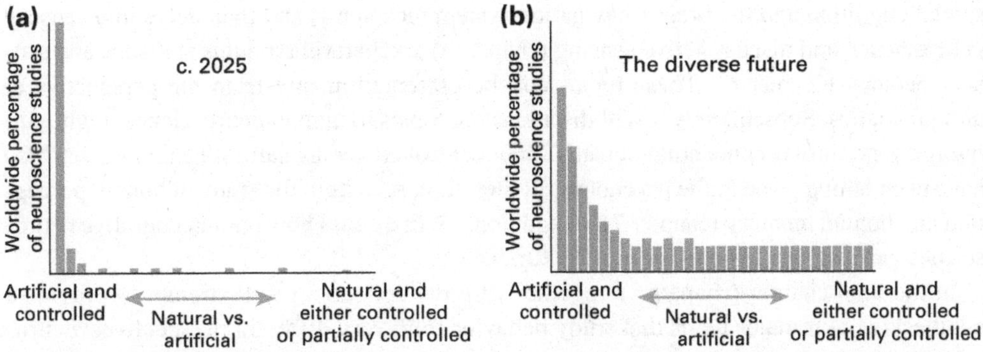

Figure 1.4
Schematic illustration of the fraction of neuroscience experiments conducted worldwide in controlled but artificial setups versus natural but less controlled setups. The x-axis here is roughly equivalent to the diagonal dashed line in figure 1.2 (or to a somewhat higher diagonal line, which goes from the top-left to middle-right of that figure). (a) Until the early 2000s, the vast majority of experiments in neuroscience were conducted in highly controlled setups, which were also highly artificial and impoverished (see the left bin of the histogram). This was still the situation to a large degree circa 2025, although the picture is starting to change. (b) Proposed future diversification in the styles of neuroscience experiments. This book argues that the histogram of experimental styles should become flatter in the future, representing increased diversity—and in particular, many more naturalistic experiments.

setups, as illustrated by the leftmost bin of the histogram in figure 1.4a, while the desired diversification is illustrated by the more uniform histogram in figure 1.4b. I submit that only through a true diversity of approaches do we stand a chance to truly understand brain and behavior.

1.2 The Scope of This Book

This book will strive to develop and promote a new naturalistic approach to conducting research in behavioral neuroscience and sensory neuroscience, with two main aims. The first aim is to point out, using multiple concrete examples of recent data, the importance of studying the brain under as natural conditions as possible. This will be the focus of chapters 3–7, where I will show that, in many cases, the details of natural behavior and natural context can have key importance for understanding both the computational/algorithmic levels of behavior and the governing neural circuits. The second aim is to outline a way forward for elucidating the neural basis of natural behavior. This will be the focus of chapter 8.

More specifically, the structure of the book is as follows. The first three chapters are of an introductory nature: this introduction (chapter 1) is followed by a comparison to related approaches such as neuroethology and comparative neuroscience (chapter 2) and then by a discussion of the recent technological and methodological advances that have enabled the natural neuroscience approach (chapter 3). These new technologies allow for the recording and manipulation of neural circuits in freely behaving, freely moving animals, together with evermore detailed quantification of the complexities of natural behavior. The next four chapters (chapters 4–7) discuss specific topics that provide particularly good examples of how a detailed consideration of the natural world and animals' natural behaviors has fundamentally changed our understanding of the relevant brain functions. These chapters discuss spatial cognition and the brain's navigation system (chapter 4) and then delve into sensory neuroscience and discuss active sensing (chapter 5) and natural stimulus statistics and sensory ecology (chapter 6). These topics will be presented mostly from the perspective of animal studies. Subsequently, I will discuss some topics in human neuroscience, highlighting the very similar conceptual debates about controlled versus natural behaviors that also have been taking place in the psychological literature, such as in the study of human perception and human memory (chapter 7). In addition, I will discuss how human cognitive neuroscience can and should be made more naturalistic.

In the final chapter (chapter 8), together with the epilogue, I will attempt to provide a synthesis of four major fields that study behavior from very different perspectives: neuroscience, psychology, ethology, and behavioral ecology. I will suggest that natural neuroscience should blend these four influences. Specifically, we ought to apply the mechanistic "How?" approach of neuroscience and psychology in order to understand natural behaviors as studied by ethology, which are illuminated by functional "Why?" explanations of behavioral ecology. My hope is to bridge some of the gaps between these four major scientific disciplines, bringing them all closer together and closer to real-life behaviors. Such a four-way combination can generate novel questions and new research directions on the neural basis of behavior, and chapter 8 will outline a few examples of such new research directions. I will finish chapter 8 by presenting a future outlook for natural neuroscience.

As already discussed, the book will focus primarily on the standard mammalian model species in neuroscience, such as mice, rats, and primates. The topics covered will come from behavioral neuroscience (in particular, spatial cognition and the brain's navigation system) and from sensory neuroscience (in particular, active sensing, natural stimuli, and sensory ecology). The selective choice of these topics comes from the fact that these disciplines offer some of the best examples of how our understanding of brain function has deeply changed by applying the natural neuroscience approach—studying the brain and behavior under naturalistic conditions. In sensory neuroscience, both the active-sensing approach and the natural-stimulus approach, while emphasizing different aspects of sensory perception, have suggested that to understand perception, it is crucial to study natural movements and natural sensory stimuli. In behavioral neuroscience, there is a growing realization of the importance of studying complex natural aspects of behavior: this development was facilitated by recent technological advances that enabled neuronal recordings and neuronal manipulations in freely behaving animals (e.g., miniaturized headstages, head-mounted microscopes, wireless electrophysiology systems, and wireless optogenetics), as well as technologies for measuring complex behaviors (such as multicamera systems, microphone arrays, motion sensors, and global positioning system [GPS] data loggers), together with sophisticated algorithms for behavioral data extraction and classification. Importantly, similar approaches can be applied—and in many cases are already being applied—to essentially all subfields of neuroscience. The purpose of this book is to review these directions and discuss their conceptual, empirical, and technological underpinnings with the aim of synthesizing a future research program for this emerging field of natural neuroscience.

The book will not dwell much on classical approaches in neuroscience or on mechanisms. A lot of the focus of neuroscience has been to dig into ever more detailed mechanisms at the circuit, cellular, and molecular levels. The mechanistic "zoom in" approach to neuroscience is certainly important and can unravel basic building blocks of brain function—but my worry is that in some cases, we might have zoomed in too early. In particular, by focusing on unnatural behaviors, we may have been studying the detailed mechanisms of the wrong things. Instead, I propose in this book to zoom out and better understand the basic operations of the brain in more naturalistic scenarios, to elucidate which of the naturalistic details are important and which aren't. Zooming in on detailed mechanisms will then come in due time.

Finally, some clarifications are due. First, the terms "natural" and "natural behaviors" may mean different things to different readers. As already mentioned, in this book it would mean mostly a conjunction of two things: (1) the experiments should be conducted with unrestrained, freely moving, and freely behaving animals; and (2) the behavior itself should bear strong resemblance to the same behavior in the wild. A corollary of the latter requirement is that, in some cases, the experimental setup should be made larger or more enriched/naturalistic than usual to allow animals to exhibit their natural, wild-like behavior. For example, the foraging behavior of animals looks completely different in large versus small environments (chapter 2), and likewise for navigation behavior (chapter 4), social behaviors (chapter 8), and other kinds of behaviors, so often a larger environment is a must. But this does *not* mean that we should necessarily conduct experiments outdoors or reproduce every detail of natural environments indoors—that's not the point. Rather, we should make the environment *as large or as naturalistic as required to allow animals to express fully the wild-like behavior of interest.* Thus, if the animal behaves naturally in a small, simple environment, that's fine. From my perspective, a richer/more naturalistic environment is a means to an end—the end goal is to recapitulate faithfully behaviors expressed by the species in the wild. The important aspect is not the naturalness of the environment, but the naturalness of the behavior.[9]

Second, although this book calls for natural experiments, it does *not* claim that classical, controlled experiments are not useful at all. When we say "useful," the question is: Useful for what? Clearly if we want to understand the basic biophysics of channels or the molecular cascades leading to synaptic transmission, then we should study them in a dish. The natural behavior of animals has little relevance here, in most cases. Similarly, if we want to understand basic neuronal computations in V1 for simple visual stimuli, then it is clear that Hubel and Wiesel's discovery of orientation tuning in V1 was very important.

The key question is: Is that all? For example, is our understanding of orientation tuning useful when going beyond the simple stimuli that were used to characterize it? Again, the question is: Useful for what? If we want to understand neuronal representation of simple stimuli, then yes, it is extremely useful. But if we want to understand neural codes for natural visual scenes, then the answer is arguably no (or perhaps "a little bit useful"). The answer is no both because responses to simple stimuli usually fail to explain responses to natural images (see chapter 6), and because it fails to integrate the importance of self-locomotion and eye movements (chapter 5). Therefore, my claim is that *in addition* to continuing to study basic biophysical and cellular mechanisms and mechanisms of neural responses to simple controlled stimuli, the neuroscience community must start devoting much more effort to studying the effects of natural stimuli and natural behaviors. In the case of V1, we should study neuronal responses to natural visual scenes and natural objects and investigate the dramatic effects of natural body movements and eye movements on visual perception and sensory coding in the visual cortex. Likewise, the same argument goes for every aspect and every subfield of systems neuroscience and behavioral neuroscience—we should always pursue *both* controlled studies and more naturalistic studies, in parallel.

Third, I do *not* claim that we should avoid any experimental manipulations or shun any controls. Yes, I do think that crucial insights about brain function will be attained from neural recordings performed in totally natural experiments, especially if we focus on natural behaviors that the animal is voluntarily repeating many times (i.e., the top-right corner of figure 1.2). But if the animal is *not* repeating its behaviors in a reasonably reproducible manner, then there is a limit on what can be learned. Here comes the importance of performing controlled experimental manipulations—either manipulations of the behavior or direct manipulations of the brain—to elucidate the system according to Marr's three levels of understanding (see chapter 8 for a discussion of these levels). Importantly, however, because of the twelve reasons outlined in this chapter, these manipulations—especially on the behavioral side—should be done more subtly than the dramatic behavioral manipulations used classically in behavioral neuroscience. Throughout the book, and in particular in chapter 8, I will try to outline such naturalistic yet controlled experimental manipulations, which aim to achieve understanding in the classical sense. In fact, I claim that the understanding that will be reached through such natural neuroscience experiments would be much more complete and real.

Fourth, related to the previous point, I do not propose here that we should abandon methodological reductionism—that is, have zero reductionism in our experiments. I believe that for every question, you must employ some level of methodological reductionism. Even when studying natural behaviors, it doesn't make sense to reproduce the animal's environment in full, down to the last twig and gust of wind. The key question is what the "right level of reductionism" is—that is, whether we allow too much reductionism or too little. I posit that within the range of relevant levels of reductionism (between too much and too little reductionism), we need to utilize the entire range—that is, we need *diversity*. We should not do *only* experiments that are on the "very reductionist" end of the scale. Instead, we should use a *broad* range of reductionism levels, including an increased emphasis on "less reduced and more natural."

Finally, promoting diversity of approaches to studying the brain is also related to my belief that neuroscience is in its pre-Newtonian era—namely, that we still lack a grand theory of the brain. We are in a phenomenological era of brain research. On the one hand, this is good because it means that there is a lot to discover about the brain: the future is wide open, and it's a great time for students to become neuroscientists and to join the quest to understand the brain. On the other hand, this is also not so good, because this means that it is unclear which experimental data we should collect. Mature post-Newtonian sciences, like physics, have grand theories that make it relatively easy to understand which data would be important to collect. This is not the case in neuroscience, where we lack such theories. And therefore, my claim is that we should collect *diverse* sets of data, including those collected under natural behaviors, as well as under reductionist conditions. This is paramount because having the right data is crucial—Newton likely couldn't have generated his theory of gravity and celestial mechanics without the right kind of data from Johannes Kepler. If Kepler didn't collect the right kind of data, Newton's efforts likely would have failed. The analogy for neuroscience is that it's crucial that we collect the right kind of data—as Kepler did—for the benefit of our own future Newton of neuroscience. But because we don't know what the "right kind of data" are, we'd better collect diverse data. If we collect only narrow kinds of data, we may find out in retrospect that we have spent the last century collecting the wrong data and elucidating the detailed mechanisms of the wrong phenomena. Diversity is an insurance policy against such an outcome.

My own personal belief, of course, is that "the right kind of data" in brain research means data that are close to the natural behavior of the animal. This belief is motivated by all the twelve arguments surveyed in this introductory chapter, including the *relevance* argument (that natural behaviors are most relevant for the animal) and the *evolution* argument (that brains and neural circuits evolved to generate natural behaviors). I thus hope to convey in this book the importance of studying the neural basis of natural behaviors.

2

Related Approaches: Neuroethology and Comparative Neuroscience

The quest to understand the neural basis of natural behavior is not new: it runs through several disciplines, including neuroethology, comparative and evolutionary neurobiology, and behavioral neuroscience. Here, I will discuss some of these approaches, aiming to put natural neuroscience into historical context. In particular, I will discuss the similarities and differences between these related approaches and the natural neuroscience approach proposed in this book.

2.1 Neuroethology

The neuroethological approach is the most closely related to natural neuroscience. The word *neuroethology* merges two words encompassing very different research traditions: *neurobiology,* with its typical focus on laboratory studies in controlled conditions, and *ethology,* with its emphasis on the study of animal behavior in its natural habitat outdoors. Until the 1970s, neurobiology and ethology differed markedly not only in their research style, but also in the type of questions they asked. Neurobiology usually focused on the "how" questions, seeking a mechanistic understanding of neuronal processing at the cellular and circuit levels, while ethology focused mostly on "why" questions, seeking to elucidate ultimate causes underlying behavior—both functional and evolutionary causes—although some efforts were also directed at understanding mechanisms (Tinbergen 1951; Camhi 1984); see chapter 8 for further discussion of ultimate "why" questions versus proximate "how" questions in neuroscience. The field of neuroethology arose in the 1960s and 1970s from the merger of these two traditions by several researchers who opposed the nearly dichotomous separation of early neurobiology and ethology and sought instead to understand natural animal behavior more broadly, across the different levels—from molecular, cellular, and circuit mechanisms all the way to functional and evolutionary understanding (Camhi 1984; Hoyle 1984; Zupanc 2019).

2.1.1 Achievements of Neuroethology

The focus of neuroethology on understanding the neural basis of natural behaviors has yielded a sequence of amazing discoveries. Rather than providing a comprehensive account of neuroethology (see Camhi 1984; Carew 2004; Zupanc 2019), I will mention here briefly only some of the notable achievements of neuroethology: this list will serve as examples that illustrate the neuroethological approach, which I will then compare to the natural neuroscience approach. Therefore, this list is partial—and it is also biased toward vertebrate rather than invertebrate models of neuroethology, primarily due to my own background and expertise in vertebrates.

The birdsong system provides one major example of neuroethological studies. Birdsong was originally studied behaviorally by ethologists, and then in the 1970s, Mark Konishi and Fernando Nottebohm, two pioneers of neuroethology, showed that birdsong is learned (Konishi), and identified a set of brain nuclei that are responsible for song production and song learning (Nottebohm). Juvenile male birds learn their song from a tutor—usually the father—in a process of *vocal learning* (Bolhuis & Gahr 2006; Zeigler & Marler 2012). Very few vertebrates apart from humans show true vocal learning—the ability to learn new vocalizations on the basis of experience (e.g., through imitating others)—and hence songbirds became a powerful model for vocal learning in vertebrates. The process of vocal learning occurs in several stages. In zebra finches, for example, the bird has to initially hear his father's song repetitively—if he is isolated during this critical period, then he will develop an aberrant song. Subsequently, he starts practicing his song; at first, it is imperfect ("babbling"), but over time, it crystallizes into a perfect adult song, which is similar to (but also distinct from) his father's. This process of practicing and crystallization does not require the presence of the father (Bolhuis & Gahr 2006; Zeigler & Marler 2012). Thus, by listening to his father early on, the juvenile zebra finch creates a template of the father's song in his brain, and then subsequently, he starts to practice singing, trying to match his performance to the memorized song-template. The process of singing and song learning involves a whole set of specialized brain nuclei (Bolhuis & Gahr 2006), some of which contain auditory neurons that exhibit exquisite selectivity to the bird's own song (Doupe & Konishi 1991; Margoliash 1997; Doupe et al. 2004)—one of the most striking examples of stimulus selectivity in any brain system.

In the 1980s, Fernando Nottebohm showed that the birdsong system exhibits *neurogenesis* (formation of new neurons) in juvenile birds that are learning to sing, which possibly contributes to song learning. Importantly, Nottebohm also demonstrated *adult neurogenesis*—formation of new neurons in adult animals—particularly in canaries that continue to learn new songs throughout adulthood (Nottebohm & Liu 2010). Nottebohm's discovery of adult neurogenesis came at a time when the dogma in neuroscience held that adult neurogenesis does not exist in the mammalian brain. Its discovery in the birdsong system paved the way in the 1990s for the acceptance of the idea that adult neurogenesis does in fact occur in mammals: in the dentate gyrus, subventricular zone, and other brain regions (Gross 2009); however, it is noteworthy that while adult neurogenesis is very prominent in rodents, it is much less prominent in humans, which has led to controversy over the functional significance of adult neurogenesis across mammals (Spalding et al. 2013; Sorrells et al. 2018).

Another striking contribution of neuroethological research in songbirds was in the field of motor sequence generation: Michale Fee and colleagues have demonstrated that neurons in the high vocal center (HVC) of zebra finches burst in an ultrasparse yet highly precise manner, with each neuron firing just one burst of spikes throughout the duration of the song, at a very precise timing, such that different neurons fire sequentially, one after the other (Hahnloser et al. 2002). Subsequent experiments, using both temperature-cooling and intracellular recordings in singing birds, provided evidence that this precise **temporal code** is consistent with a synaptic chain model of sequence generation—a theoretical neural-network model that was proposed in the 1980s as a mechanism for generating precisely timed sequences (Abeles 1982; Long & Fee 2008; Long et al. 2010). Thus, the study of the birdsong system provided crucial pioneering discoveries in several domains of neuroscience: vocal learning, neurogenesis, sequence generation, and neural coding.

Barn owls provide another key example of neuroethological studies. Barn owls are night hunters that rely on both hearing and vision to hunt their prey—primarily rodents—and they have an outstanding capability for auditory localization of sound sources in space (Payne 1971), as well as good visual acuity (Wathey & Pettigrew 1989). In a series of studies, Eric Knudsen and Mark Konishi made a number of important discoveries on the neural basis of the barn owl's ability to localize prey. First, they found that two regions in the owl's midbrain—the external nucleus of the inferior colliculus (IC) and the optic tectum (the bird homolog of the mammalian superior colliculus)—contain an ordered map of auditory space in head coordinates, with different directions in space being mapped topographically onto the brain surface (Knudsen & Konishi 1978; Knudsen et al. 1987). This map is a *computational map*: it is not inherited from the sensory periphery (the cochlea), where there is in fact no information about sound direction, and hence this map has to be computed in the brain (Knudsen et al. 1987). Catherine Carr and Mark Konishi then showed that the azimuth of the auditory source is computed primarily through a calculation of the time difference of sound arrival to the two ears (the "interaural time difference"), which is performed by a set of delay lines and coincidence detectors in the nucleus laminaris of the barn owl (Carr & Konishi 1988, 1990). Interestingly, this experimental finding corresponded precisely to an early theoretical model of sound localization, the Jeffress model, which also relied on delay lines and coincidence detectors (Jeffress 1948). Second, Knudsen and Michael Brainard demonstrated that there is a visual map and an auditory map in the optic tectum, and these two maps are coregistered: They experimentally shifted the visual map of young barn owls by rearing them with prisms placed on their two eyes, and showed that this leads to a subsequent slow shift of the auditory map, whereby after several weeks the auditory map comes into register again with the visual map (Knudsen & Brainard 1991). This landmark paper was among the studies that laid the foundation for modern research on multisensory integration in vertebrates (Knudsen & Brainard 1995).

Bats are another classical model in neuroethology—in particular, their sensory system of echolocation, or biosonar. Echolocation is an example of an *active sensing* system, in which the animal produces energy or movement to actively collect information from the environment. Specifically, echolocating bats emit brief acoustic calls, typically at ultrasonic frequencies, and from the timing of the returning echoes, they estimate the distance to objects in their environment (Griffin 1958; Schnitzler et al. 2003; Jones 2005; Ulanovsky & Moss 2008). They can also compute precisely the direction of objects, assess their identity, and even discriminate the surface properties of objects (e.g., texture) based on the echoes. Further, when an insect-eating bat attacks an insect, it increases the rate of echolocation calls (i.e., increases the sensory acquisition rate), and shortens the duration of its calls (to avoid masking of the incoming echo by the outgoing call). In the 1970s, Nobuo Suga reasoned that the bat therefore provides a very useful model to elucidate auditory processing because unlike generalist species such as rats or cats, where it is unclear what the auditory cortex is trying to compute, in the bat we do in fact know at least one essential sound stimulus that is particularly important for the animal's natural behavior and for its survival—namely, the bat's own echolocation calls and the ensuing echoes. Therefore, Suga decided to move away from simple, artificial stimuli, such as pure tones or broadband white noise—the classical sounds that are commonly used to this day in auditory neuroscience, but bear little behavioral relevance for bats. Instead, Suga started using bat echolocation calls and their echoes to dissect the bat's auditory system. What Suga then discovered was quite striking. First, he found neurons that were matched exactly to the behavior of the animal in exquisite ways. Specifically, he discovered delay-tuned neurons, which were strongly tuned to a particular pulse-echo delay—these neurons may thus represent the physical distance to the object (Suga et al. 1978; O'Neill & Suga 1979; Suga & O'Neill 1979). Moreover, Suga found neurons that specifically preferred sequences of pulse-echo pairs in which the time delay between pulse and echo became progressively shorter, matching the behavioral increase in call rate and decrease in time delay as a bat is tracking and approaching an insect outdoors. Suga called these cells "tracking neurons" (O'Neill & Suga 1979). These neurons thus provide an amazing matching of sensory processing in the brain to the motor behavior of the animal—the first such dramatic example of sensorimotor matching in any mammalian species. Second, Suga discovered a series of auditory cortical areas (fields) that were specialized for different functions (Suga 1989, 1990; Ulanovsky & Moss 2008). One cortical area contained an abundance of delay-tuned neurons, which specialize in measuring time differences between pulse and echo, thus representing the distance to the target. Moreover, the preferred distances of these neurons were mapped progressively along the surface of the brain. Another area contained neurons that specialized in measuring frequency differences between pulse and echo, thus representing the flight velocity of the bat relative to the target based on the Doppler effect. (The Doppler effect is the change in sound frequency due to movement, and it is proportional to the movement velocity—which allows directly inferring the movement velocity from the frequency change). Here again, the preferred velocities were mapped progressively along the surface of the brain. Both of these functional anatomical maps—the target-distance map and the target-velocity map—are computational maps, which have to be computed in the brain and are not inherited in any simple way from the animal's sensory periphery.[1] The discovery of these specialized auditory areas, with each area computing a single behaviorally relevant variable, was the first demonstration of such sensory cortical specialization in any mammalian species (two specialized areas in the primate visual cortex—the movement-selective medial temporal [MT] area and the face-selective areas in the inferior temporal [IT] cortex—were discovered at about the same time). I will discuss bat echolocation further in chapter 5.

Weakly electric fish are another classical neuroethological model of active sensing. These fish inhabit murky rivers in the tropical regions of Africa and South America, where the sense of vision is very limited. They produce weak electric fields, and by measuring the received distortions of the field via electroreceptors on their skin, the fish can estimate the presence and location of tiny objects in the water, such as small prey (Heiligenberg 1977, 1991b). This remarkable sensory system is called *electrolocation*, and it is mediated through the fish's electrosensory system. In particular, electrolocation was studied extensively in relation to a behavior called *jamming avoidance response*—the change in frequency of the electric fields produced by one fish when another fish is approaching. This frequency shift is thought to minimize the jamming (interference) of one fish's electrolocation by the electrical signals of the other fish. The jamming avoidance response is perhaps the best-understood sensorimotor behavior in any vertebrate, with decades of mechanistic research at the cellular and systems levels, conducted across all the brain regions involved in this behavior (Heiligenberg 1991b; Krahe & Maler 2014). The study of the electrolocation system has provided key insights into the mechanisms by which the fish can detect the small signals from its prey on the background of the large electric fields that the fish itself is generating. It was found that cerebellum-like structures in the fish brain create a "negative image" that allows subtracting the large yet predictable self-generated electric fields of the fish, which then enables the fish to detect the very small external sensory inputs from the environment (Bell et al. 1997, 2008; Requarth & Sawtell 2011). A similar kind of adaptive subtraction was subsequently demonstrated in the auditory system of mice, where a cerebellum-like structure, the dorsal cochlear nucleus, allows the animal to hear external sounds over the background of self-generated sounds (Singla et al. 2017); and similar results were found in the mouse auditory cortex (Schneider et al. 2018). This is a characteristic example of neuroethology, where the study of one system in one species (electrolocation in fish) may inform the understanding of another brain system in another species (hearing in mice). This cross-system transfer of knowledge is yet another reason why neuroscience research needs to have more diversity—diversity of behaviors, diversity of brain systems, and diversity of model species. Walter Heiligenberg, who pioneered the study of the circuit mechanisms of electrolocation, stressed the importance of studying the neural circuits of behavior across multiple species. He emphasized how exquisitely skilled these fish are in detecting small prey, and in particular, how accurate they are in detecting tiny temporal deviations in the incoming signal, which makes these fish a champion species for the study of temporal coding, active sensing, and sensorimotor behaviors (Heiligenberg 1991a, 1991b). I will come back to the concept of "champion species" later in this section.

Aplysia californica, the Californian sea slug, is a marine invertebrate that provided invaluable insights into the cellular mechanisms of learning and memory. Eric Kandel chose to study *Aplysia* because it exhibits a number of simple yet reproducible natural behaviors, such as the gill withdrawal reflex, which undergo learning. These behaviors include nonassociative learning (habituation and sensitization) and associative learning (classical conditioning). In addition, as is the case for many other invertebrates, *Aplysia* offers the distinct advantage that it has a small number of neurons (about 20,000 in total), many of which are very large, permitting easy electrophysiological recordings. Further, quite a few of these neurons are *identified neurons*, which recur precisely from individual to individual, occupying the same node in the neural network across individuals. Kandel and coworkers have elucidated the cellular and molecular bases of these forms of learning (Carew 2004; Squire & Kandel 2008). For example, the sensitization behavior is driven by serotonin-mediated *synaptic facilitation* (short-term plasticity that enhances synaptic strength), which involves the activation of a cyclic adenosine monophosphate (cAMP) cascade in the sensory neuron terminals. Classical conditioning is driven mostly by mechanisms that are similar to sensitization, but with an additional component of activity-dependent presynaptic facilitation, which acts effectively as a coincidence detector between the unconditioned and conditioned stimuli. *Long-term memory* of sensitization was shown to require protein synthesis, which is elicited by cAMP that activates transcription factors such as CREB in the sensory neurons, which in turn activates a cascade of immediate early genes and late-response genes, that ultimately produce proteins that are shipped to the synapses of the sensory neurons, where these proteins cause the structural and functional synaptic changes that underlie long-term memory. Thus, the study of *Aplysia* provided the first rather complete cellular and molecular elucidation of learning and memory mechanisms.

Finally, John O'Keefe's discovery of place cells in the rat hippocampus in the early 1970s was largely a neuroethological endeavor. In the 1960s, several researchers recorded from the hippocampuses of rats that were performing highly artificial, constrained, well-controlled tasks, but these researchers were unable to elucidate a coherent view of what the hippocampal neurons were doing during these artificial experiments (O'Keefe & Nadel 1978). Then came O'Keefe, who employed a very different approach. He reasoned that it is better to let the animal exhibit its natural behavior, and since rats "love" exploring space, O'Keefe simply let them perform curiosity-driven spatial exploration. This led to the discovery of hippocampal place cells (O'Keefe & Dostrovsky 1971; O'Keefe & Nadel 1978). O'Keefe called his approach the "neuroethological approach," and he contrasted it with the more artificial "neuropsychological approach" that preceded his experiments. Here is O'Keefe and Nadel's account of the discovery of place cells (O'Keefe & Nadel 1978, pp. 190–195):

> There are two fundamentally different approaches to the study of single-unit activity in the freely moving animal. . . . The first can be called *neuropsychological* and the second *neuroethological*. The neuropsychological approach draws its inspiration from the methodology of behavioral psychology. . . . The experimenter addresses himself to a limited number of preordained questions (usually one) about unit activity, and seeks to design a well-controlled study which will answer them. . . . The advantages of the neuropsychological approach are clear: it affords a degree of control over the experimental situation which will eventually be necessary if the laws governing the operation of the nervous system are to be discovered. Its drawbacks result from its premature application. When we first venture into the unknown, we need not the incisive beam of the proud penetrating laser, but the gentle diffuse illumination of the humble torch. There will be plenty of time later for detailed investigation of the nooks and crannies; first we must find the mountains. . . . The neuroethological approach differs from the neuropsychological one in several respects. First, it seeks to study the activity of single units in as naturalistic a setting as possible, in the belief that an animal's behavior in its natural environment maximizes the possibility of producing changes in unit activity that are meaningfully related to that unit's function. . . . A second principle of neuroethology is that, unlike the neuropsychological approach, it seeks not to ask a preordained set of questions about a population of units, but instead treats each unit as an individual and attempts to describe the full range of its behavior in as many different situations as possible. . . . The strategy used in our own work on single-unit activity in the hippocampus of the freely moving rat leans toward the neuroethological, rather than the neuropsychological, approach.

O'Keefe's discovery of place cells in 1971 laid the foundation for the neuroscientific study of animal navigation (chapter 4). More generally, this was likely the first example of a major discovery about brain function that was revealed by letting animals move and behave freely. In this sense, the discovery of place cells can be thought of as an early forerunner of the natural neuroscience approach advocated here. Indeed, O'Keefe and Nadel's description of the "neuroethological approach" bears similarity to the approach described in this book. However, as will be discussed in chapter 4, most studies of hippocampal place cells still remain far removed from truly natural navigation behaviors. In fact, once researchers started to test animals in more complex and natural situations, it became clear that the neural codes for space are very different in complex, large-scale environments (chapter 4). Moreover, hippocampal neurons turned out to encode more than just space. Thus, O'Keefe's study was an extremely important step in the direction of natural neuroscience—but only the first step.

The same can be said more generally about most neuroethological models, which, as I will argue next, rarely went far enough in their examination of the neural basis of truly natural behaviors. Most neuroethological studies were done either in vitro, under anesthesia, or using highly constrained behaviors. In what follows, I will discuss these limitations and also describe the fundamental differences between the scope of neuroethology and the scope of the natural neuroscience approach proposed here.

2.1.2 Neuroethology versus Natural Neuroscience

Neuroethology has had several historical characteristics. First, most neuroethology research focused on a "tandem approach" (Camhi 1984): studying the natural behaviors of animals in the field to characterize their behaviors in detail and get inspiration for laboratory studies, and then conducting laboratory studies in reduced preparations (e.g., under anesthesia or head fixation, or even in vitro). Neuroethologists thus take very seriously the animal's natural behaviors and study them in detail, which is a great advantage. However, this approach has the disadvantage that most neuroethologists do *not* record brain activity during the *actual* execution of natural behaviors by freely moving animals (with a few notable exceptions, such as birdsong, O'Keefe's place cells, and more recently, studies in bats). This sharp separation into two phases of research—a natural in vivo approach in freely behaving animals for studying behavior, and a separate phase of using reduced preparations for studying the neural basis of that behavior—primarily arose from the technical difficulty of recording and manipulating neural activity in the brain while the animal was actually performing its natural behavior. However, these technical challenges have largely been overcome in recent years, as will be detailed in chapter 3: today, it is possible to record and manipulate neurons in many animal species while they are actually freely moving and behaving. This "simultaneous approach" (as opposed to the tandem approach) was enabled by technological advances in miniaturization of electronics and optics, which now allow conducting such dream experiments in ever-smaller species. In vertebrates, these species include mice, bats, small birds, lizards, and fish, while in invertebrates, such experiments are being performed in octopus and even in some large insects. However, most neuroethologists today are still adhering to the tandem approach, carrying out behavioral experiments and observations on whole animals in the field, under natural conditions, and then studying reduced preparations in the lab.

One example of the tandem approach is the study of bat echolocation. Neuroethologists have studied for years the complex echolocation behaviors of flying bats in the field; but then, when turning to laboratory experiments, they *passively* presented *simulated* pairs of pulses and echoes to head-fixed bats while these restrained bats were *not* actually echolocating. In fact, all the fascinating findings described above on the neural processing of bat echolocation were measured in head-fixed, passively listening bats that did not emit any echolocation calls, except for only three studies that recorded auditory neurons in vocalizing bats (Kawasaki et al. 1988; Kothari et al. 2018; Wohlgemuth et al. 2018). Similarly, when studying auditory localization in barn owls, neural responses have been recorded in the brains of head-fixed owls. Thus, during recordings, the owl was not *actually* performing the highly precise orienting response, which was the reason why barn owls were chosen for studies of auditory localization in the first place. Such examples of a tandem approach abound across neuroethological animal models and brain systems. This current state of affairs in neuroethology, whereby the neural basis of behavior is studied mostly in nonbehaving animals, is clearly unsatisfactory. In fact, the same concerns and arguments that were raised in chapter 1 against overcontrolled experiments in neuroscience in general are directly applicable also to neuroethology—namely, that it is highly limiting to study the brain of any animal under reduced, overcontrolled laboratory conditions. Instead, neuroethology should be expanded to incorporate the principle of natural neuroscience—studying brain mechanisms in animals that are *actually* performing complex natural behaviors.

Despite this critique of the tandem approach, it is also crucial to highlight the importance of behavioral field studies, which have always been the first step in any neuroethological study. Fieldwork reveals to the researcher the true richness of the behavior at hand, and it uncovers the full natural context in which this behavior is displayed, including the environmental problems that it may have evolved to solve, the possible relations to other behaviors, and the interactions with other individuals. This focus on behavior is a major advantage of the neuroethological approach, and it stands in stark contrast to mainstream neuroscience, where quite often there is very little known about how the behavior at hand is expressed in the real world. For example, studies of place cells in rats have suffered from the absence of such fieldwork. Compared to the enormous amount of laboratory work on spatial cells in the hippocampal formation carried out in the past fifty years since place cells were first discovered (O'Keefe & Dostrovsky 1971), there is very little known about natural navigation outdoors by wild rats and mice. This lacuna is partly due to the technical difficulty of tracking rodents outdoors. However, this difficulty has been recently overcome by new technological developments, such as miniaturized GPS data loggers and other tracking devices, which now allow ameliorating this situation (see chapters 3 and 4).

Second, as already noted in chapter 1, another historical characteristic of neuroethology was the focus on studying innate *fixed action patterns,* which are relatively simple, stereotyped behavioral sequences that are reliably elicited by a particular stimulus, called a *sign stimulus* (Camhi 1984; Hoyle 1984). Focusing on innate fixed action patterns enabled studying natural behaviors that are nevertheless highly reproducible, allowing internally generated experimental control of the animal's behavior without the need to enforce control by the experimenter externally. This in turn allowed the double benefit of studying natural yet controlled behaviors (figure 1.2, top-right quadrant). However, not all animal behaviors are innate, nor are they fixed action patterns—in fact, many behaviors are neither. Thus, if one wishes to understand the neural basis of behavior more broadly—including complex learned behaviors and behaviors that require complex cognitive processing—it is simply too restrictive to focus solely on fixed action patterns. Research must be conducted on the entire rich spectrum of animal behaviors. And indeed, many neuroethologists have started moving in these directions in recent years.

Third, neuroethologists have often focused on expert species (sometimes called "champion species")—animals that have superior sensory or motor capabilities that may yield advantages in elucidating the neural basis of these capabilities. This concept was advocated by Walter Heiligenberg, who studied one such champion species, the weakly electric fish *Eigenmannia virescens*. These fish exhibit an extraordinary ability to detect temporal differences between outgoing and incoming electrical signals. These temporal differences are behaviorally important because they indicate interference from the presence of another fish. Heiligenberg's dictum concerning the use of expert champion species had the following rationale. He suggested that those species that have superior sensory or motor capabilities evolved neural solutions that are more conspicuous or more optimal than in nonexpert species, and this would allow better dissection of the neural basis of that behavior (Heiligenberg 1991a). For example, monkeys have exquisite visual acuity, making them an excellent model for studying visual perception (and, of course, monkeys are evolutionarily close to humans, which is another motivation to study them). Electric fish and barn owls have exquisite capabilities in measuring time differences (in their electrolocation and auditory systems, respectively), making them excellent models for studying temporal coding in the brain (Heiligenberg 1991a). Another advantage of using expert species is the high reproducibility of their behaviors, which makes it easier to reach the natural-and-controlled corner of figure 1.2 (top-right corner).

However, a critique of the "champion species" approach is the question of *generalizability*. The neural structures that these expert species have evolved are often highly specialized. On the one hand, this proved to be a key advantage: perhaps counterintuitively, the brains of some expert animals (e.g., HVC in birds, V1 in monkeys and cats, and the barrel cortex of rodents) turned out to be relatively easy to understand, even though the hardware is presumably much more complex than the simpler brains of nonexpert animals. The biology seems to be much clearer in expert brains than in nonexpert brains. This is why studies on cats and monkeys have revolutionized our understanding of vision since the work of Hubel and Wiesel (1959, 1962, 1968)—while research on mouse visual cortex has not led to anything remotely as revolutionary (even though the methodologies applied in mice are much more powerful). Paradoxically, mouse visual cortex is harder to understand than monkey visual cortex. On the other hand, it remains unclear how well the insights gleaned from these expert species can be generalized to nonexpert species. In some cases, it does generalize. For example, in bats—a species expert in navigation (Yovel & Ulanovsky 2017)—the neuronal representation of small-scale spaces by hippocampal place cells and entorhinal grid cells is very similar to rodents (Las & Ulanovsky 2014; Geva-Sagiv et al. 2015). In other cases, it does not generalize. For example, in barn owls, an expert species in sound localization (as previously discussed), the mechanism of computing sound direction, which is based on arrays of delay lines and coincidence detectors (Carr & Konishi 1990), turned out to be very different from mammals, where the computation of sound direction is based on precisely timed inhibition (Brand et al. 2002). And in the case of the visual cortex, the picture is mixed: some features do generalize between monkeys and mice, and some do not (see more on this shortly). Thus, the question of cross-species generalizability is complicated and seems to vary between brain systems. Of course, if we are interested in understanding the wonderfully interesting systems of electrolocation in fish or sound localization in barn owls—understanding them in their own right without looking for any cross-species generalizations—then the generalizability critique is irrelevant. We clearly cannot study electrolocation in mice. Further, one may argue that evolution has acted to find unique solutions in different species, so looking for generalizations is meaningless. However, research shows that in some cases, neural solutions *do* generalize across species, as in the above-mentioned case of hippocampal place cells, which are found across mammalian species. Therefore, if we want to look for general principles across species, then we must study multiple species. Just as we should not study only one species of a standard animal model (e.g., the mouse), we should not study only expert champion species. Both of these extremes yield a view of brain function that may be too narrow, limited by the specifics of a particular species. Instead, we should diversify and study a *variety* of species—experts and nonexperts, and everything in between.

Fourth, neuroethologists often focused on brain systems and animal models that provide easy experimental accessibility, such as invertebrate animal models that have particularly large neurons that are easy to record from[2] or vertebrate systems that have orderly anatomical map organization (Knudsen et al. 1987; Fitzpatrick & Ulanovsky 2014). However, many interesting and important behaviors are not necessarily governed by very large neurons or by highly organized circuits, but we would still like to understand them. As Jeffrey Camhi wrote in his classic *Neuroethology* textbook: "And yet, by focusing on neuroanatomical order, large neurons, and stereotypic movements, neuroethologists have tended towards behaviors that, to most ethologists, are just too simple to be really interesting" (Camhi 1984, p. 375). However, recent years have seen a movement toward studying the neural basis of more complex behaviors, in both neuroethology and mainstream neuroscience.

This book proposes to conduct neural recordings and circuit manipulations of *complex natural behaviors*—both learned behaviors and innate ones—and crucially, these recordings and manipulations must be performed while the animal is unrestricted and *actually* performing those natural behaviors. To this, I would add that we must also simultaneously record from *large populations of neurons* in the freely behaving animal, for three reasons. First, the brain operates at the network level, so it is evidently important to understand circuit dynamics and collective behaviors of neuronal ensembles beyond the single neuron (Mante et al. 2013; Pfeiffer & Foster 2013; Rigotti et al. 2013; Rubin et al. 2019; Gardner et al. 2022). Second, population recordings arguably are a better way to understand the level of *neural computations,* which in turn builds a crucial bridge between the level of biological circuits, cells and synapses and the level of animal behavior (Carandini 2012). Third, as suggested by the theoretical work of Gao and Ganguli (Gao & Ganguli 2015; Gao et al. 2017), already mentioned in chapter 1, more complex behaviors require a larger number of recorded neurons to correctly describe the dynamics of the neuronal population (see also chapter 1, endnote 5). This may indicate a fundamental need for recording larger numbers of neurons when conducting more naturalistic experiments with complex behaviors. Therefore, we should strive to develop a combination of large-scale recordings of *thousands of neurons simultaneously* (a point to which I will return in chapter 3), while the animal performs *truly complex natural behaviors* (described in chapters 3–8). In addition, we must develop *new tools for data analysis,* which will allow understanding single-neuron selectivity and ensemble activity during natural behavior (chapter 8). This combination will facilitate understanding the neural basis of behavior in its full richness: the richness of the neuronal mechanisms, circuit dynamics, and the richness of behavior.[3]

2.2 Comparative Systems Neuroscience

An approach related to neuroethology is that of *comparative neuroscience*. Indeed, as already discussed, neuroethological research is done on a variety of animal species. But the comparative approach is not limited to neuroethological investigations. In fact, the entire history of neuroscience is, to a large degree, a history of the diverse study of nonstandard organisms (Marder 2002; Brenowitz & Zakon 2015; Yartsev 2017; Laurent 2020). Some of these organisms are shown in figure 2.1.

Thus, Lord Adrian discovered the action potential by recording from the nerve fibers of frogs. Hodgkin and Huxley elucidated the mechanism of the action potential by studying squid. Eccles and Katz elucidated synaptic transmission by studying frogs. Hartline discovered lateral inhibition in the visual system of the horseshoe crab *Limulus*. Wilson found the first central pattern generator (CPG) in a locust. Kandel elucidated some of the basic cellular mechanisms of learning and memory in *Aplysia*. Young has shed light on the molecular and neural bases of social preference by studying voles. Hubel and Wiesel discovered orientation selectivity in the visual cortex of cats. And O'Keefe and the Mosers discovered place cells and grid cells in rats (Adrian 1928; Hodgkin & Huxley 1952; Hartline et al. 1956; Hubel & Wiesel 1959; Wilson 1961; Eccles 1964; O'Keefe & Dostrovsky 1971; Kandel & Schwartz 1982; Lim et al. 2004; Hafting et al. 2005). To this, we can add the discoveries, in diverse species, of some key molecules that we now know are crucial for brain development and brain function, including the discovery of nerve growth factor in chicks by Rita Levi-Montalcini, and the discovery of the inhibitory neurotransmitter gamma-aminobutyric acid (GABA) in crabs by Ed Kravitz—as well as the discovery of a whole host of useful molecular tools, from green fluorescent protein (GFP) in jellyfish to channelrhodopsins in algae (Kravitz et al. 1962; Levi-Montalcini 1998; Nagel et al. 2003; Shimomura 2006).

Why did neuroscientists and biologists study so many species? The answer is threefold. First, curiosity-driven research leads scientists to explore the wonderful diversity of the natural world, simply for the sake of understanding nature in all its richness. Second, since variability across species seems to be the rule rather than the exception—including both variability in brains and in behaviors—it is of fundamental biological importance to understand mechanistically how variability in brain structure and function underlie variability in behaviors across species. Third, another reason for studying diverse species comes from a key principle of biological research—one that was too often forgotten in recent years in mainstream neuroscience—the so-called *Krogh's principle*. This idea was proposed by Danish physiologist August Krogh, who wrote, "For such a large number of problems there will be some animal of choice or a few such animals on which it can be most conveniently studied" (Krogh 1929). In other words, Krogh's principle proposes to choose the model organism based on the biological problem rather than choosing the problem based on the model organism. Thus, Hodgkin and Huxley used the squid's giant axon because of its large diameter, which allowed inserting a macrowire into the axon to create voltage clamp and space clamp, which was essential for their experiments. Hartline studied the visual system

Figure 2.1
Some of the species that were used in neuroscience research since Lord Adrian discovered the action potential in the 1920s using the frog's nerve-muscle preparation.

of the evolutionarily ancient horseshoe crab, *Limulus,* because horseshoe crabs have long optic nerves that can be split to record from individual fibers. Moreover, the visual system of horseshoe crabs is relatively simple, with a compound eye whose optical units (ommatidia) each possesses a single photoreceptor or a small number of photoreceptors, and a single output neuron—an organization that eventually facilitated the discovery of lateral inhibition. Kandel studied *Aplysia* because of the size of its neurons, many of which are also identified (i.e., they can be located repeatedly among individuals). Moreover, these animals display very robust behaviors, in particular highly reproducible forms of learning and memory, such as sensitization and classical conditioning of the gill withdrawal reflex. O'Keefe chose to study rats to elucidate hippocampal function despite the fact that in the 1960s and 1970s, other species, such as rabbits, were equally popular for hippocampal studies. He chose rats primarily for a behavioral reason: namely, that rats are curious and like to explore space. All these neuroscientists specifically chose the species that was best suited to study the biological question that interested them. That is, they followed Krogh's principle.

Today, however, a typical study of the neural basis of behavior would be conducted on a standard inbred (i.e., domesticated) strain of the laboratory mouse, performing a very simple behavioral assay in an impoverished laboratory setup.[4] The focus of this book is to propose that we should go beyond the highly controlled laboratory setup, to more natural behaviors, even if we are studying mice. However, it is still worthwhile to consider how it came to be that a few inbred mouse strains became the dominant model species in neuroscience—and what price we might be paying for this.

2.2.1 Inbred Mouse Strains: Benefits and Limitations

The most common laboratory mouse strain, C57BL, was developed in 1921 by Clarence Little from female no. 57 on Abbie Lathrop's farm (Zilkha et al. 2016). This and other strains—all of which were derived from a small number of founders—have become increasingly popular models for laboratory studies in biology and neuroscience, for several reasons. First, mice are small, easy to maintain, mature quickly, and reproduce rapidly with large litters, making them a convenient species for the confined spaces of research laboratories. Second, there is a rich body of biological knowledge about mice, which serves as a platform for formulating new research questions using mice—or using any standard model organism, for that matter—which then drives a rapid increase in the corpus of knowledge about the mouse. This in turn encourages the further use of mice. Third, an ever-growing arsenal of molecular tools developed for the mouse have enabled detailed studies of biological mechanisms, which for many years were impossible in other mammalian species. Fourth, breeding of specific strains decreased the genetic variability between individuals, which was considered beneficial because it reduced the "noise" (i.e., variability) inherent in wild populations, thus allowing better statistics.

The first two reasons—convenience and knowledge base—are certainly genuine advantages, which ensure that the mouse will remain the dominant mammalian model in the foreseeable future. The third reason—molecular tools—while historically a key advantage, has gradually eroded in recent years with the advent of virus-based gene targeting methods, as well as CRISPR-based gene-editing methods, which allow editing the genomes of a wide variety of nonstandard organisms (Heidenreich & Zhang 2016; Lander 2016). These virus-based gene-targeting methods and revolutionary gene-editing techniques will likely encourage an increase in the fraction of neuroscientific studies of nonstandard organisms (Brenowitz & Zakon 2015), thus somewhat eroding the dominance of the mouse model. The fourth reason—the improved experimental control and increased statistical power of inbred strains—has some important merits, but also major drawbacks. It is useful to consider this issue through the statistical notions of variance and bias: studying inbred mouse strains reduces the variance, but at the cost of introducing large systematic biases—most notably biases in behavior—which makes such mouse strains particularly worrisome for neuroscientists interested in the neural bases of behavior. Behavioral traits became highly biased in inbred mouse strains because the early mouse breeders selected the animals for particular behavioral phenotypes—as will be explained shortly—thus creating an accelerated evolution of behavior.[5]

Social behaviors constitute one class of behaviors that has become highly biased by the domestication and artificial inbreeding of laboratory mice (Jakubowski & Terkel 1982; Zilkha et al. 2016). This is very understandable, as one of the first tasks of the early mouse breeders was to get rid of the natural avoidance of wild animals to mate with close relatives, which led to the disappearance of the natural social behavior of inbreeding avoidance (Sherborne et al. 2007). Likewise, selective breeding led to the reduction of aggression in laboratory mice, which nowadays are much more docile than wild mice (Blanchard et al. 1998). This difference is particularly strong in female mice: whereas laboratory females never attack intruders, it has been shown that both wild and wild-derived female mice exhibit dramatic rates of adult-directed attacks and pup-directed attacks (figures 2.2a, left; and 2.2b, left) (Chalfin et al. 2014). In fact, the dogma that female mice are much more

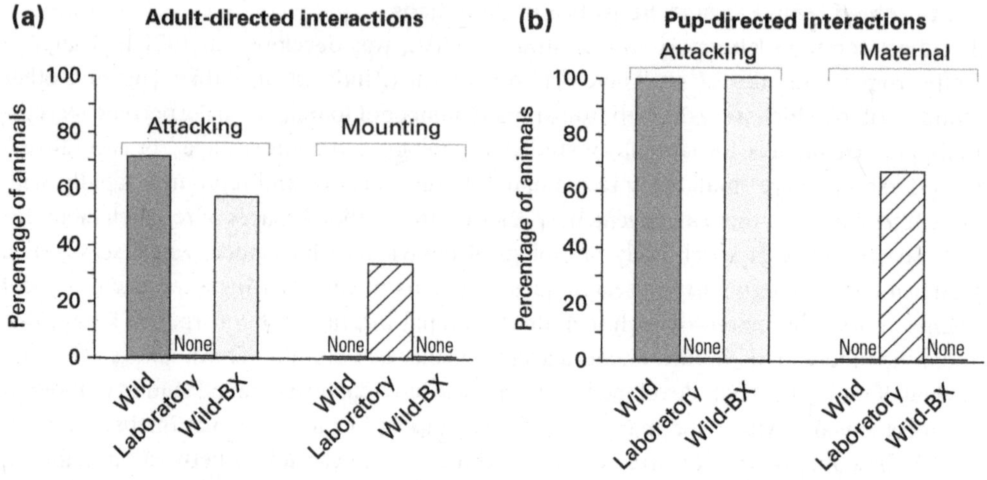

Figure 2.2
Social behaviors differ dramatically between laboratory mice and wild mice (as well as wild-backcrossed mice, Wild-BX). (a) Adult-directed social interactions. Left, attacks by females on alien intruder adult females; right, same-sex mounting. (b) Pup-directed social interactions. Left, attacks by adult females on alien intruder pups; right, spontaneous maternal behaviors toward alien intruder pups. Data replotted from Chalfin et al. (2014). Laboratory mice were from the C57 × 129 strain. Note that in all cases, the wild mice and wild-backcrossed mice exhibited similar social behaviors, but both behaved completely differently from laboratory mice. Thus, social behaviors in this standard laboratory mouse strain, C57 × 129, are not representative of the natural social behaviors of wild mice—let alone of other mammalian species.

docile than male mice is incorrect: it is true only for laboratory strains, not for wild mice. Further, sexually naive female laboratory mice exhibit unnaturally high maternal behaviors toward alien pups that are introduced into the cage, which never happens in wild or wild-derived mice (figure 2.2b, right). However, some of these differences may result from using small laboratory cages for testing the mice: in fact, in nature, wild mice form large, complex communal nests and sometimes nurse pups communally—and these communal behaviors reduce both female aggression toward pups and male infanticidal behaviors (Manning et al. 1995). This reinforces the need to use large, complex, seminatural environments for studying social behaviors in mice, because such seminatural environments encourage much more natural social behaviors compared to the standard laboratory assays that usually involve highly impoverished small setups with minimal or no physical contact (Shemesh et al. 2013; Weissbrod et al. 2013; Zilkha et al. 2016). Another major problem is that results collected in artificial laboratory assays for social behaviors in mice are notoriously sensitive to minutia such as the details of daily cage cleaning, lighting conditions, or feeding (Gray & Hurst 1995; Reardon 2016). This is not the case in complex seminatural environments, which allow for testing social behaviors not only more naturally, but also more robustly (Zilkha et al. 2016). In summary, there is a clear need to apply the natural neuroscience approach to social neuroscience.

Spatial movement behaviors of individual mice also differ markedly between wild mice and domesticated laboratory mice. For example, wild and wild-derived mice exhibit wild behaviors, such as spontaneous high jumping and prolonged freezing, to an extent never observed in laboratory mice (Chalfin et al. 2014). Further, in a series of studies, Ilan Golani and colleagues have found remarkable differences in spatial exploration patterns between wild mice and two laboratory strains, C57BL/6 and BALB/c, that were tested in an open-field test (Fonio et al. 2009, 2012a; Benjamini et al. 2011). They started by introducing a more naturalistic spatial-exploration paradigm, in which they extended the arena size by 10-fold compared to arenas used in classical open-field tests, and prolonged the experimental time 100-fold compared to typical laboratory studies in mice (2.5-meter circular arena, 45-hour experiment). They also provided food inside the home cage that was connected to the arena, thus allowing for testing of the intrinsic exploratory drive of mice independent of food foraging. They found that both C57BL/6 and BALB/c mice, when first introduced to this arena, followed a very stereotyped sequence of ontogeny of movements that was comprised of twelve movement stages, which were repeated almost exactly in all individual mice of both strains. This sequence consisted first of local movements near the home cage (stages 1–4); then one-dimensional (1D) shuttling along the arena borders, with shuttle amplitude increasing over time (stages 5–8); then two-dimensional (2D) forays into the arena center, with foray amplitude increasing over time (stages 9–11); and finally, three-dimensional (3D) jumping (stage 12) (Fonio et al. 2009). By contrast, wild mice often entirely skipped stages 1–9 and went directly to stages 10–12—the wild mice immediately darted into the center of the arena and also quickly exhibited 3D jumping behavior (Fonio et al. 2012a; Ilan Golani, personal communication).

Some of these differences in spatial behaviors are illustrated in figure 2.3 (Fonio et al. 2012a). Here, the exploratory behavior of wild mice and laboratory mice (BALB/c) is shown for the first 5 hours of the 45-hour experiment. From the very first moment, wild mice spent a large time within the center of the arena (figure 2.3a, dark gray) and moved a lot (figure 2.3b), while BALB/c laboratory mice avoided the center initially (figure 2.3a, light gray) and moved much less than the wild mice did (figure 2.3b). Interestingly, after a few hours, the picture *reversed* and the wild mice stopped moving and started freezing a lot (figures 2.3b and 2.3c), while the BALB/c mice exhibited a stable calm behavior, including repeated visits to the center of the arena (figures 2.3a and 2.3b). This reversed behavior was then stably maintained for the duration of the 45-hour experiment.

It is noteworthy that BALB/c laboratory mice are considered a model of anxiety because they are thought to avoid the center of open arenas; but this notion was based on standard 15-minute or 30-minute testing sessions in small arenas (dashed vertical lines in figure 2.3 denote the maximal standard laboratory testing duration: 30 minutes). By contrast, this study showed that at longer timescales (> 2 hours), the picture reverses almost completely, and BALB/c mice do *not* seem to exhibit chronic anxiety, but rather a stable, calm behavior (figures 2.3a and 2.3b). In other words, at short timescales of < 30 minutes, BALB/c mice exhibit anxiety and neophobia (fear of novelty), but at long timescales of > 2 hours, these mice exhibit calmness and neophilia (love of novelty). This leads to two conclusions. First, we need to reexamine the notion that BALB/c is a good mouse model for chronic anxiety—and more generally, we must reexamine very thoroughly the behavioral validity of mouse models of psychiatric and neurological disorders (Ennaceur et al. 2008; Nestler & Hyman 2010; Fonio et al. 2012b; Brenowitz & Zakon 2015). Second, it is quite remarkable that after a century of experiments on mice in the open field, a simple naturalistic extension of the test space by 10-fold and of the test time by 100-fold has revealed a completely new picture of the ontogeny of natural exploratory behaviors. This finding highlights once more the need to study animals under more naturalistic conditions.

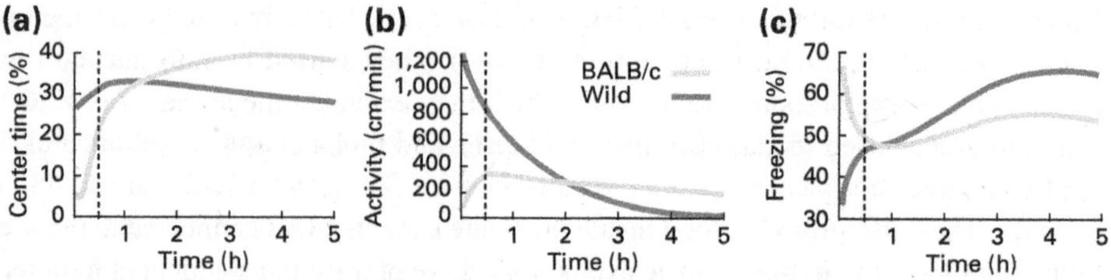

Figure 2.3
Spatial behaviors differ dramatically between laboratory mice (BALB/c) and wild mice. Data replotted from Fonio et al. (2012a). Here, the spatial exploration behavior of mice was tested in a very large, open arena (2.5-meter diameter circle) for a long time (5 hours; for comparison, the vertical dashed lines show the maximal duration of standard open-field tests: 30 minutes). (a) Percent of time that the mice spent at the center of the arena. (b) Activity levels (distance traveled per minute). (c) Freezing. Note that BALB/c mice exhibited anxiety at short timescales (see high freezing at < 30 minutes in [c], light-gray), but showed calmness at long timescales (see the high center-time at >2 hours in [a], light-gray). By contrast, wild mice exhibited almost the opposite tendencies—that is, anxiety at long timescales (see low activity and high freezing at >2 hours: [b] and [c], dark gray).

Behavioral differences between wild species and their inbred laboratory-strain counterparts are not particular to mice.[6] Rats also exhibit large differences between laboratory strains and wild animals. An interesting example is related to spatial movements—and is thus close to my heart. This example comes from a study of Norway rats, in which wild rats were shown to exhibit excellent climbing capabilities even when denied any early climbing experience, unlike laboratory rats from the Long-Evans strain, which cannot climb well unless they have had early climbing experience (Huck & Price 1976). Interestingly, the deficient 3D-climbing of Long-Evans rats was much more pronounced in males than in females, with Long-Evans males spending 350-fold less time 3D-climbing than wild males, whereas in females, this deficiency was "only" 19-fold (Huck & Price 1976). The deficient 3D locomotion of male laboratory rats may have implications for studies of the neural basis of 3D navigation in rats, particularly because many studies on the neurobiology of navigation are done in male Long-Evans rats. Notably, wild mice also exhibit much better 3D climbing abilities than laboratory strains (Ilan Golani, personal communication). I will come back to 3D locomotion in chapter 4.

There are many other behavioral differences between wild rodents and laboratory inbred strains, as well as among different laboratory strains. Thus, some of the most common strains of laboratory mice differ dramatically among themselves in various tests of anxiety, fear, acoustic startle, and in a host of learning and memory tests—including the Morris water maze, contextual fear conditioning, radial-arm maze, and spatial alternation—in addition to major interstrain differences in locomotion and in social behaviors (reviewed by Crawley et al. 1997). Large behavioral differences were found even between extremely close substrains coming from different providers (Crawley et al. 1997). Some of these differences might perhaps be linked to the very large sensory differences that exist between certain mouse strains and between inbred strains and wild mice (Wong & Brown 2006) because sensory abnormalities may affect behavioral patterns. However, many of these differences likely stem from true high-level cognitive differences. Surprisingly, even a seemingly basic behavioral notion—that mice are nocturnal animals—was shown to be primarily a laboratory finding: while laboratory mice are indeed strictly nocturnal (and hence should be studied behaviorally during the dark phase), wild mice were found to also exhibit substantial daytime activity from spring until late autumn (Robbers et al. 2015). It is thus clear that much more work must be done to characterize the natural outdoor behavior of wild mice (*Mus musculus*).

In addition to major differences in behavior, some laboratory mouse strains differ markedly from wild mice on a whole host of other parameters, from endocrinology to basic cell biology (Harper 2008). These differences arise from the highly deviant genetic composition of many inbred strains, which led to well-documented problems for studies of genetics, molecular biology, and biomedicine—as reviewed for example by Austad (1993) and Harper (2008). Importantly, gene-expression profiles in certain brain regions may differ up to 100-fold (or more) between laboratory mice and wild mice (Chalfin et al. 2014). These problems are of particular importance for molecularly oriented studies of the brain mechanisms of behavior. However, almost no attempts have been made to characterize possible electrophysiological differences—or differences in neural coding—between wild mice and laboratory mice. Since wild mice and laboratory mice differ in so many behaviors, this must be manifested in differences in brain circuits. Further, wild-born mice are experiencing a completely different behavioral history in the wild, outdoors, compared to laboratory mice that have been raised in small impoverished cages; this difference may again influence brain circuits and brain function. Therefore, a serious concerted effort is needed to study systems neuroscience in wild mice.

What can be done to overcome the systematic behavioral biases (and likely neural biases) that rapidly evolved in inbred mouse strains during the twentieth century and that may bias the study of the neural basis of behavior? One possibility is to study wild-caught mice. A second possibility is to use wild-derived mice (mouse strains that were recently derived from wild-caught mice)—an approach that has become increasingly popular in recent years in a number of fields in biology (Guénet & Bonhomme 2003), although it still accounts for only a small minority of mouse experiments. Wild-derived mice do not have natural experience outdoors as real wild mice do, which is a disadvantage; but on the other hand, this approach allows combining the power of mouse genetic tools with the reduced bias and normal natural variability of wild populations (Chalfin et al. 2014; Zilkha et al. 2017). A third possibility is to diversify the choice of species and to increase the worldwide fraction of brain studies that use nonstandard model species for studying systems neuroscience and behavioral neuroscience—an approach which, as I will argue next, offers some fundamental conceptual advantages.

2.2.2 Examples of Comparative Studies: Cortical Columns and Hippocampal Oscillations

Before I discuss in more detail why it is important to diversify the set of species used in neuroscience, I want to start with two examples of comparative neuroscientific studies. These examples illustrate how comparative studies allow drawing *general* conclusions about brain function, by elucidating which aspects of a brain system are invariant across species—and are thus likely important for its function—and which aspects are not invariant.

The first example concerns cortical columns and cortical maps. **Cortical columns** were discovered by Mountcastle in the cat somatosensory cortex and by Hubel and Wiesel in the cat visual cortex (Mountcastle 1957; Hubel & Wiesel 1963b)—and can be summarized by the maxim: Nearby neurons have similar tuning properties (and hence, most likely, similar functions). Moreover, the tuning properties of neurons in cortical columns often change smoothly along the cortical surface, creating orderly **cortical maps** of the relevant parameter. Cortical maps and columns have been found across a wide range of cortical regions, including sensory, motor, and high cognitive areas (Fitzpatrick & Ulanovsky 2014). Many possible functions have been proposed over the years for the prevalence of maps and columns in many brain areas, including (1) economy of wiring; (2) sharpening of neural tuning via lateral inhibition; (3) implementation of a "**labeled-line code**," which may allow more efficient and rapid computations; (4) simple readout by downstream neurons; (5) facilitation of plasticity mechanisms; and (6) allowing innate wiring of the neuronal network in pups, prior to any experience (Knudsen et al. 1987; Fitzpatrick & Ulanovsky 2014; Kössl et al. 2014; Krahe & Maler 2014).

In the primary visual cortex, V1, a number of stimulus parameters were found to be mapped along the cortical surface and to exhibit cortical maps and columns. These parameters include stimulus retinal position (retinotopy), ocular dominance, and orientation (Nauhaus & Nielsen 2014). Such orderly maps and columns in V1 were found across a range of species, including multiple species of primates, tree shrews, and carnivores such as cats and ferrets (Ibbotson & Jung 2020). However, studies of V1 in rodents have suggested that, at least with respect to orientation tuning, this picture may need reconsideration. Clay Reid and colleagues have conducted two-photon imaging of neuronal responses to oriented gratings in V1 of rats and cats (Ohki et al. 2005). In cat V1, they found beautifully organized orientation maps and columns, with nearby neurons having similar preferred orientations (figure 2.4a; the preferred orientation of each neuron is marked by a line). They also found in cat V1 "orientation pinwheels": distinct points on the cortex where different preferred-orientations converge, with the preferred orientation being organized as spokes on a wheel around these pinwheels (Ohki et al. 2006). This confirmed many previous findings using other methods, such as electrophysiology and optical imaging, all of which reported orientation maps, columns, and pinwheels in V1 of cats and monkeys. However, when the same two-photon imaging method was applied to rat V1, they found no apparent local structure: nearby neurons had completely different preferred orientation—a "salt-and-pepper" organization (figure 2.4b; note the random arrangement of preferred orientations for nearby neurons) (Ohki et al. 2005). One may suspect that this fundamental interspecies difference in functional architecture—between cats and monkeys that exhibit orientation columns versus rodents that do not—might stem from studying a rodent species with relatively poor vision: the rat. However, this explanation does not seem to hold because of another study that has examined the functional architecture of orientation selectivity in V1 of a highly visual rodent, the gray squirrel, which has high visual acuity and also heavily relies on vision for its natural behavior. Surprisingly, this study found in V1 of squirrels basically the same result as in rats: a salt-and-pepper organization of orientation tuning (Van Hooser et al. 2005). So the difference is not about visual abilities or about reliance on vision, but is a real phylogenetic difference.[7] Moreover, in both rats and squirrels, the tuning of the individual neurons to orientation was as sharp as in cats (Ohki et al. 2005, 2006; Van Hooser et al. 2005), so it is not the case that V1 neurons in rodents are poorly tuned to orientation—they are in fact strongly tuned. Nevertheless, there is a fundamental interspecies difference in the functional organization of orientation tuning along the cortical surface. Rodents are truly different.

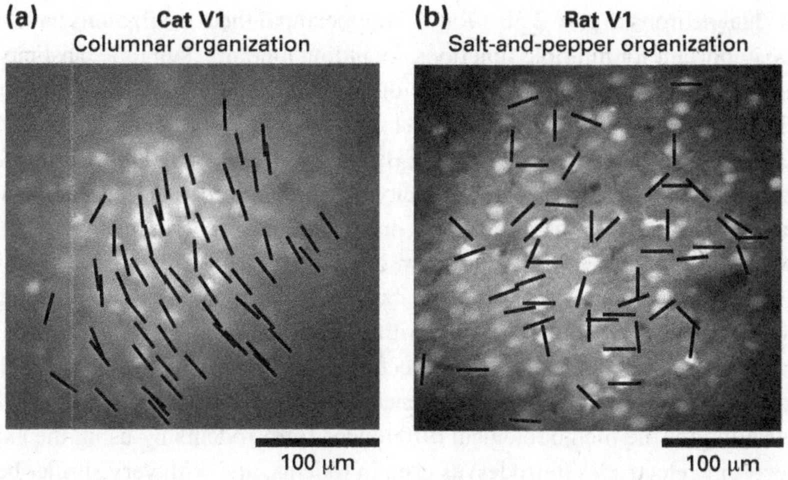

Figure 2.4
Fundamental differences in V1 functional organization between cats and rodents. Adapted from Ohki et al. (2005). Here, the orientation tuning of individual V1 neurons was measured using two-photon imaging of a calcium indicator in cats and rats. (a) Cat V1. Bright spots show imaged neurons. The orientation of the lines indicates the preferred orientation of the underlying neurons; the orientation was remeasured from Ohki et al. (2005). Clear orientation columns are seen in cat V1, with nearby neurons having similar preferred orientations; note that all the lines in this patch of V1 have rather similar orientations. (b) Rat V1, with the same graphical convention as in (a). No orientation columns are seen; rather, in rat V1, nearby neurons can have very different preferred orientations; note the haphazard orientations of the lines: a "salt-and-pepper organization."

These results created somewhat of a mystery. Why are there functional columns in V1 of cats and monkeys, but not in rodents? One possibility is that in terms of topology (connectivity-wise), neurons in rodents may perhaps be arranged just like in cats and monkeys but simply lack the topographical organization along the cortical surface. This possibility is supported by the finding that in mouse V1, neurons with similar orientations are preferentially connected (Ko et al. 2011), similar to previous findings in cat V1 (Malach et al. 1993). Another possibility is that the function of rodent V1 does not correspond to that of cat V1. For example, the function of V1 in rodents may correspond to the function of higher visual areas in the cat or primate—hence the different functional architecture. In any case, these comparative studies of V1 functional architecture have taught us that cortical columns—in particular, orientation columns and pinwheels—are *not* necessary for creating sharp orientation tuning in individual neurons. Nor are they necessary for high-acuity visual perception (as demonstrated by the lack of columns in the highly visual squirrel). This conclusion is not entirely satisfactory because this interspecies difference still remains somewhat enigmatic, and it calls for future studies aimed to understand more deeply the importance (or lack thereof) of cortical map topography. Importantly, this conundrum brings modesty to older claims about the pervasive importance of cortical columns. This modesty was achieved through comparative studies.

The second example of a comparative systems neuroscience approach concerns *hippocampal oscillations*—in particular, the theta oscillation. Hippocampal theta oscillation is an 8-Hz rhythm observed in the hippocampus and entorhinal cortex of rodents as the animal moves through the environment. The theta oscillation is clearly visible both in recordings of the local field potential (LFP; figure 2.5a, "Rat")—and in **autocorrelations of spike trains** from individual neurons (figure 2.5b, "Rat"). Hippocampal theta oscillations have been proposed to be important for multiple functions, including memory, synaptic plasticity, temporal coding of position, communication with other brain regions, and generating the hexagonal grid patterns of grid cells (Huerta & Lisman 1993; Skaggs et al. 1996; Mehta et al. 2002; Huxter et al. 2003; Buzsáki 2006; Burgess et al. 2007). All these proposals, however, were based on the strong theta rhythmicity observed in rodents. Reports from monkey and human hippocampus revealed much less prominent theta, and in many cases, theta was almost absent (Stewart & Fox 1991; Ekstrom et al. 2005). However, this major difference between rodents and primates was often attributed to the fundamental methodological differences between electrophysiological recording methods used in rodent versus primate hippocampal studies—both differences in electrodes and differences in behavioral tasks.

To shed light on these interspecies differences, we conducted a series of studies in bats, while controlling for the methodological differences from rodents by using the exact same types of recording electrodes (tetrodes) as used in rodents, and with very similar behavioral tasks as in rodent studies. We found an absence of continuous theta oscillations in the bat hippocampal formation, both in crawling bats and in flying bats (Ulanovsky & Moss 2007; Yartsev et al. 2011; Yartsev & Ulanovsky 2013; Eliav et al. 2018). This lack of theta oscillations was manifested in two ways: LFP in the bat were nonrhythmic, in contrast to the highly rhythmic LFP in rodents (figure 2.5a: compare rat and bat)—and in addition, autocorrelations of spike-trains from single neurons in the bat were completely nonrhythmic, in contrast to the highly rhythmic autocorrelations in rodents (figure 2.5b: compare rat and bat) (Eliav et al. 2018). Importantly, gamma oscillations and high-frequency ripple oscillations *were* present, and were very similar to rodents (Ulanovsky & Moss 2007; Yartsev et al. 2011; Eliav et al. 2018): thus, the interspecies difference was specific to the lower-frequency theta oscillations. The lack of continuous theta oscillations in bats has created a major controversy regarding the role of theta across mammals (Ulanovsky & Moss 2007; Yartsev et al. 2011; Barry et al. 2012; Heys et al. 2013) because we did find grid cells, but without theta oscillations, which argued strongly against many computational models of grid cells, which posited that temporally periodic theta oscillations are required for generating the spatially periodic hexagonal pattern of grid cells (Yartsev et al. 2011). Furthermore, the lack of theta oscillations in bats seemed to argue against other proposed roles of theta in learning, memory, and neural coding.

However, a recent study suggested a surprising resolution to the controversy (Eliav et al. 2018). In this study, we found that hippocampal neurons in bats *do* exhibit surprising similarity in their neural coding to hippocampal neurons in rodents, despite the lack of theta oscillations (or any other oscillations in the range of 1–20 Hz). Specifically, similar to rodents, bat neurons exhibited pronounced phase-locking—that is, locking their activity to the nonrhythmic (broadband) fluctuations of the LFP, such that neurons were synchronized together despite the absence of oscillations: this suggested that synchronicity is important for hippocampal function across species. Further, as in rodents, some bat neurons exhibited "phase precession," wherein the phases of spikes relative to the LFP shifted earlier as the animal moved through the place field (figure 2.5c), resulting in phase coding of the bat's position, whereby the phase of spikes relative to the LFP encoded the animal's position in space. In rodents, this phase coding is found relative to the periodic theta oscillation, while in bats, this phase coding was relative to the bat's nonoscillatory, nonperiodic LFP. This

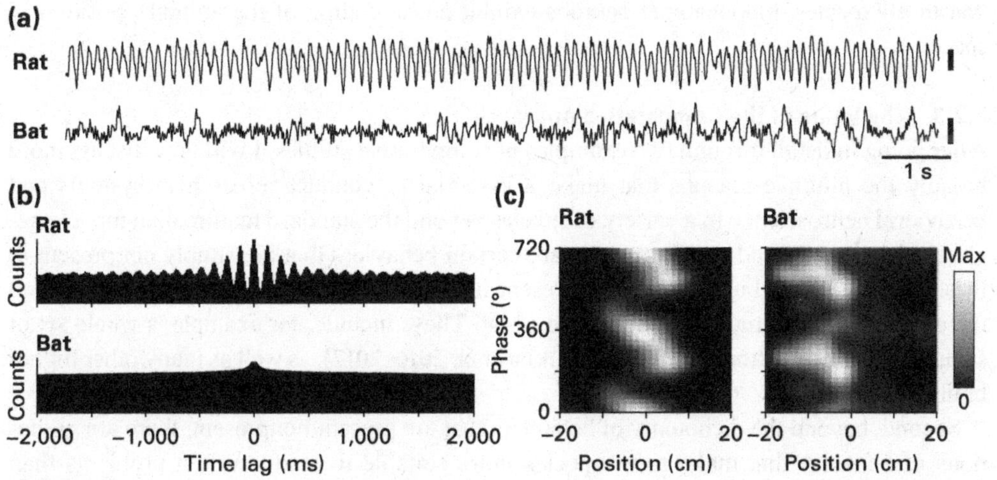

Figure 2.5
Fundamental differences in hippocampal oscillations between bats and rodents. Bat data are from Eliav et al. (2018). Rat data are from www.crcns.org, courtesy of G. Buzsáki. (a) LFP recorded in dorsal hippocampal area CA1 of a rat moving along a horizontal linear track (top) and from an Egyptian fruit bat moving in a horizontal arena (bottom). Note the strong theta oscillation in the rat LFP, and the lack of theta oscillation in the bat LFP. Vertical scale bars, 300 μV. (b) Spike-train autocorrelations from an interneuron in rat CA1 during locomotion on a horizontal linear track (top) and an interneuron in bat CA1 during locomotion on a horizontal surface (bottom). Note the strong theta oscillation in the rat interneuron, which illustrates why CA1 interneurons in rats are often called "theta cells," compared to the lack of theta oscillations in the bat interneuron. Counts go from 0–6,000 in the rat and 0–8,500 in the bat interneuron. Bin size, 10 ms. (c) Phase-by-position firing-rate maps (white color: high firing rate), showing clear phase precession (change in the phase of firing as a function of position) in a rat CA1 place cell (left) and in a bat CA1 place cell (right) during locomotion on a horizontal linear track (rat) or horizontal 2D arena (bat). Position 0 denotes location relative to the center of the place field (for the bat neuron, this is the linearized position along 2D trajectories passing through the place field: see details in Eliav et al. 2018). Two cycles of the phase are shown for clarity. The phase in both cases was computed by identifying the troughs of the LFP fluctuations and assigning a 0–360-degree phase to each cycle by linearly interpolating between adjacent troughs. This revealed the classical oscillatory phase coding of position in the rat neuron, based on the theta oscillation (left), and a surprising nonoscillatory phase coding of position in the bat neuron, based on nonrhythmic LFP fluctuations (right). Analyses courtesy of Tamir Eliav.

study provided the first demonstration of nonoscillatory phase coding in the brain of any species—a unique type of temporal coding. Similar results were also recently found in human hippocampus and entorhinal cortex (Qasim et al. 2021). Taken together, the results of this bat study (Eliav et al. 2018) suggest that, surprisingly, a similar kind of neuronal coding (phase coding) may be common to hippocampal and entorhinal neurons in both bats and rodents—and also in humans. Importantly, these results dissociated between theta oscillations, neuronal-assembly synchronicity, and phase coding of position—three phenomena that in rodents are tightly coupled together, and that were believed for many decades to be fundamentally related to each other—whereas our study showed that only the latter two phenomena generalize across species.

This result highlights the value of the comparative approach. Only through studying hippocampal physiology across species was it possible to offer a new synthesis. Specifically, we proposed a shift of focus from rhythmicity (periodic oscillations) to synchronicity and neural coding, which may reconcile some of the debates about the roles of hippocampal oscillations across different species. Our results argue that although there are clear differences in hippocampal oscillations, there is in fact deep similarity in *neural coding* across species—namely, that in all species, hippocampal neurons exhibit phase coding of the animal's position in space.

2.2.3 The Value of the Comparative Approach

After going in detail through two examples of comparative studies, I will now discuss more broadly the multiple reasons that make it essential to conduct research in systems and behavioral neuroscience in a variety of species beyond the standard mammalian model species such as mouse and rat. First, there are certain behaviors that are simply not present in these rodent species but are strongly present in other mammals, including humans—and therefore cannot be studied in the mouse or rat. These include, for example, a whole set of social behaviors (Clutton-Brock 2016; Zilkha et al. 2016, 2017), as well as many other higher brain functions.

Second, beyond the dichotomy of behaviors that are present/nonpresent, there are gradations of behavior that make some species more suitable to study certain problems than others—which is exactly the point of Krogh's principle. For example, bats fly very fast and may exhibit flight speeds of about 10 meters/second in large laboratory setups, which allows them to fly many kilometers during a single recording session of 1 or 2 hours. Therefore, as described in chapter 1, this provided a unique opportunity to study hippocampal neural codes for very large spaces, on the order of hundreds of meters or even kilometers (Eliav et al. 2021). Laboratory rodents, which run much slower, cannot cover such distances many times per hour—as bats do—and hence rodents do not permit studying the encoding of such large spaces within the time-limits of a typical experiment. Therefore, the one or two orders of magnitude higher speed of locomotion in bats, as compared to rats, has allowed tackling a unique question that was not approachable in rodents, namely: How are very large-scale naturalistic spaces encoded in the brain? The answer turned out to be very different than hippocampal codes found in small laboratory setups (Eliav et al. 2021), as mentioned in chapter 1 and as will be discussed in detail in chapter 4.

Third, the comparative approach allows a reality check for testing whether specific findings or insights that were derived from one species also generalize to other species. For example, as detailed previously, orientation columns in V1, which are found in primates, carnivores, and tree shrews, were *not* found in rodents (Ohki et al. 2005)—including not in a highly visual rodent, the gray squirrel (Van Hooser et al. 2005). This means that, contrary to long-held beliefs, cortical maps and columns are not an essential prerequisite for high-acuity visual perception. This example illustrates the point that the only way to test whether *any* finding or principle of brain function generalizes across species is to conduct direct comparisons of several carefully chosen species. Comparative studies provide the only way to validate the generality of neuroscientific results and concepts across species—or, conversely, they allow for discovering that different species may have evolved different brain mechanisms and distinct computational algorithms. In short, comparative studies are absolutely necessary if we want to understand diversity versus generality in brain function.

Fourth, comparative systems neuroscience allows using a contrast-and-compare approach to glean *totally new general insights* about certain brain functions. One classical example comes from the studies by Larry Young and colleagues on the roles of oxytocin and vasopressin in social behaviors. Young powerfully utilized the ability to compare closely related species of voles (prairie voles and montane voles) that strongly differ in their social affiliative behaviors, which allowed him to illuminate the molecular, genetic, and neural bases of affiliative behaviors (Walum & Young 2018; Froemke & Young 2021). Another example is the case of hippocampal oscillations: as detailed previously, these oscillations turned out to be very different in bats and primates as compared to rodents, which argued strongly against oscillation-based theories of grid cells (Yartsev et al. 2011). This study of bats allowed for testing and refuting some computational theories of grid cells—thus constraining the space of possible theories of hippocampal and entorhinal function. Importantly, however, we also showed phase locking and phase coding of position in the bat's hippocampus, in the absence of oscillations (figure 2.5) (Eliav et al. 2018). This latter study generated two new insights: that phase coding without oscillations can exist in the brain; and that phase coding and neuronal synchronization are likely essential features of hippocampal function across species, while periodic oscillations are not. Further, this serves to illustrate a more general point: the comparative systems neuroscience approach can be used very effectively to narrow down the space of possible theories of a certain brain circuit, computation, or function.

Fifth, by comparing brain properties across many species, it is possible to find interesting correlations that reveal a new story. For example, a number of studies that compared hippocampal volume across a set of closely related bird species have found that hippocampal volume increases for species with larger demands on navigation and spatial memory, such as migrating species or food-storing species that rely heavily on spatial memory (Krebs et al. 1989; Healy et al. 1996). Similar results were also found in closely related species of bats, revealing again a correlation between hippocampal volume and navigational load (Safi & Dechmann 2005). These early findings in birds (Krebs et al. 1989) later led to a search for similar principles in humans, yielding the famous studies of London taxi drivers that found enlarged hippocampal volume in taxi drivers, who rely heavily on spatial memory to navigate (chapter 7). Thus, cross-species comparisons can spawn new insights.

Sixth, there are the unknown unknowns. As detailed previously, nonstandard model species have led to a glowing trail of discoveries in neuroscience, ranging from elucidating the mechanisms of the action potential and synaptic transmission, through the discoveries of lateral-inhibition and axonal delay lines, and ending with key molecules such as GABA. All these advances were made possible because specific nonstandard species were particularly suited to discover them—in line with Krogh's principle—either for technical reasons (e.g., large size of axons or synapses, permitting easy accessibility) or because the phenomenon was particularly pronounced in those species (e.g., adult neurogenesis in songbirds). Who knows which other phenomena in neuroscience are awaiting to be discovered in nonstandard species—phenomena that might never be discovered in mice? Prudence mandates that we should cast a broad net as a research community—that is, we must utilize biological diversity to enhance the diversity of our scientific studies.

Seventh, there is the general argument in favor of diversity in curiosity-driven research, which was beautifully summarized by Eve Marder (2002):

> As we revel in the sometimes outrageous forms that species take, we remember that species diversity was an outcome of survival in disparate environments. The peculiarities of body plan and nervous systems that allowed animals to live in strange ecological niches remind me that the most important findings in science often result from individual scientists' foibles, genius, insight and personal taste. Although brute-force science

has its place, we risk an incalculable loss of individual creativity and imagination if we work only on consensus problems and consensus preparations. We should value and protect those who dare to be fascinated by animals that have evolved nervous systems to best carry out a specific task.[8]

Finally, it is important to stress that—contrary to a common misconception—the comparative approach is *not* an endeavor of "stamp collection" or "butterfly collection" where we collect ever more disparate and disconnected details about ever more species, and thus increase confusion about the brain. No! The exact opposite is true: the comparative approach allows the identification of *general principles* and *invariant functional properties* in systems neuroscience and behavioral neuroscience—those principles and functional properties that generalize across species. The comparative approach may also help identify neuronal *computations* and *algorithms* that are invariant across species, independent of biological implementation (Borst & Helmstaedter 2015; Hemberger et al. 2016)—meeting David Marr's distinction between levels of understanding (Marr 1982). Further, the comparative approach can yield new insights by utilizing the relative advantages of each species (i.e., Krogh's principle). It is true that one can, and should, conduct curiosity-driven research on the neurobiology of nonstandard animals for its own sake—and many researchers do exactly that—but at its best, the comparative approach is a key gateway to elucidating general principles in neuroscience. At its best, the comparative approach does not increase confusion; rather, it *reduces* confusion by eliminating the biases and misconceptions that might be caused by a too-narrow focus on too-few model species (as demonstrated by the examples of cortical columns and hippocampal oscillations). The inherent diversity of the comparative approach can thus lead to a more unbiased identification of broad insights and *general principles* of brain function.

2.3 Summary

In this chapter, I compared *natural neuroscience* to some related ideas. Natural neuroscience promulgates the notion that we must study the neural basis of natural behaviors in freely behaving, freely moving animals—or at least we should strive to approach as closely as possible the natural behaviors of the animal, even if we perform indoor laboratory studies.

A related notion is that of *neuroethology,* another branch of neuroscience that seeks to understand the neural basis of natural behaviors—but it classically focused mostly on *innate* behaviors (e.g., fixed action patterns), as studied by the fathers of ethology, Konrad Lorenz and Niko Tinbergen (Tinbergen 1951; Lorenz 1982). Therefore, neuroethology usually focuses on champion species—those species that excel in executing a particular innate behavior and are thus very well suited to address specific research questions, according to Krogh's principle. Consequently, neuroethology usually studies nonstandard species.

Another related notion is that of *comparative systems neuroscience,* the branch of neuroscience that seeks to unravel general principles of brain function by studying the brains of multiple species in a comparative manner. I have listed in this chapter several reasons why it is absolutely essential that—in parallel to research efforts on standard model species such as the mouse, zebrafish, or *Drosophila*—the neuroscience community should indeed maintain highly active research efforts on nonstandard species (often called "nonmodel species"). In

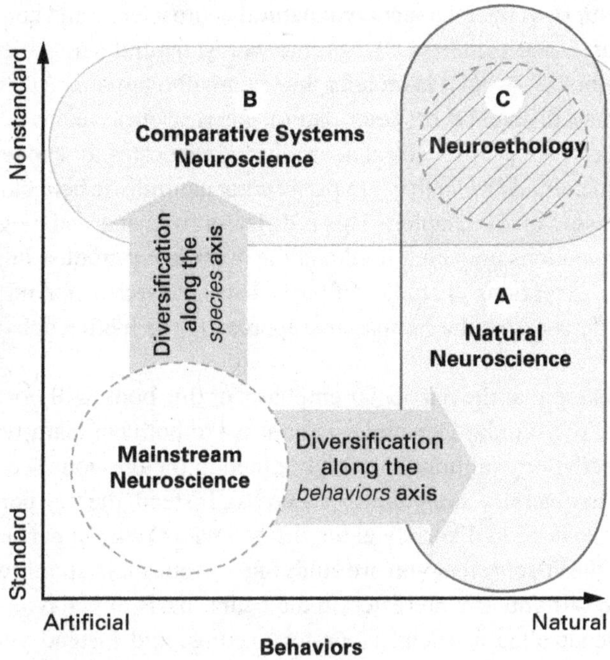

Figure 2.6
A schematic graph showing the range of possible behaviors on the x-axis (Artificial → Natural), and the range of possible model species on the y-axis (Standard → Nonstandard). Mainstream neuroscience typically addresses standard animal models (e.g., the mouse) performing artificial behaviors. Neuroethology typically addresses nonstandard animal models conducting natural behaviors. Comparative systems neuroscience calls for studying a wider variety of model species, including nonstandard models, *regardless* of behavior (i.e., the upper half of this 2D-plane: ellipse B). The natural neuroscience approach, which is the focus of this book, calls for studying more natural behaviors, *regardless* of the animal species used (the right half of this 2D-plane: ellipse A). In other words, natural neuroscience posits that for both nonstandard species and standard species—such as the mouse, zebrafish, or *Drosophila*—there is an urgent need to diversify along the behaviors axis.

fact, as genomic sequences, viral transfection methods, and gene-editing tools become rapidly available for nonstandard organisms, the relative advantage of these genetic models becomes at least partially eroded—and we now see increasing species diversification as neuroscience emerges from the "model-species bottleneck" (Brenowitz & Zakon 2015).

The differences between natural neuroscience, neuroethology, and comparative systems neuroscience are schematized in figure 2.6. Both comparative systems neuroscience and natural neuroscience call for diversification in the field of neuroscience, but each of them emphasizes diversification along a different dimension. Comparative systems neuroscience stresses the need to diversify along the dimension of *model species* and seeks to increase studies of nonstandard species regardless of behavior (figure 2.6; diversification along the y-axis—see ellipse B). Thus, in comparative systems neuroscience, you could, in principle, study nonstandard species that perform completely artificial behaviors—or no behaviors at all. Natural neuroscience, by contrast, emphasizes the need to diversify along the dimension of *behaviors* and seeks to incorporate more studies on the neural basis of natural behaviors—regardless of the species (figure 2.6; diversification along the x-axis—see ellipse A). Thus, natural neuroscience and comparative systems neuroscience seek to diversify the research in neuroscience along two orthogonal dimensions. Neuroethology, in many respects, comprises the intersection of natural neuroscience and comparative systems neuroscience (figure 2.6—see circle C). In other words, natural neuroscience encompasses the field of neuroethology—but it is broader than neuroethology.

An additional key difference between natural neuroscience versus neuroethology and comparative neuroscience is this: natural neuroscience advocates for conducting neuroscientific experiments in animals *while* they are performing naturalistic behaviors (the simultaneous approach discussed in this chapter). This is different from neuroethology, where in many cases reduced preparations are being used and the brain is *not* studied in a freely behaving animal (with some exceptions in studies of birds, bats, and rodents). And certainly, natural neuroscience is different from the comparative approach, which often does not even focus on behavior.

I wish to stress here that the particular emphasis of this book will *not* necessarily be to promote the use of nonstandard species—in contrast to both comparative systems neuroscience and neuroethology (although, as highlighted in the previous section, I do support strongly the need to diversify along the *species axis*). Instead, the key point emphasized in this book will be the need to diversify along the *behaviors axis*. In particular, I will argue that regardless of the species that you are studying—even if it is an inbred mouse, zebrafish, or *Drosophila*—if you are interested in the neural basis of behavior, then you should try to avoid overcontrolled artificial laboratory setups, and instead you should seek to make at least some of your experiments closer to the natural behavior of your subjects. This is the only way to ensure that your results will be relevant to the real world.

3

Technological Advances Enabling Natural Neuroscience

Sydney Brenner (2002) famously said: "Progress in science depends on new techniques, new discoveries and new ideas, probably in that order" (the original sentence was slightly different, but this is the version that has been consolidated in scientists' collective memory). While some may argue with this particular order, it is clear that new techniques enable major breakthroughs in science, as has been demonstrated time and again in the history of every scientific field, including neuroscience. The idea advocated in this book—that is it crucial to study the brain in freely moving animals under evermore naturalistic conditions—has been around for many decades, as discussed in chapter 2. However, it was difficult to fulfill this idea due to limited techniques. Therefore, with a few exceptions, this idea remained mostly a long-term dream, while most neuroscientists continued to study the brain in anesthetized and/or head-fixed animals, or in vitro. However, this situation has changed dramatically over the last decade, with the advent of a series of new and powerful methods and technologies, which finally have allowed this dream to come true. This chapter will survey some of these new methods.

To study the neural basis of behavior, we need four classes of methods, which can be arranged in a 2 × 2 matrix:

Measuring brain activity	Manipulating brain activity
Measuring behavior	Manipulating behavior

The rubric "manipulating behavior" has two distinct meanings. One entails *indirectly* affecting the animal's behavior by changing its environmental conditions or its ecology—and in chapter 8, we will discuss some useful examples of such indirect manipulations, which may help connect neuroscience to behavioral ecology. The other meaning is *direct* training of the animal to perform a particular task—and as argued in chapter 1, when studying the neural basis of *natural* behaviors in animals, we would like to minimize training as much as possible to avoid overtraining and its associated problems. This is not to say that we should avoid any training whatsoever: even when studying totally natural behaviors, it is often necessary to conduct some amount of training to familiarize the animal with humans and with the experimental environment. However, there is a fundamental difference between such short familiarization and conducting extensive focused training for a particular task, which may be quite unnatural to the animal. Such extensive training can dramatically alter brain circuitry and physiology and also cause stress—both of which should be avoided (as argued in chapter 1). Therefore, this chapter will focus on the remaining three classes of methods: measuring brain activity, manipulating brain activity, and measuring behavior. Notably, this chapter will focus on methods for natural neuroscience in animals, while chapter 7 will discuss novel methods for natural neuroscience in human subjects. Although there are major similarities in the conceptual aims of natural neuroscience in both animals and humans, the technical approaches differ quite substantially, which is why we are discussing the methods for studying animals and humans in two separate chapters.

All three of these methodological domains—measuring brain activity, manipulating brain activity, and measuring behavior—underwent major revolutions in recent years. There was an explosion of new methodologies for recording large populations of neurons simultaneously, for manipulating neural activity with great specificity, and for quantifying the detailed nuances of animal behavior—and all of this in completely unrestrained animals that are freely moving and behaving. Consequently, after decades of dreaming about studying the neural basis of truly natural behaviors, but not being able to carry it out due to technological limitations, we can now stop dreaming and start doing—because we now finally possess all three classes of technologies that enable conducting natural neuroscience research.

The survey of these three classes of technologies will necessarily be brief. I will not attempt to describe the methodological details of particular technologies and subtechnologies, as such a technical manual is tedious, and in any case it would become rapidly obsolete as newer technologies emerge. Instead, I will try to provide a big picture survey, and in particular, I will highlight the *why*: why certain features of the technology are important for understanding the brain.

3.1 Measuring Brain Activity in Freely Moving, Freely Behaving Animals

When studying the neural basis of natural behaviors, there are arguably two key requirements for conducting neural recordings: first, to record in a manner that does not limit the animal's behavior, ideally using a lightweight wireless recording device; and second, to record from large neuronal populations. I will start with the second requirement.

Recording from large populations of neurons. Recording from *individual* neurons is necessary because the neuron is the basic processing unit of the brain. Recording from *large populations* of individual neurons is needed for four main reasons. First, the brain is a dynamical system that operates at the level of populations, so ultimately we need to measure from neuronal populations. In particular, population recordings arguably can allow a better understanding of neural computations that underlie behavior (Stevenson & Körding 2011; Carandini 2012; Saxena & Cunningham 2019; Vyas et al. 2020; Urai et al. 2022). Second, as discussed in chapter 1, theoretical work by Gao and Ganguli (Gao & Ganguli 2015; Gao et al. 2017) suggested that more complex behaviors require a larger number of recorded neurons to describe the dynamics of the neuronal population correctly (see chapter 1, endnote 5). In other words, recording large neuronal populations is particularly important when studying natural, complex behaviors. Third, recording from large populations of neurons is necessary for discovering potential distributed neural codes, which are not easily visible at the level of individual neurons. Fourth, increasing the number of recorded neurons also allows for identifying rare types of neuronal responses (Urai et al. 2022), which again might be important when studying natural behaviors.

Technological innovations over the past few decades have steadily miniaturized the recording devices and increased the density and count of electrophysiological recording probes—and consequently increased the number of neurons that can be recorded simultaneously at single-spike resolution. These improvements stemmed from two sources:

1. *Miniaturized electronics and denser electrodes.* This thrust included tinier headstages, as well as improved lithography, which allowed manufacturing silicon probes and multiwire probes with ever-increasing number of recording sites, and with integrated electronics for signal amplification, filtering, and digitization. These include Neuropixels probes (Jun et al. 2017; Steinmetz et al. 2021; Gardner et al. 2022), Neuralink probes (Musk & Neuralink 2019), and others (Raducanu et al. 2017; Hong & Lieber 2019). Currently, using the Neuropixels 2.0 probe, more than 2,500 spike-sorted neurons were recorded simultaneously from a freely moving rat, and hundreds of neurons were recorded simultaneously in mice (Steinmetz et al. 2021; Gardner et al. 2022). Figure 3.1 summarizes the advances in the number of simultaneously recorded neurons in both freely moving animals (right) and head-fixed animals (left), revealing exponential growth over time: "Moore's law for electrophysiology" (Stevenson & Körding 2011). These data suggest that the number of recorded neurons doubles every 6.4–6.8 years on average. Extrapolating this law for the freely moving case (figure 3.1, right) suggests that by the year 2030, we shall routinely reach typical populations of almost 1,000 neurons, with some studies likely achieving populations of >10,000 simultaneously recorded neurons in freely moving rodents and other small animals.

2. *Miniaturized mechanics.* This was enabled by better machining capabilities, and also by the rapid development of high-resolution, 3D printing technologies. These improvements allowed manufacturing ever-smaller and denser microdrives (e.g., Voigts et al. 2020). This has two distinct implications. First, it allows advancing much larger numbers of independently moveable electrodes or tetrodes into the brain, thus recording ever-increasing numbers of neurons. Currently, up to about 300 spike-sorted neurons can be recorded simultaneously from a freely moving rat using a 64-tetrode microdrive, which features independent mechanical control of all the tetrodes (Widloski & Foster 2022); and the numbers keep growing. Second, if one wishes to keep the weight down and allow a more natural behavior for the animal, then 3D printing technology has now allowed this by dramatically reducing microdrive weights. For example, the weight of multitetrode microdrives (e.g., 16 tetrodes) has gone down almost tenfold over the last decade. This transformational weight reduction now allows for studying animals that behave much more naturally.

Such massive population recordings, of hundreds or even thousands of neurons simultaneously, come with distinct computational challenges. First is the spike-sorting challenge: Manual spike sorting is not feasible for such large neuronal data sets. Therefore, a major useful development in recent years was the advent of reliable automatic spike-sorting algorithms, which substantially reduce the need for human curation (Hennig et al. 2019; Buccino et al. 2020; Steinmetz et al. 2021). Second, the data analysis challenge: What shall we do with a huge ensemble of simultaneously recorded neurons? This remains a major open question in neuroscience (Stevenson & Körding 2011; Kohn et al. 2016). Some advances in recent years include using nonlinear dimensionality reduction and tensor component analysis to reveal the internal structure and dynamics of neuronal ensembles (Williams et al. 2018; Chaudhuri et al. 2019; Rubin et al. 2019; Gardner et al. 2022). However, much more work needs to be done, and the question of how to analyze neuronal big data in relation to the animal's behavior is a very active area of research and a key objective for computational and theoretical neuroscience. I shall return to some of these questions in chapter 8.

Alongside the dramatic advances in electrophysiology, calcium imaging techniques have undergone a parallel revolution of miniaturization (Helmchen et al. 2001; Aharoni & Hoogland 2019; Aharoni et al. 2019). Several generations of tiny, one-photon epifluorescence microscopes (known as "miniscopes") were developed over the years. These enable using genetically encoded calcium indicators to perform calcium imaging of large populations of neurons, at single-cell resolution, from identified cell types (Flusberg et al. 2008; Ghosh et al. 2011; Cai et al. 2016; Liberti et al. 2017; Guo et al. 2023). The easy access to cell types is a major advantage of miniscopes; such access is also possible with electrophysiology (Cohen et al. 2012), but it is somewhat more difficult to achieve. Additional advantages of miniscopes are the ability to record from the same neurons over many days (Ziv et al. 2013) and the ability to reveal the fine spatial organization of neuronal populations with particular functional properties (Yang et al. 2022). On the other hand, calcium imaging with miniscopes has also several notable drawbacks. First, calcium transients are related nonlinearly to spiking and also miss many of the underlying spikes; hence, calcium imaging can give only a partial picture about the true underlying spiking activity (Greenberg et al. 2014; Lecoq et al. 2019; Huang et al. 2021). Moreover, these nonlinearities can sometimes lead to *qualitatively* different conclusions about functional properties of neurons and circuits, when using calcium imaging versus gold-standard electrophysiology: these qualitative differences apply both to single-neuron tuning properties and to population-level coding (Wei et al. 2020). Second, because of the slow dynamics of calcium, these miniscopes forgo the single-spike temporal resolution of electrophysiology (note that even though calcium indicators are becoming faster over the years, the physiological calcium signal itself is inherently slow, thus strongly limiting the temporal resolution of calcium imaging). Consequently, miniscopes cannot be used to study very fast population phenomena such as replay/preplay of spike sequences (Foster & Wilson 2006; Diba & Buzsáki 2007; Pfeiffer & Foster 2013), neuronal oscillations (Buzsáki 2006), and the fine temporal coordination of neuronal activity with behavior (Joshi et al. 2023). Moreover, calcium-based miniscopes cannot be used to study neuronal responses to very fast behaviors—including many natural behaviors—in which the timescale of behavioral changes is much faster than the timescale of calcium dynamics. To overcome this key problem, there are now substantial efforts to develop new voltage indicators that yield millisecond-level temporal resolution (Adam et al. 2019; Abdelfattah et al. 2023). If successful, these indicators will allow developing miniscopes capable of direct voltage imaging at single-spike resolution instead of the sluggish temporal resolution of calcium indicators. Third, miniscopes are much more invasive than electrophysiology—particularly when imaging deep structures such as the hippocampus, striatum, amygdala, or hypothalamus, where implanting the miniscope's bulky lens requires removing large portions of the brain overlying the recording site (Aharoni et al. 2019). Nonetheless, the use of miniscopes has become very widespread in recent years.

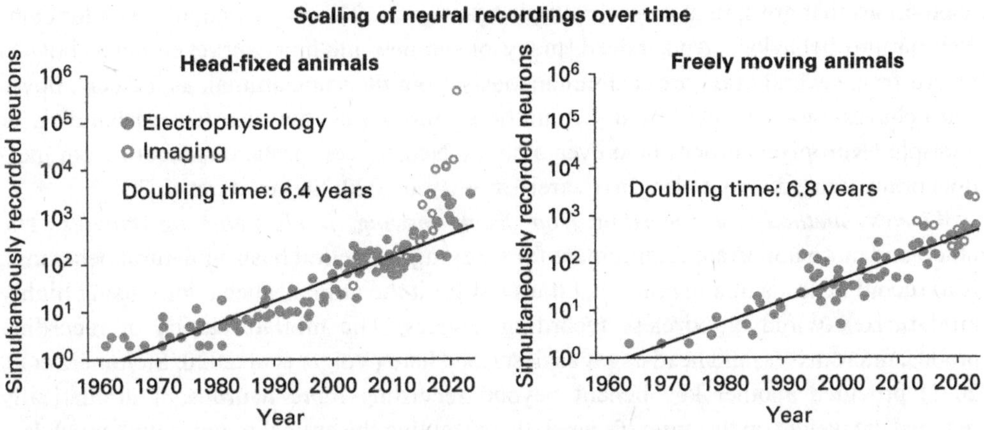

Figure 3.1
Rapid advancement in electrophysiology over the years. Shown is Moore's law for electrophysiology: the number of simultaneously recorded neurons in various studies (filled circles: electrophysiology, open circles: imaging) is plotted as a function of publication-year, separately for recordings from head-fixed animals (left) and freely moving animals (right). These plots reveal doubling of the number of neurons every 6.4–6.8 years (regression lines are based on electrophysiology only). Data from Stevenson and Körding (2011) and Urai et al. (2022), with some corrections and additions by the author; shown are only in vivo data from adult mammals (excluding nonmammals such as zebrafish). Note that for head-fixed recordings (left), the yield of imaging methods is growing even faster and has surpassed the neuronal yield of electrophysiology (Demas et al. 2021; Urai et al. 2022)—but this is *not* the case for freely moving animals (right), where both imaging and electrophysiology give similar neuronal yields.

Miniscopes allow imaging populations of a few hundred neurons and up to about 2,500 neurons simultaneously in the freely moving mouse (Ziv et al. 2013; Scherrer et al. 2023)—quite comparable to the hundreds to about 2,500 neurons that can be recorded simultaneously with electrophysiology using the Neuropixels 2.0 probe[1] (Steinmetz et al. 2021; Gardner et al. 2022). In addition to the one-photon microscopes, there have been parallel efforts to produce miniaturized two-photon and three-photon microscopes that can be carried by freely moving rodents (Helmchen et al. 2001; Zong et al. 2017; Wallace & Kerr 2019; Klioutchnikov et al. 2020). These microscopes offer a much superior resolution as compared to one-photon microscopes, and this high resolution also allows imaging of dendrites and even dendritic spines. Two-photon and three-photon microscopes also allow for imaging much deeper inside the brain tissue.[2] Currently, miniature multiphoton microscopes have a relatively small field of view, which enables imaging of up to about 300 neurons simultaneously (Zong et al. 2021), which is less than the numbers of neurons that can be recorded using one-photon miniscopes or Neuropixels probes (Ziv et al. 2013; Steinmetz et al. 2021; Gardner et al. 2022; Scherrer et al. 2023). In upcoming years, we can expect to see a continued evolution and improvement of both types of miniscopes (one-photon and multiphoton), as well as of miniaturized electrophysiology probes and microdrives: all of these will become increasingly smaller. Each of these methods offers distinct advantages for recording large populations of neurons in freely moving animals, and together, these diverse approaches open the way to studying the neural basis of evermore natural behaviors.

Finally, it is noteworthy that the same arguments that were listed here for the importance of recording from large neuronal populations also suggest the need to record from *several brain areas simultaneously* because most behaviors involve multiple neuronal populations that are spread across multiple brain areas. This is particularly true for complex natural behaviors. And indeed, many of the new methods surveyed here allow to record from several brain areas simultaneously from the same animal, as it freely moves and behaves—for instance, by using multiple miniscopes, multiple tetrode bundles, or multiple Neuropixels probes (and even a single Neuropixels probe can record from multiple brain areas, due to its length) (Varela & Wilson 2020; Steinmetz et al. 2021).

Wireless methods for recording from freely moving, freely behaving animals. The other, and more important requirement for studying the neural basis of natural behaviors, is to record neurons in a manner that does not limit the animal's behavior—using highly miniaturized wired or wireless recording devices. The miniaturization of recording probes, microdrives, and headstages as described here (Voigts et al. 2020; Steinmetz et al. 2021) provided another key benefit beyond recording more neurons: it dramatically reduced the weight on the animal's head, thus enabling the animal to move with much less effort and behave much more naturally. But one key problem remained: recording wires (tethers) impose severe limitations on the naturalness of animal movement and on the naturalness of experimental setups that can be used, because tethers pull on the animal's head and can also get entangled in nonempty environments that contain objects or other individuals. However, in recent years, there has been a true revolution on this front, with the introduction of *wireless-electrophysiology* methods, to which our lab also contributed (Schregardus et al. 2006; Vyssotski et al. 2006; Szuts et al. 2011; Yartsev & Ulanovsky 2013; Schwarz et al. 2014; Finkelstein et al. 2015; Sarel et al. 2017; Omer et al. 2018; Eliav et al. 2021). Miniaturized wireless electrophysiology devices allow recording from untethered, unconstrained animals, performing natural behaviors that were impossible with classical tethered methods. Examples include the following:

1. *Large environments*: Recording from animals moving in very large environments, such as bats flying in a 200-meter-long tunnel (Eliav et al. 2021; Sarel et al. 2022) or foraging in a large 3D flight room (Yartsev & Ulanovsky 2013; Sarel et al. 2017; Ginosar et al. 2021)—or even bats flying outdoors, in the wild (see chapter 4); and rodents or monkeys running in large arenas (Harland et al. 2021; Mao et al. 2021).
2. *Complex environments*: Recording from animals in seminatural environments that are more complex than standard empty laboratory setups, such as rodents running inside burrow systems, hiding within roofed boxes (Reinhold et al. 2019), or climbing inside complex 3D structures (Grieves et al. 2020, 2021)—all of which would be impossible without the wireless technology used in these studies because any wires would get entangled.
3. *Social neuroscience*: Conducting social-neuroscience experiments in multiple animals simultaneously (Rose et al. 2021; Ray et al. 2022), where standard tethers would tangle, making the experiment impossible.

Thus, wireless electrophysiology opens the door to the emergence of a true natural neuroscience research program across species.

Wireless electrophysiology comes in two flavors: wireless telemetry devices, which transmit the data to a nearby receiver via a radio-frequency (RF) transmitter; and wireless loggers, which store the data on-board the animal. Both types of devices are very useful for

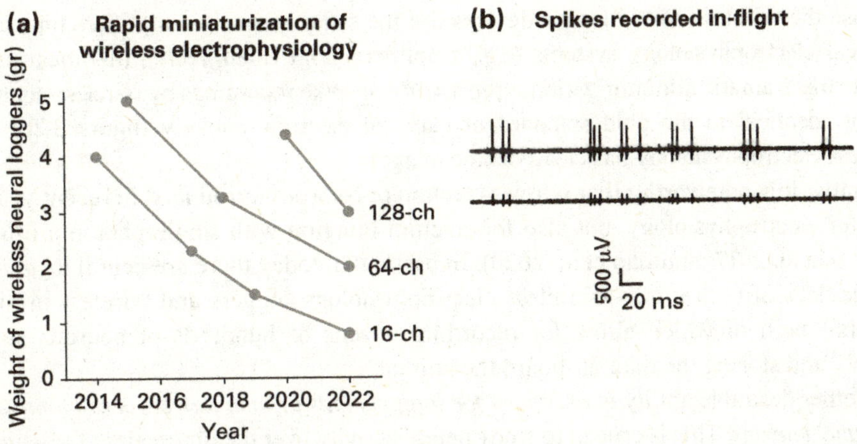

Figure 3.2
Rapid improvement in wireless electrophysiology. (a) Wireless neural loggers have decreased dramatically in weight over the years and increased in channel (ch) count. Shown are various neural loggers from Deuteron Technologies: these wireless neural loggers were used in a variety of species, including bats (Finkelstein et al. 2015; Sarel et al. 2017, 2022; Omer et al. 2018; Zhang & Yartsev 2019; Eliav et al. 2021; Ginosar et al. 2021; Forli & Yartsev 2023), rodents (Reinhold et al. 2019; Bagi et al. 2022), fish (Vinepinsky et al. 2020), birds (Ben-Yishay et al. 2021; Takahashi et al. 2022; Agarwal et al. 2023), and monkeys (Mao et al. 2021; Testard et al. 2024). Other types of neural loggers were likewise markedly miniaturized in recent years and have been used in multiple species. (b) Spikes recorded using a wireless neural logger from hippocampal area CA1 of a flying bat. Shown here are two of the four channels of a tetrode. The bat was flying inside a 200-m tunnel at a speed of about 8 m/s (Eliav et al. 2021).

natural neuroscience experiments, but the telemetry type has three key limitations. First, it is usually limited to transmitting within a few meters from the receiver, thus precluding the use of very large, naturalistic setups of the kind that will be discussed in chapters 4 and 8. By contrast, neural-logger devices allow recording essentially at unlimited distances (Eliav et al. 2021). Second, telemetry-type wireless devices suffer sometimes from RF noises and interference, which is not the case with loggers, which store the data immediately at the source and thus provide consistently high quality. Third, logging the data locally requires much less energy per gigabyte of data than transmitting it via RF telemetry, which means that telemetry systems necessitate a much larger battery, and hence more weight on the animal, leading to less-natural behavior. Therefore, while my lab initially started to work with telemetry devices (Yartsev & Ulanovsky 2013), we very quickly switched to wireless neural loggers (Finkelstein et al. 2015; Sarel et al. 2017, 2022; Omer et al. 2018; Eliav et al. 2021; Ginosar et al. 2021).

The size and weight of wireless neural-logger devices went down dramatically over the years, while the channel count and number of recorded neurons went up (figure 3.2a). Further, two key factors in wireless devices are the battery size and power consumption of the device, and on both of these fronts, there have been steady improvements over the years, driven by the general technology market, which keeps improving both battery technologies[3] and the power consumption of electronic components. Wireless neural loggers are increasingly used in a variety of vertebrate species, including bats, rodents, fish, birds, and monkeys. Soon, their ongoing miniaturization also will allow recording neuronal activity in some freely moving *invertebrates,* such as large beetles, locusts, crabs, and octopus. Importantly, because the wireless neural-logger devices use the same low-noise amplifier chips used in classical electrophysiology systems (e.g., amplifiers from Intan Tech.), this means that—despite the dramatic miniaturization—the *quality* of spike recordings by wireless neural loggers is identical to the gold standard of classical electrophysiology (figure 3.2b). Thus, wireless electrophysiology has clearly come of age.

Finally, it is noteworthy that wireless technologies were introduced in recent years not only for electrophysiology, but also for calcium imaging with single-photon miniscopes (Liberti et al. 2017; Shuman et al. 2020). In particular, today there are neural logger wireless devices of both kinds—wireless electrophysiology loggers and wireless miniscope loggers—both of which allow for recording dozens or hundreds of neurons simultaneously and storing the data on-board the animal.

Another desirable ability is to *record for long periods of time (days/weeks/months) from the same neurons.* This is crucial to study neural activity over the timescales of learning and plasticity. Here again, there has been a technological revolution over the last decade. Head-mounted one-photon miniscopes now allow for performing calcium imaging of the same neurons over many weeks, thanks to improved mechanical stability and efficient algorithms for motion correction and image registration (Ziv et al. 2013; Sheintuch et al. 2017). Head-mounted two-photon miniscopes can similarly track the same neurons over weeks, in freely moving mice (Zong et al. 2021). Similar advances occurred recently in electrophysiology as well: Neuropixels 2.0 combined silicon-probe technology with an algorithm that allows efficient motion-correction by utilizing the high density of recording sites on the probe, which ensures that upon brain motion, a neuron is usually not lost but is recorded by another recording site: this approach enables reliable tracking of the same individual neurons over many days, and even weeks, in freely moving animals (Steinmetz et al. 2021). Together, these recent methodological developments, in both calcium imaging and electrophysiology, now enable to record the activity of large neuronal populations in freely moving, freely behaving animals, while tracking the same individual neurons over timescales of weeks and even months (and in the two-photon miniscopes, potentially tracking individual dendritic spines as well). Stable recordings over these long timescales are crucial for elucidating the neural mechanisms of slow behavioral processes such as long-term memory and learning, the dynamics of complex social interactions in animal groups, or seasonal changes in natural behavior.

In summary, a clear picture emerges from the steady improvements in wireless neural-logger technology (figure 3.2) and battery technology, together with the steady increase in the number of channel counts in both tetrodes and silicon probes (figure 3.1). We predict that within the next decade, it will become possible to *wirelessly record the activity of hundreds or even thousands of neurons simultaneously, in animals performing completely natural behaviors* in large, complex, seminatural environments, and in the presence of many socially interacting individuals.

3.2 Manipulating Brain Activity in Freely Moving Animals

The second major technology (or set of technologies) that emerged in recent years allows reversible manipulations of brain activity in freely moving animals with ever-increasing precision and specificity. This is due in large part to developments in optogenetics (Bernstein & Boyden 2011; Adamantidis et al. 2015; Deisseroth 2015; Rost et al. 2017), but also in other methods, such as chemogenetics and pharmacology. Classically, optogenetics uses light to synchronously activate or inactivate many neurons in a particular brain region, belonging to a specific molecular class, and then to observe the effect of this perturbation on the animal's behavior. As discussed briefly in chapter 1, classical optogenetics has several important limitations. First is the *complex connectivity of the brain*: the neuronal class and brain area of interest are usually highly interconnected with other neuronal populations, and hence the activation or inactivation of the target neurons may also perturb other neuronal populations—for example, in downstream brain regions. If this downstream population is crucial for generating the behavior, then this unintended and unaccounted perturbation may in fact underlie the observed changes in behavior. Thus, in a highly interconnected dynamical system like the brain, it is not valid to attempt determining causality via a transient, massive focal perturbation. This problem of unintended activation of downstream brain regions is sometimes called the problem of *off-target effects* (Otchy et al. 2015). See further discussion of this problem in chapter 1 and in Pearl (2009), which discusses in detail the fundamental difficulties of making causal inferences in complex networks.

Second is the *spatiotemporal pattern of activation*: the neuronal class of interest may itself contain millions of neurons, whose natural complex spatiotemporal dynamics of activity is very different from the synchronized unison that is imposed by classical laboratory-based optogenetic activation. Such synchronized activation treats all these millions of neurons as one "neuron" that is turned on and off. In other words, the rapid simultaneous activation is creating extremely artificial brain activity, which imitates very poorly the normal activity of brain networks. This may lead to another problem called *off-manifold perturbations*: perturbation away from the manifold of normal brain activity (Jazayeri & Afraz 2017).

Third, the firing rates evoked by optogenetic activations are often much higher than the normal firing rates observed in the same circuits during natural behaviors (Headley et al. 2019).

Despite these caveats, optogenetics has been widely used to shed important light on the neural mechanisms of behavior, perception, and cognition. However, as discussed in chapter 1, these mechanisms must be studied in animals that perform behaviors that are as naturalistic as possible, or else the relevance of these mechanisms to normal brain function remains limited. Therefore, it is crucial that circuit manipulations—by optogenetics or other methods—be performed in naturally behaving animals. And indeed, already from the first days of optogenetics, methods were developed for delivering light stimulation via flexible optical fibers, which allow relatively free movement and free behavior (Aravanis et al. 2007). Subsequently, methods were developed to combine calcium imaging with spatially selective holographic photoactivation in freely moving and freely behaving rodents (Szabo et al. 2014; Accanto et al. 2023). The newest developments allow for performing a combination of *two*-photon imaging and optogenetic holographic photoactivation in freely moving animals, reaching very high spatial resolutions (Accanto et al. 2023). Future work is likely to further increase the field of view and the speed of these systems. Importantly, the holographic photoactivation allows for stimulating ensembles of neurons in more naturalistic spatiotemporal patterns rather than stimulating all the neurons synchronously, thus partially overcoming one of the key problems of classical optogenetics.

However, despite these advances, tethering animals to optical fibers substantially limits the naturalness of behaviors that can be studied using optogenetics. This tethering prohibits the use of large environments; it precludes using complex environments; it does not allow animals to go under things (e.g., to enter into burrow systems or home boxes); and it limits the ability of performing social experiments in several animals simultaneously, since their multiple fibers would get entangled. In short, wired optogenetics allows for using only small, empty, and impoverished behavioral setups, and therefore it suffers from all the problems that were discussed in section 3.1 in relation to wired electrophysiology and imaging—problems that have led to the development of wireless electrophysiology and wireless microscopy. This problem was recognized by multiple researchers, and recent years have seen the development of a number of *wireless optogenetics* devices (Dagnew et al. 2017; Emara et al. 2019; Qazi et al. 2019; Yang et al. 2021; Wu et al. 2022; Stuart et al. 2023). Some of these studies combined wireless optogenetics with wireless sensing of neurohormone levels, or with wireless microfluidic delivery of various drugs into the brain, or with wireless release of caged glutamate (Wu et al. 2022; Stuart et al. 2023). These combinations promise versatile approaches for the experimental manipulation of animal behavior in large naturalistic environments.

The development of wireless optogenetics has already enabled some interesting discoveries. Thus, a study by Yamanaka and colleagues (Izawa et al. 2019) has used wireless optogenetics to show that a subclass of hypothalamic neurons, which project densely to dorsal hippocampus, are involved in active memory forgetting. Another study, by Rogers and colleagues (Yang et al. 2021), used wireless optogenetics to stimulate the brains of pairs or triads of animals simultaneously. They induced interbrain synchronization between the prefrontal cortices of these mice and showed that this manipulation modulated the dynamics of social interactions within the group. Interestingly, the latter study used an RF-powered device without batteries, which allowed dramatic miniaturization. In a third study, Alon Chen and colleagues (Anpilov et al. 2020) have constructed a miniature, head-mounted, wireless optogenetic device for mice. The device consisted of a switchable light-emitting diode (LED) that was coupled to a short optical fiber, which was implanted into the target region in the brain. This device allowed them to conduct optogenetic experiments in complex seminatural environments that contained many objects, shelters, and multiple socially interacting animals. They stimulated oxytocin neurons in the hypothalamus, and by using detailed behavioral analysis in the seminatural environment, they discovered that activation of oxytocin neurons could lead to diverse behavioral effects—to enhancement either of prosocial behaviors or of aggressive chasing behaviors, depending on the social context (Anpilov et al. 2020). In other words, they discovered that the effects of oxytocin on social behaviors are more complex and nuanced than found in standard, small, artificial laboratory environments, where most studies reported that oxytocin enhances prosocial effects and reduces aggressive behaviors. This finding highlights again the importance of studying social behaviors in more natural conditions. It also emphasizes the need to measure and quantify behaviors in more detail than is typically done in neuroscience—as will be discussed in the next section.

3.3 Measuring and Quantifying Behaviors

The third dramatic methodological revolution that took place over the last decade is a revolution in detailed measurement and quantification of animal behaviors. But before discussing this revolution in more detail, one may ask: *Why* is there a need to measure the details of animal behavior? Perhaps this is too much information and is irrelevant for understanding the brain?

I will discuss here six answers to this question. First, a crucial point is that many behaviors are complex and cannot be quantified by a one-dimensional (1D), simple variable such as the animal's position or head direction, or the timing of its vocalizations. In particular, many natural behaviors—including spatial exploration, foraging, navigation, predator avoidance, and social behaviors—are very complex and multidimensional, involving multiple degrees of freedom, and therefore require many behavioral variables to describe them properly. If we use partial, simplified descriptors of behavior, we may lump together behaviors that are in fact completely different from the perspective of the animal. This may create great confusion if we wish to understand the neural bases of behavior. Therefore, we should measure in detail many more behavioral variables than we usually do—in particular, we must measure at least those variables that we think may be relevant for fully capturing the behavior as the animal sees it.

The discrepancy between the true dimensionality of the behavior and the dimensionality of behavioral measurements is very prominent in laboratory-based, standardized behavioral tests, which are used routinely today in brain research. These tests are typically based on measuring very simple metrics, which quantify aggregate average behavior over an entire session into one number along a 1D scale. For example, in the Morris water-maze memory test, the standard metric measures the percentage of time that the animal spends in the quadrant of the maze with the submerged platform (Andersen et al. 2007). Similarly, in social behaviors, one of the classical tests—the three-chamber test of sociability—measures the average time that the mouse spends in one chamber or another (namely, near another mouse or not). However, by measuring the average time that one animal has spent in each chamber, one completely misses any fine temporal dynamics of behavior. Further, by averaging behavior into one number (one metric), such as the average time spent in one chamber versus another—you miss many other aspects of social behaviors, such as social vocalizations and social sniffing. In other words, because behavior is complex and multidimensional, it is impossible to summarize it all by one number or one dimension. To comprehensively characterize social behavior, it is crucial to measure and quantify in detail the multidimensional complexity of social behaviors, including the time spent near other animals, but also the angle of the head and the body toward them; you need to quantify aggressive chasing versus social grooming; you need to measure and quantify sniffing behaviors, and social whisking; and you need to record ultrasonic social vocalizations, as well as emitted pheromones, and much more. In short, when studying social neuroscience, one needs to measure and quantify many complex details of animal behavior. This is clearly true for social behaviors, but it is likely true for many other behaviors, and perhaps for all animal behaviors.

Importantly, as will be discussed next, although natural behavior is not 1D, it is also not extremely high-dimensional and complex, since natural behaviors occupy only a small subspace of the entire possible movement space (more on this to come). Thus, the dimensionality and complexity of natural behaviors are, in some sense, middle range. Further, certain natural behaviors tend to repeat multiple times in a relatively stereotypical manner. Together, this mid-dimensionality and repeatability make natural behaviors amenable to investigation.

Second, while measurements of brain activity have undergone a revolution over recent decades, with ever-finer levels of granularity, animal behavior was classically measured at a very coarse level of granularity. This mismatch between the levels of granularity in measuring brain and behavior seems inappropriate if we wish to understand the brain basis of behavior. We must therefore measure animal behavior at a much higher level of detail, as we do for brain measurements. Furthermore, it is likely that if we stop treating the animal as a dot moving through space, and instead we increase the level of detail in analyzing behavior, we will be in a better position to understand the *algorithms* for generating behavior (Krakauer et al. 2017).

Third, there is mounting experimental evidence, from a variety of brain areas, that demonstrates that neurons encode subtle behavioral variables that classically were not thought to be encoded in those areas. For example, a recent study by Lisa Giocomo and colleagues in freely moving mice (Mallory et al. 2021) found that neurons in the medial entorhinal cortex, which are classically thought to represent the animal's position in space (e.g., grid cells, border cells) or its azimuthal direction (azimuthal head-direction cells), also encode the pitch and roll angles of head direction—that is, they encode 3D head direction. So far, this is perhaps not very surprising, as this is just an extension of the known azimuthal head-direction tuning in the rodent entorhinal cortex (Sargolini et al. 2006); and in addition, our previous studies in bats already showed 3D head-direction coding in both entorhinal cortex and in the adjacent presubiculum (Finkelstein et al. 2015; Ginosar et al. 2021). What was much more surprising in the Giocomo study was that it found a substantial fraction of neurons in mouse medial entorhinal cortex, which encoded the *eye* gaze in 3D (albeit the eye-gaze tuning was measured separately in a virtual-reality setup). While it is known that rodents move their eyes substantially during free behavior (Wallace et al. 2013; Michaiel et al. 2020), the finding of such strong neuronal selectivity to eye movements in the mouse entorhinal cortex came as a surprise. This selectivity never would have been discovered without measuring unconventional details of behavior—in this case, eye-gaze behavior.

Another example comes from our studies in flying bats, which showed that many neurons in hippocampal area CA1—which classically are thought to represent the animal's absolute position in space (place cells)—surprisingly encode also the egocentric direction of the animal's movement relative to its navigational goal (Sarel et al. 2017), as well as the directions and positions of other bats (Omer et al. 2018; Ray et al. 2024). Here again, if we wouldn't measure the animal's movement direction relative to the goal or the positions of other individual bats—both of which are quite unconventional behavioral variables in the context of hippocampal literature—we wouldn't have discovered these surprising types of coding.

Another example demonstrating that neurons encode subtle behavioral variables comes from a study by Jonathan Whitlock and colleagues (Mimica et al. 2018), who studied neurons in the posterior parietal cortex of freely moving rats. They measured six postural features of the rat's body—the azimuth and pitch of the animal's back, the elevation of the neck, and the azimuth, pitch, and roll of the head—and they found that about 50 percent of the neurons in the parietal cortex represented some or all of these six variables of the animal's posture. While previous ideas on the function of the rodent posterior parietal cortex included involvement in navigation, in action coordination, or in decision-making (Nitz 2006; Whitlock et al. 2008, 2012; Harvey et al. 2012), it was never considered that this brain area may encode the detailed body posture of the animal. Here again, if the detailed posture hadn't been measured, this coding wouldn't have been discovered.

Yet another example comes from several recent studies, from a number of groups, which have videotaped mouse faces and discovered that neurons across several brain regions encode mouse facial expressions (Musall et al. 2019; Stringer et al. 2019; Dolensek et al. 2020), reflecting very high-dimensional behavioral variables. These facial expressions were task-irrelevant, having no direct influence on the mouse's performance in the experimental tasks used in these studies; and yet the facial expressions were strongly encoded by many neurons.

These examples are just a few of the many recent studies demonstrating that neurons in a variety of brain regions encode much more diverse behavioral variables than traditionally thought. Crucially, to discover neuronal tuning to such unconventional behavioral variables, we must *measure* in our experiments a much more diverse and detailed set of behavioral variables.

A fourth reason for measuring the animal's behavior in great detail is to counter a central critique that is directed at the freely moving approach—namely, that no two behavioral trials are the same. Specifically, this critique posits that any observed differences in neural activity between different trials may be due to unobserved differences in behavior, because during free movement, the animal's behavior during different trials is likely to differ, whether to a large extent or in subtle ways. For example, if the experimenter had not measured the eye gaze of the animal, then perhaps the neuronal differences are due to some unobserved (unmeasured) gaze differences. To control for such differences in behavior, a first step is to *measure* the behavior, including its subtle details. The next step is to use sophisticated data-analysis methods to disentangle the contributions of different behavioral variables to neural activity. Such analysis methods are discussed in chapter 8, but they require first measuring all the relevant behavioral variables that we suspect to potentially affect neural activity. Of course, one could take this critique to an extreme and argue that perhaps the neuron is responding to a very subtle movement (e.g., to the acceleration of the second joint of the third toe of the front right foot), so does this mean that we should measure the movement of each and every joint in the body? Perhaps not. But at least one should measure behavior in more detail than was done classically—where, instead of a detailed pose with complex motion, the behavior of the animal was typically reduced to the movement of a single dot (namely, its center-of-mass). Any detail is surely better than this highly reductionist description.

Fifth, it has been counterargued that acquiring behavioral "big data" (e.g., massive videos of behaving animals) is not enough because eventually what we need are not the raw behavioral data, and not even their parsing into units of movement. Instead, goes the argument, we need to parse the behavior into units of *meaning*. That is, parsing the behavior to different segments and what they *mean* for the animal, from the animal's perspective (Gomez-Marin et al. 2014). Classically, this has been done by ethologists through careful observation of the animal's behavior and parsing it into ethograms of specific meaningful behaviors. However, to scale up this approach to much larger amounts of data, one needs to parse behaviors automatically. Such parsing can likely be done much better if one measures more details of the animal's behavior.

Sixth and finally, there is the argument of "unknown unknowns." In brain research, we have operated for many years under specific assumptions about the behavioral role of particular brain areas—roles that were elucidated via highly reductionist, simplified experiments, which were quantified using highly simplified behavioral metrics. But what if we are wrong? What if neural circuits care about much richer aspects of behavior? We would be able to answer this question only if we *measure* richer, more detailed aspects of behavior. In other words, what I am proposing here is *not* an overemphasis on behavioral details—because, in most cases, it is not clear at all which details are the important ones and which details we can neglect. Therefore, at this stage, it is prudent to measure as many behavioral details as possible.

After discussing *why* we should measure the details of behavior, let us now turn to *how* to do it. Recent years have seen a methodological revolution on this front—or actually three revolutions. The challenge of measuring and quantifying the details of behavior was met in the last decade by two different *experimental* approaches, and also by developing a flurry of *computational* tools that utilized the deep-neural-network revolution, which provided a means to automatically quantify many aspects of behavior.

The two experimental approaches are virtual reality (VR) experiments and freely moving experiments. Let us start with animal VR—a methodology that started to rapidly expand about fifteen years ago. VR in head-fixed animals offers several advantages. First, all the sensory stimuli can be completely controlled. Second, head fixation allows for measuring the animal's behavior in great detail, more than would be possible in freely moving scenarios—for example, videotaping the facial expression of the mouse at high resolution, as has been done in a number of studies (Stringer et al. 2019; Dolensek et al. 2020). Third, and perhaps most important, head fixation makes it easier to conduct brain recordings, either via two-photon imaging, intracellular electrophysiology, or extracellular electrophysiology (e.g., Neuropixels). On one hand, these are indeed substantial advantages. On the other hand, as was discussed in chapter 1, head fixation is very artificial and creates major problems:

1. It eliminates many normal inputs, such as vestibular inputs, and also substantially distorts motor and proprioceptive signals. The elimination of vestibular inputs is particularly important when studying the neurobiology of navigation (e.g., place cells, grid cells, or head-direction cells), where vestibular inputs are known to be crucial for normal function. Moreover, the strong mismatch between vestibular inputs and visual inputs, which is experienced by the animal in the VR environment, makes VR navigation quite different from real-world navigation. This problem has led to strong arguments against the use of VR in both animal navigation and in human navigation experiments with functional magnetic resonance imaging (fMRI; see Taube et al. 2013; Minderer et al. 2016). The lack of vestibular inputs is important also when studying sensory systems—even vision—because, for example, it was shown that most layer 6 cells in V1 respond to vestibular signals (Vélez-Fort et al. 2018). Since in VR there are no normal vestibular inputs, we handicap the study of V1 by using head-fixed VR.
2. Head-fixed setups may stress the animal.
3. In most cases, animal VR does not feel very realistic.

4. Very minor differences in the VR setup may lead to very different neuroscientific results. This was demonstrated by two studies of spatial neurons in two-dimensional (2D) VR: one study found in VR normal-looking place cells, grid cells, and head-direction cells (Aronov & Tank 2014), while the other reported a complete disappearance of hippocampal place-tuning in VR (Aghajan et al. 2015). Thus, these studies found fundamentally different results in seemingly similar experiments. Both studies used VR setups where the animals were body-fixed (or body-restrained) rather than head-fixed. It turned out that the major differences in neural selectivity between these two VR-based studies were caused by minor differences in how much the animals were able to move: the rats were able to move their heads slightly more in the study that found normal spatial tuning (Aronov & Tank 2014), thus allowing more naturalistic vestibular inputs, as compared to the study that found a lack of normal spatial tuning (Aghajan et al. 2015). These differences are very worrying because the two studies used very similar VR setups—the differences in the setups were truly minor—and yet the neuronal results were diametrically opposite. This suggests that VR setups should be used in neuroscience with great caution. These differences between the two studies also suggest that one possible way to retain some of the advantages of VR (and in particular retain the ability to fully control the sensory inputs) while avoiding most of the pitfalls of VR is to avoid altogether any head fixation or body fixation, and instead to use freely moving VR.

A freely moving VR system was developed by Andrew Straw and colleagues and was implemented for flies, fish, and mice (Stowers et al. 2017). They used computer-vision algorithms to track the animal in real time, and they used computer-graphics methods to project visual images that are calibrated to show the world correctly *from the animal's perspective* (i.e., from the animal's position) as the animal walks, flies, or swims. This VR environment avoids the problem of vestibular-input elimination, which characterizes head-fixed VR, because the animal moves completely freely, but it still maintains two of the main advantages of VR: tight control over the sensory inputs and precise manipulations of the environment. This immersive VR allowed even introducing artificial agents: virtual animals whose behavior is fully controlled by the experimenter, thus allowing interesting perturbational studies of collective animal behaviors and social behaviors (Stowers et al. 2017; Sridhar et al. 2021). However, for the latter direction to be viable, it needs to be shown that the real animals perceive the virtual animals as real—a formidable challenge indeed. The problem of believable VR in animals is accentuated by the fact that all the current VR systems are vision-based, which may be reasonable for animals that rely primarily on vision, such as humans—but many animal species rely strongly on other senses, such as olfaction, hearing, and touch. For such species, it would be important to develop multisensory VR systems that combine vision, olfaction, hearing, and/or touch, depending on the **Umwelt** of each species.

* * *

The second experimental approach for measuring the details of behavior, which has boomed over the last decade, is to employ freely moving experiments in a real environment, as opposed to virtual environments (figure 3.3). Here, recent technological developments allow evermore detailed measurements of animal behavior. In addition to videotaping the behavior, often from multiple cameras simultaneously, many experimenters now record the animal's acoustic vocalizations using microphones; measure the animal's olfactory sniffing patterns using nasal cannulas; measure its whisking using dedicated high-speed cameras; and measure the details of the animal's movement using motion sensors. For example, in our wireless neural loggers for bats, we now have on-board—in addition to 64 neural recording channels—a host of behavioral sensors: an ultrasonic microphone for recording the bat's social vocalizations and echolocation calls; a "nine-axis" motion sensor (three-axis accelerometer, three-axis gyroscope, and three-axis magnetometer), which allows measuring the animal's 3D head direction, as well as measuring fine movement patterns of the head; a GPS module, for measuring the animal's 3D position; and an altimeter, for measuring the animal's altitude more precisely than GPS. It is also possible to add a small video camera on-board, as well as many additional sensors. All these behavioral sensors are highly miniaturized and are mounted on-board the neural logger—on the bat's head—all flying with the animal wherever it goes while we perform neural recordings. This approach—putting all the behavioral sensors on the animal's head, together with the neural recordings—enables studying the neural basis of behavior while the animal is behaving freely in large, complex environments (Omer et al. 2018; Eliav et al. 2021; Sarel et al. 2022). This allows much more natural behaviors by the animal. The ability to put all these sensors in one miniaturized electronic board, which is small enough to be carried by a flying bat, is due to rapid technological developments in recent years. For example: only in the recent few years did GPS modules appear that combine improved spatial precision and high temporal sample rate of 25 Hz or more, thus allowing positional measurements with high temporal resolution and high spatial resolution; only in recent years did altimeters become miniaturized enough to be carried by small animals; and only recently did motion sensors become small enough and high-quality enough to enable precise measurements of animal movements and 3D head direction. These and similar technological developments now allow wirelessly measuring the details of animal behavior while performing wireless neural recordings in multiple freely moving animals simultaneously. Similar systems have now been developed for smaller animals, such as mice (Meyer et al. 2018, 2020), albeit this system is wired, not wireless. In chapters 4 and 8, I will return to these on-animal systems and discuss in more detail the results that can be obtained using these new methodologies for behavioral measurements.

Another technological advance that took place in recent years allows measuring, in freely moving animas, what the animals is seeing, and what it is looking at—*from the animal's perspective*. This relies on a combination of several technologies. First, technologies for 3D scene reconstruction, such as 3D laser scanning using light detection and ranging (LIDAR) systems, or camera systems, which allow mapping the 3D structure of the environment in great detail (Stürzl et al. 2015; Couzin & Heins 2023; Koger et al. 2023). These cameras or lasers can be ground-based or they can be carried by drones and map the 3D structure of the landscape from the air (Couzin & Heins 2023; Koger et al. 2023). This type of mapping is often used in behavioral studies outdoors, but similar methods can also be used for performing 3D reconstruction of complex seminatural environments indoors, in neuroscience experiments. Second, there are new technologies for measuring the animal's gaze direction. In animals that do not move their eyes much, such as birds, one can measure the 3D head direction as a proxy for the animal's gaze direction, using nine-axis motion sensors similar to the one used in bats (Kano et al. 2018), as well as on-board tiny cameras. For animals that do

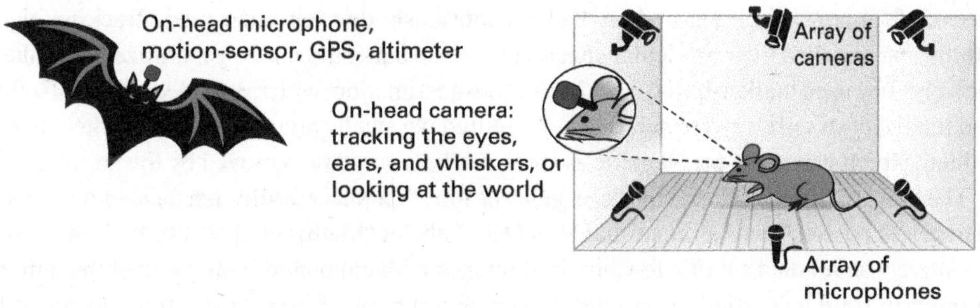

Figure 3.3
Multiple behavioral sensors are used today in freely moving animals. Experimenters document the animal using arrays of cameras, allowing measuring its 3D pose and detailed behavior (see also figure 3.4), and also using arrays of microphones, allowing for documenting animal vocalizations and their directionality in space. Furthermore, experimenters place on the animal's head a whole slew of sensors, including on-head microphone, piezoelectric sensor, motion sensor (accelerometer, gyroscope, and magnetometer), sniffing sensor, GPS module (and other localization tags), and altimeter. They also place miniature head-mounted cameras that can document the world from the animal's point of view, videotape the animal's eyes for computing the gaze direction, or videotape its ear movements and whisking.

move their eyes, such as mammals, several labs developed approaches for tracking the eye gaze using tiny, head-mounted cameras (Wallace et al. 2013; Meyer et al. 2018; Wallace & Kerr 2019). Combining the various types of information—namely the gaze-direction measurements, the animal's measured position, and the reconstructed 3D structure of the environment—allows computing, in freely moving animals, what exactly the animal is looking at and which visual information is available to the animal at every moment. Measuring what the animal is looking at might be important not only for questions of sensory processing, but also for understanding the animal's behavioral goals, including where it intends to go (i.e., assessing its planning), as well as addressing questions of decision-making and more.

Importantly, the head-mounted cameras can be also directed to the ear's pinna to measure how freely moving animals direct their pinnae during sound localization. The cameras can also be directed at the whiskers to measure whisking patterns during free movement (Meyer et al. 2018). In short, we are at the beginning of an era of evermore detailed behavioral measurements in freely moving animals.

* * *

The third, and perhaps the biggest, methodological revolution in quantifying animal behavior is the advent of machine-learning-based methods for data extraction and data analysis of the massive behavioral data that are being acquired by neuroscience laboratories. These approaches focus mostly on automatic analysis of behavior from video data and rely on a number of machine-learning tools, including heavy use of deep neural networks. These methods have been recently described in detail by several excellent reviews (Egnor & Branson 2016; Datta et al. 2019; Mathis et al. 2020; Mathis & Mathis 2020; Pereira et al. 2020; Goodwin et al. 2022; Luxem et al. 2023), so I will not describe them in length here. Instead, I will try to give a very brief overview of the main methods and what they are used for.

First are methods for *automatic pose estimation and tracking of body parts.* This is a difficult problem because body parts look very different in different species, and even in different experimental setups for the same species—creating a challenging generalization problem.

Indeed, for many years, researchers had to laboriously develop tailor-made tracking algorithms that matched their particular species and setup but could not be generalized—or alternatively, they used marker-based tracking and pose estimation, which provided a generalizable and relatively straightforward solution but required physically attaching many markers to the animal, which was often problematic as these markers could be removed by the animal.

The original and most widely used *general* tool that successfully performed automatic markerless pose estimation in animas was DeepLabCut (Mathis et al. 2018). It allows using a relatively small data set of a few hundred images with annotated body parts of the animal in order to train a (partially pretrained) deep neural network to robustly track those body parts in new video data, thus solving the generalization problem. DeepLabCut has been used on a variety of species, including on videos of wild animals outdoors, and a recently developed version allows pose estimation in videos of multiple interacting animals (Mathis et al. 2018; Nath et al. 2019; Mathis et al. 2020; Mathis & Mathis 2020; Lauer et al. 2022). Other tools, such as LEAP and SLEAP (Pereira et al. 2019, 2022), allow for going beyond pose estimation to analyzing the *gait* of multiple socially interacting animals. Similar tools, for 3D pose estimation in multiple socially interacting mice, were developed by the lab of Robert Froemke (Ebbesen & Froemke 2022). Their study revealed that many neurons in the somatosensory barrel cortex encoded multidimensional combinations of social variables (such as nose-nose distance and nose-tail distance between the recorded mouse and another mouse), postural variables (angles of the head, the hip, and the spine), and movement variables (velocity, angular velocity, rate of posture changes). Many neurons were tuned to five or more behavioral variables, usually to combinations of social, postural, and movement variables. Such a high-dimensional neuronal representation in the barrel cortex could be discovered only via detailed 3D pose estimation of multiple animals. Thus, there has been a true revolution in recent years, evolving from a coarse description of the animal as a moving dot (centroid tracking) to finer and more detailed pose estimation (figure 3.4a).

The second are methods for *tracking the identity of multiple animals within large animal groups*. This is a difficult problem because different individuals often look very similar to each other and their paths may cross, making it nontrivial to determine who's who. A very successful tool that does this is idTracker.ai (Romero-Ferrero et al. 2019), which can track up to 100 individual animals over long periods of time without putting any markers on them. The idTracker.ai tool uses pretrained convolutional neural networks to identify animal crossings and to track individual animals between and across these crossings. This method works with very high accuracy, provided that the background is uniform, that all the animals stay within the camera's field of view, and that there are no complex occlusions/overlaps of different individuals. In situations where the background is complex and nonuniform, or if occlusions are expected, one can use other methods, which rely on marking different individuals with unambiguous markers, such as giving them unique colors (Shemesh et al. 2013) or placing barcodes or QR codes on them (Crall et al. 2018; Ray et al. 2024).

The third of these are methods for automatic *classification and segmentation of an animal's behavior into behavioral motifs* (figure 3.4b). One method, VAME (Luxem et al. 2022), relied on pose estimation using DeepLabCut, followed by an autoencoder and hidden Markov model (HMM), to segment the continuous behavior of the animal into discrete behavioral motifs, and to characterize the transition matrix between the various behavioral motifs. These motifs captured the elemental repetitive movements that comprise distinct

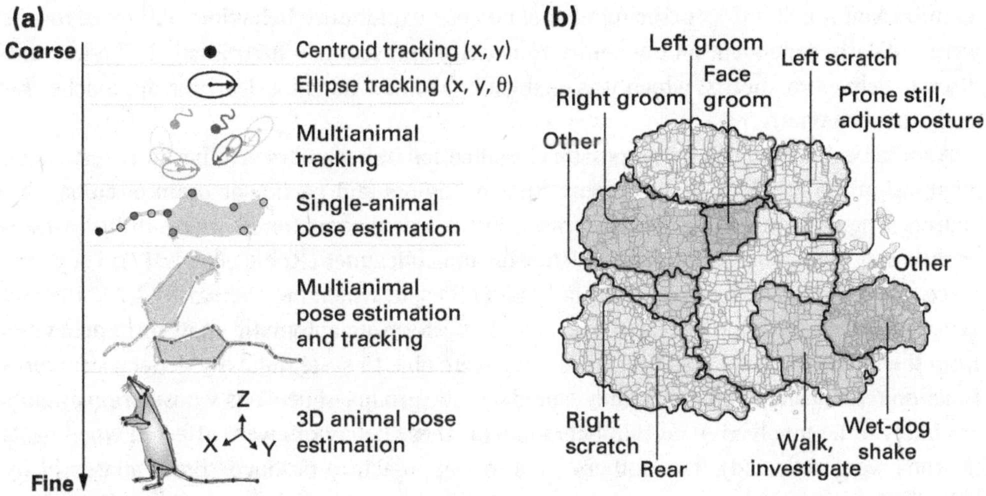

Figure 3.4
Quantifying animal behavior. (a) Progression of pose estimation over the years, from coarse description of the animal as a moving dot (x, y) or a dot with a head direction (x, y, θ), to 2D pose estimation, to full 3D pose estimation. Adapted from Pereira et al. (2020). (b) Unsupervised low-dimensional embedding of rat postures: computed using *t*-SNE dimensionality reduction of a very large data set of postures from one animal. The behaviors and subbehaviors were then classified based on the density representation of the behaviors in *t*-SNE space. Adapted from Dunn et al. (2021).

behaviors, which in rodents include walking, turning, rearing, and grooming. A related method, called MoSeq, was developed by Bob Datta and colleagues and used in a series of studies (Wiltschko et al. 2015; Markowitz et al. 2018; Weinreb et al. 2024). They found that mouse behavior was composed of a few discrete and stereotyped behavioral motifs (also called "behavioral modules" or "syllables"), with durations of a few hundred milliseconds each. These motifs can be flexibly concatenated, forming highly structured behaviors on longer timescales. To analyze this, they took video data from depth cameras and used an AR-HMM model, where each stereotyped behavioral motif was modeled by an autoregressive (AR) process describing a stereotyped trajectory through pose space, and the switching dynamics between the behavioral motifs was captured by an HMM. They showed that the short behavioral motifs were reused by the mice repeatedly to form different behavioral sequences: this reuse strategy forms an interesting modular-hierarchical outlook on the generation of behavior (Wiltschko et al. 2015). This outlook is in line with early ethological ideas (Tinbergen 1951), and also with a classical notion in motor neuroscience, whereby movements are composed of "motor primitives" (behavioral motifs) that are concatenated into sequences and are reused to produce a diversity of complex movements (Flash & Hochner 2005; Anderson & Perona 2014).

In another study (Markowitz et al. 2018), Datta and colleagues showed that the striatum is important for behavioral sequence generation during free behavior. They found that striatal neurons encode the identity of the short behavioral motifs ("syllables"), and these neurons also represent the ordering of the motifs, thus encoding behavioral sequences. Lesions in the dorsolateral striatum created disorganization of motif sequences, supporting a causal role for this brain region in the generation of behavioral sequences. A subsequent study (Markowitz et al. 2023) showed that dopaminergic inputs into the dorsolateral striatum play an important role in the structuring of spontaneous exploratory behaviors. All these results were critically dependent on the ability to classify and segment naturalistic behaviors into discrete behavioral motifs, which was enabled by the new machine-learning approaches for analyzing behavior.

Another interesting use for automatic classification of behaviors is to perform systematic mapping of the relations between particular behaviors and particular brain circuits. This approach may be too naive in vertebrates, but it was demonstrated successfully in *Drosophila melanogaster* flies by Kristin Branson and colleagues (Robie et al. 2017). They analyzed videos collected from 400,000 individual flies, in which they activated 2,204 distinct genetically targeted populations of neurons. By performing automatic analysis of behaviors from the videos in these 2,204 fly lines, they were able to systematically deduce structure-function relationships—namely, they elucidated the groups of neurons whose brain circuits are likely to be involved in certain behaviors (as these behaviors were affected when these neurons were activated). The authors used this approach to produce "brain atlases of fly behavior," which can now be used to test in depth the mechanistic bases of specific behaviors. Here again, this work was possible only through the new machine-learning approaches for automatic analysis of behavior.

The fourth set of methods is for *characterizing the dynamics of behavior*. Natural behavior can be thought of as a time-evolving dynamical process, so it is clearly important to develop methods for automatically describing the rich dynamics of behavior. This can be achieved by either supervised methods or unsupervised methods. Supervised methods rely on an annotated data set that is used to train classifiers, which can then automatically label new behavioral data. Supervised methods have been very successful in generating automatic ethograms of complex behaviors (Kabra et al. 2013). However, they also suffer from two main drawbacks: first, it is often difficult to generate large enough annotated training sets that would allow achieving high performance of the classifiers in capturing truly rich and complex behaviors; and second, supervised methods do not allow for discovering new behaviors beyond those used a priori in the training set. These limitations can be overcome by using unsupervised methods, which automatically classify clusters of behaviors and extract their dynamics (Todd et al. 2017; Goffinet et al. 2021).

An early example of the unsupervised approach came from the work of William Bialek, William Ryu, and colleagues, who analyzed the behavior of the worm *Caenorhabditis elegans*, and showed that the worm's behavior can be fully captured by low-dimensional (four-dimensional) descriptors of the worm's body shape, called "eigenworms" (Stephens et al. 2008; four is a very low number of dimensions compared to the dozens of dimensions required a priori to fully describe a worm's body shape). They illustrated the power of this approach by analyzing the behavior of mutant strains of *C. elegans*, which did *not* differ in their classical behaviors as defined via traditional ethograms, but they did differ substantially in the detailed statistics of their behavior, as quantified by projecting their behavior onto a 4D eigenworm space (Stephens et al. 2010a). Another study used this low-dimensional representation to discover a previously unknown behavior in *C. elegans*, the delta turn (Broekmans et al. 2016). These results illustrate the importance of quantifying the rich details of behavior (in this case, worm behavior).

A study by Joshua Shaevitz and colleagues (Berman et al. 2014) developed a different method for unsupervised clustering of behaviors based on dimensionality reduction of spectrograms of postural modes in *Drosophila* flies. Because the dimensionality reduction was done on the spectrogram, this approach extracted repeatable *periodic* behaviors with specific frequencies, such as various types of walking, grooming, and wing waggling. This study showed that flies exhibit about 100 distinct and stereotyped periodic behaviors. A subsequent study (Berman et al. 2016) showed that fly behavior consists of hierarchically organized actions forming behavioral sequences on multiple timescales. Similar results also apply to mammals: unsupervised behavioral analysis revealed distinct and stereotyped behaviors with hierarchical structures in the behavior of mice (Wiltschko et al. 2015) and rats (Marshall et al. 2021).

Overall, a large body of work revealed several principles of behavior, which are certainly not universal but do apply in many instances (Stephens et al. 2011; Brown & de Bivort 2018): First, behavior is relatively *low dimensional* (or medium dimensional) and does not cover the entire space of possible postures. Second, behavior is often *discretized* into multiple behavioral modes that are distinct from each other. Third, behavior is *hierarchical*, with long and complex behavioral sequences assembled hierarchically from simpler behaviors. Fourth, while *many behaviors are complex and nonstereotypic,* reflecting a complex transition matrix between the many different behavioral modes—there are also some behaviors that are relatively *stereotypic,* reflecting repeated occurrences of particular behavioral modes or particular transitions between modes. It is clear that future efforts to derive general rules of natural behavior will benefit from further development of automatic approaches for quantifying behavior.

Fifth, there are *combinations of these behavioral methods.* For example, a recent study (Weinreb et al. 2024) introduced Keypoint-MoSeq, a method that combines DeepLabCut's pose estimation, which tracks key points on the animal's body, with MoSeq's classification and segmentation of behavior into behavioral motifs. In the future, it would be useful to also combine this with methods for multianimal tracking, such as idTracker.ai: this will yield a very rich and structured characterization of natural social behaviors in large animal groups.

In sum, recent years have seen a true revolution in the automatic analysis of the details of animal behavior. However, multiple challenges remain (Datta et al. 2019). One challenge is that behavior occurs in 3D, while most of the methods that have been developed thus far act on 2D videos, although substantial progress has been made in automatic 3D pose estimation (Mathis & Mathis 2020; Dunn et al. 2021; Marshall et al. 2021). A second challenge is to determine at which level of granularity is it correct to describe behavior. This refers both to the granularity of the behavior itself (i.e., the level of *detail* used to describe the behavior) and to the question of what the relevant timescales are. There is no obvious answer here, and it remains an important challenge for the future. A related problem is that the animal's behavior on long timescales provides *context* for its behavior on short timescales, and it remains unclear how to incorporate context into the study of the neurobiology of animal behavior. This is important because the same sequence of movements that an animal performs might mean different things depending on context. A third challenge is that animals often perform several behaviors at the same time (e.g., predator avoidance during navigation; or running, flying, or swimming during foraging), which raises the question of what are the correct labels for such multifaceted behaviors.

A fourth challenge is that truly natural behaviors that occur outdoors take place on multiple spatial scales, from centimeters to thousands of kilometers. While videotaping the behavior of a mouse running in a small box allows for using advanced machine-learning tools to analyze these limited behaviors, one may ask: How are we going to videotape the behavior of animals that navigate many kilometers outdoors and *also* engage in complex behaviors on much smaller spatial scales? A recent study by Iain Couzin and colleagues (Koger et al. 2023) has addressed this problem by developing a drone-based approach to track and videotape animals outdoors, in the African savanna. They used a drone with a camera to track large animals such as baboons and zebras, over relatively long distances of hundreds of meters, while concurrently recording close-up videos of their behavior. This study then used machine-learning methods to analyze the drone's videos and automatically extracted the poses of multiple animals from these videos. Thus, at least in some cases, one can clearly overcome the challenge that behavior occurs over multiple spatial scales. But what about animals for which many of the key behaviors occur underground, such as wild mice and rats? How would one videotape them underground? Here, using a lab-based system of transparent tunnels may yield a practical solution. Generally speaking, this remains a major open challenge. In any event, these examples serve to illustrate the problems facing anyone trying to extend the methods of behavioral measurements and automatic behavioral analysis to *actual* natural behaviors in the wild—or even to complex seminatural setups in the lab.

A fifth and final challenge is that, while there has been dramatic progress in machine-learning-based, automatic analysis of behavior as captured in *video* data, much less progress has been made on automatic analysis of behavioral data captured via other sensors, such as movement data as recorded via motion sensors (accelerometers and gyroscopes) or social vocalizations as recorded via microphones—although this is beginning to change (Hammond et al. 2016; Brandes et al. 2021; Goffinet et al. 2021; Cohen et al. 2022; Peterson et al. 2023). For example, motion-sensor data can be used to classify behaviors such as walking, flying, eating, fighting, and more, by relying on the characteristic patterns of acceleration and angular acceleration of the head in each of these behaviors—even without any videos.

In closing, it is clear that despite all these challenges, the novel methods of computational ethology will likely revolutionize the study of the neural bases of natural behaviors (Anderson & Perona 2014; Datta et al. 2019).

3.4 Summary

The last decade has seen an explosion of new methods, which finally enable studying the neural basis of truly natural behaviors. These methods were briefly reviewed in this chapter and include methods for wireless neural recordings from multiple freely moving animals, using wireless electrophysiology or wireless calcium imaging; methods for wireless manipulation of neuronal circuits; methods for recording a wealth of behavioral information on-board the animal, from the animal's point of view; and methods for automatic analysis of behavior from video data (in particular, automatic pose estimation and automatic behavioral classification). To these, we should add several recently developed methods for sophisticated data analysis, which allow for disentangling the contributions of different behavioral variables to neural activity: these analysis methods are discussed in chapter 8. Together, these novel methods open the door to an exciting future for neuroscience—a future that will truly elucidate the neural basis of natural behaviors.

Some of the possible future directions that were opened by these methodological revolutions are described in chapter 8. But before discussing the future, I will focus in chapters 4–7 on discussing the past and present. In these chapters, I will describe in detail the already published, extensive experimental evidence showing that the brain deeply "cares" about natural behaviors, as neural circuits respond completely differently to natural versus unnatural behaviors and stimuli. I hope that the experimental evidence presented in these next chapters will convince the reader that it is crucial to study the brain of animals (and humans) while they conduct natural behaviors. In other words, it is crucial to do natural neuroscience.

4

Neurobiology of Navigation

The neurobiology of navigation was perhaps the first field within neuroscience where a truly fundamental discovery was made by letting animals perform their natural behaviors. As discussed already in chapters 1 and 2, when John O'Keefe allowed rats to perform their natural behavior of spatial exploration, he discovered *place cells,* neurons in the hippocampus that are activated whenever the rat passes through a particular spatial location in the environment (figure 4.1a, left) (O'Keefe & Dostrovsky 1971; O'Keefe 1976; O'Keefe & Nadel 1978). In subsequent years, additional spatial cell types were discovered in hippocampal and parahippocampal brain areas by letting rats behave freely. These included *grid cells,* which are activated whenever the animal passes through a set of locations that form a hexagonal lattice spanning the environment (figure 4.1a, middle); *head-direction cells,* which are activated whenever the animal's head points in a particular direction (figure 4.1a, right); and *border cells,* which are activated along the geometrical borders of the environment (Ranck 1985; Taube et al. 1990; Hafting et al. 2005; Solstad et al. 2008; Lever et al. 2009; Moser et al. 2017). The set of interconnected brain areas that harbor these spatial cell types was dubbed "the brain's navigation circuit." All these cell types were also found in bats that were crawling in two dimensions (figure 4.1b; also see Ulanovsky & Moss 2007; Yartsev et al. 2011; Finkelstein et al. 2015). In addition, place cells and head-direction cells were found in the hippocampal formation of birds (Ben-Yishay et al. 2021; Payne et al. 2021), and head-direction cells were found in fish (Vinepinsky et al. 2020). Collectively, place cells, grid cells and head-direction cells are thought to represent the animal's position ("map"), distance traveled ("odometer"), and direction ("compass"), respectively, and they are believed to underlie navigation.

Recent studies have provided increasing evidence that hippocampal neurons encode very rich aspects of natural behaviors—both natural navigation behaviors, and natural behaviors beyond navigation—forming a much richer neuronal representation than was classically thought. In this chapter, I will discuss some of those examples. This will *not* be a comprehensive overview, but rather a selective survey aimed to illustrate the point that the brain's navigation circuit should be studied under evermore naturalistic conditions, and focus on more naturalistic behaviors.

In particular, since the neurobiology of navigation is my own field of research, I will discuss in some detail my lab's studies on the neurobiology of navigation in bats. This will

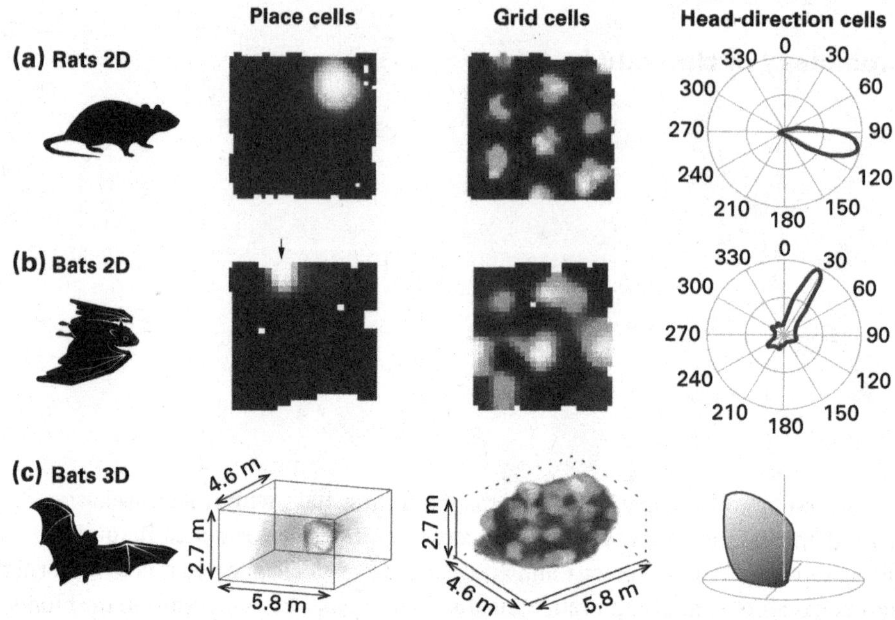

Figure 4.1
Spatial cell types in the hippocampal formation of rats and bats, in two dimensions (2D) and three dimensions (3D). The left two columns (place cells, grid cells) show the firing rates of neurons as a function of the animal's position, color-coded from zero (black) to the maximal firing rate (white). The rightmost column shows the directional tuning curve of head-direction cells: firing rate is indicated by the distance from the center. (a–b) Examples of place cells, grid cells, and head-direction cells recorded in 2D environments in rats (a) and bats (b). Recordings were made from neurons in animals exploring a 1 × 1-m square box. The arrow in (b) points to the place field. (c) Examples of a place cell, grid cell, and head-direction cell recorded in 3D in bats. The 3D place cell and 3D grid cell were both recorded in bats flying in a large flight room (5.8 × 4.6 × 2.7 m). The examples in (a) were adapted from (left to right) Whitlock et al. (2008); Boccara et al. (2010); and Solstad et al. (2008), (b) from Yartsev et al. (2011), and (c) from Yartsev and Ulanovsky (2013); Ginosar et al. (2021); and Finkelstein et al. (2016). Panels a-b were arranged similarly to Geva-Sagiv et al. (2015).

serve to illuminate my own personal journey from classical controlled neuroscience toward natural neuroscience. The bat studies will also demonstrate that if we remove some of the unnatural experimental constraints on our studies and make brain research more naturalistic, we are almost guaranteed to make surprising new discoveries about the brain.

4.1 Hippocampal Coding of Navigational Goals, 3D Spaces, and Kilometer Distances

Place cells, grid cells, and head-direction cells are typically studied in rodents running on linear tracks or in small, empty boxes of about 1 × 1 m in size (e.g., figures 4.1a and 4.1b). However, real-life navigation differs dramatically from typical laboratory setups in at least three ways:

1. The world is not empty; it contains things such as objects, areas of low or high valence, and navigational goals.
2. The world is not flat but three-dimensional (3D).
3. The world is not 1 m in size, and most animals—including rodents, bats, and humans—navigate over kilometers.

Very little was known until recently about the neural underpinnings of these three aspects of real-life navigation. Next, I will discuss these aspects, beginning with the representation of goals.

Navigation—the ability to reach a desired destination or goal—requires knowing the direction to the goal. Decades of research have focused on the representation of the current location and orientation of the animal—namely on place cells, grid cells, and head-direction cells. However, these cells are insufficient for an animal to navigate from point A (its current location) to point B (the goal) because navigation entails much more than knowing where you are and where north lies. It also requires knowledge about the goal—for example, knowing the distance and direction to the goal. Thus, a fundamental question that remained unanswered is how spatial *goals* are encoded in the brain, which is crucial for understanding the neural basis of goal-directed navigation. To address this question, we studied bats that were flying in a large room, in which—unlike in classical hippocampal studies—we introduced a navigational goal in the middle of the room (a landing platform with food) (Sarel et al. 2017). We used wireless-electrophysiology methods (chapter 3) to record the activity of hippocampal neurons in these freely flying bats. Surprisingly, we discovered a subpopulation of neurons in hippocampal area CA1, which exhibited angular tuning to the egocentric direction of the goal relative to the bat. "Egocentric direction" refers to the direction relative to the animal's orientation (hence, "ego")—for instance, whether the goal is to the right or left of the animal. By contrast, "allocentric direction" refers to the direction in the environment, in absolute coordinates, regardless of the animal's orientation—for instance, whether the goal is located east or west of the animal. Importantly, the classical spatial cell types—place cells, grid cells, and head-direction cells—are all allocentric neurons, as they represent allocentric absolute position in the environment (place cells, grid cells), or allocentric direction relative to north (head-direction cells). In contrast to this, the surprising goal-direction tuning that we found in the hippocampus (Sarel et al. 2017) was an egocentric tuning. Such type of tuning is arguably particularly useful for executing navigational actions, which are performed from the animal's first-person, egocentric perspective.

Further, we conducted another experiment in which we positioned the goal behind a curtain, through which the bat could neither see nor echolocate the goal. We found many neurons tuned to this hidden goal, suggesting that the goal-direction representation is memory-based (mental) rather than merely perception-based (sensory). In addition, we found hippocampal neurons that encoded the distance to the goal, often in conjunction with goal-direction tuning. Together, our results suggested the existence of goal-direction and goal-distance signals in the bat hippocampus—a vectorial representation that could support goal-directed navigation.

These goal-direction neurons may potentially underlie the classical findings that hippocampal dorsal-CA1 lesions impair goal-directed navigation, such as in a water maze (O'Keefe & Nadel 1978; Morris et al. 1982). The water-maze task is a widely used test, in which rodents are placed in a pool of water and need to swim and navigate to an escape platform that is submerged underwater (Morris et al. 1982). The water-maze task is a goal-directed navigation task, and therefore lesioning hippocampal goal-direction neurons or goal-distance neurons is likely to impair the underlying goal-directed navigation abilities (as discussed in Sarel et al. 2017). Further, goal-direction neurons could potentially underlie the "home vector" signal reported in multiple behavioral studies across many animal species (Etienne et al. 1998; Wallraff 2005; Mouritsen 2018; Wehner 2020). These studies suggested that animals perform vectorial computations to navigate to their home. However, the neuronal basis of these abilities was unknown. The goal-direction neurons might in fact be the neural mechanism of the home-vector signal, as well as of goal-directed navigation more generally. The findings in Sarel et al. (2017) thus filled a major gap in our understanding of the neural basis of navigation—namely, how we represent and navigate to goals.

Subsequently, multiple studies have reported similar egocentric goal-direction signals in the hippocampus, entorhinal cortex, and other brain regions (Nyberg et al. 2022). Thus, O'Keefe and Ormond found that neurons in rat hippocampus area CA1 exhibited egocentric goal-direction tuning that was very similar to the findings in the bat (Ormond & O'Keefe 2022). James Knierim and colleagues showed that neurons in lateral entorhinal cortex represented the egocentric direction to objects in the environment (Wang et al. 2018). This lateral entorhinal egocentric signal may potentially provide an anatomical input to the egocentric goal-direction neurons in the hippocampus CA1 (Sarel et al. 2017; Ormond & O'Keefe 2022), or vice versa. By contrast, in the medial entorhinal cortex of rodents, the groups of Knierim and Edvard and May-Britt Moser have found primarily the representation of allocentric absolute directions relative to objects—for example, whether an object is northeast of the animal, regardless of the animal's orientation (Wang et al. 2018; Høydal et al. 2019). Interestingly, egocentric goal-direction tuning toward a shelter box was found in neurons in the retrosplenial cortex and superior colliculus of mice during escape behavior (Campagner et al. 2023). Here again, the egocentric signal may provide rapidly accessible information that tells the animal: "This is the direction to the shelter—run for it!" Egocentric coding of goal direction is arguably particularly important during ethological behaviors, when the goal is especially salient—such as navigating toward a goal (Sarel et al. 2017) and escape to a shelter (Campagner et al. 2023). These findings thus illustrate the importance of studying the hippocampal formation—and other brain structures—during ethological, naturalistic behaviors.

* * *

The second aspect in which real-world navigation differs from typical laboratory setups is the *dimensionality* of space—namely, the fact that space is 3D. Many (albeit not all) animal species move in a 3D manner, either by flying, swimming, climbing, or burrowing underground. However, the classical studies of the "brain's navigation circuit" were all done in 1D or 2D environments. In the last decade, there have been substantial advances in understanding the coding of 3D space in rodents, bats, and primates. These studies were extensively reviewed (e.g., Jeffery et al. 2015; Finkelstein et al. 2016; Taube & Shinder 2020; Ginosar et al. 2023; Jeffery 2023). These reviews discuss in detail the properties of spatial coding in place cells, grid cells, and head-direction cells as animals move on vertical, horizontal, or angled 2D surfaces, or when they pass between two 2D surfaces. I refer the reader to these reviews for a comprehensive discussion of these issues. Here, instead, I will provide a very brief survey focusing mostly on spatial coding during 3D *volumetric* movement.

3D place cells In the first study of hippocampal spatial coding during truly volumetric 3D navigation, we used wireless-electrophysiology methods and investigated place cells in the hippocampus of bats flying in a large flight room (Yartsev & Ulanovsky 2013). We found that individual hippocampal CA1 neurons in freely flying bats were active in confined 3D volumes, creating spherical place fields (figure 4.1c, left). Thus, these 3D place cells represented all axes of space with similar resolution, forming an isotropic representation of 3D space. The 3D place fields of different neurons spanned different locations and collectively represented the entire available space in the room uniformly. These results thus suggested that the bat hippocampus represents 3D volumetric space in a uniform and isotropic manner. Similar results were subsequently found in the hippocampus CA1 of rats that were climbing on a 3D volumetric mesh: here too, place fields were spherical in shape, and created a uniform and isotropic representation of 3D space (Grieves et al. 2020).

3D grid cells In another study, we wirelessly recorded from neurons in the medial entorhinal cortex of freely flying bats in the same large flight room (Ginosar et al. 2021). We found neurons with multiple 3D firing fields. Many of these multifield neurons were 3D grid cells, whose fields exhibited local spatial order, whereby neighboring fields were separated by a characteristic distance. While this formed a local order, the field arrangement lacked any global lattice order (figure 4.1c, middle). Thus, while 2D grid cells form a global hexagonal lattice, characterized by both local and global order, 3D grid cells exhibited only local order, creating a locally ordered metric for space, but without a global lattice. A parallel study in rats, which were climbing through a 3D volumetric mesh, also revealed the absence of a 3D hexagonal lattice (Grieves et al. 2021). Together, these results from bats and rats hold important implications for our understanding of grid cells. Classical theories on the function of 2D grid cells rely on the fundamental assumption that grid-cell firing patterns are periodic (McNaughton et al. 2006; Fiete et al. 2008; Mathis et al. 2012; Stemmler et al. 2015). Moreover, multiple theoretical studies suggested that grid cells are involved in various geometric computations, such as goal-directed navigation or path integration (McNaughton et al. 2006; Fiete et al. 2008; Stemmler et al. 2015)—suggestions that were motivated by the highly geometric, periodic representation of 2D space by 2D grid cells. These suggestions now seem less likely, given the absence of global periodicity in 3D grid cells. Thus, global spatial periodicity does *not* seem to be central to the function of grid cells, which calls for developing new ideas and models to describe the contribution of grid cells to navigation in both 2D and 3D. Some possible alternative ideas for the function of grid cells were recently reviewed by Ginosar et al. (2023).

Importantly, the use of more naturalistic setups played a key role in the discovery of both the hexagonal grids in 2D (Hafting et al. 2005) and the local-only order in 3D (Ginosar et al. 2021). In particular, the original discovery of 2D grid cells was enabled by allowing rats to move in a much larger space than previously used, which uncovered the hexagonal periodicity (Hafting et al. 2005). Likewise, the discovery of the local-only order in 3D, as well as the subsequent reevaluation of the mechanisms and functions of grid cells, were enabled by using wireless-electrophysiology technologies, as described in chapter 3, which allowed one to study the brain in animals that move more freely and perform more naturalistic 3D behaviors (Ginosar et al. 2021). Nobody expected to find hexagonal grids in 2D, and nobody expected their absence in 3D: these were both completely surprising discoveries—unknown unknowns—and these surprising discoveries were both made by using more naturalistic setups, which illustrates the importance of using more naturalistic setups and studying more naturalistic behaviors.

3D border cells The same study of bats revealed 3D border cells (Ginosar et al. 2021). In 2D, border cells are activated along the 1D borders of the environment (Savelli et al. 2008; Solstad et al. 2008; Lever et al. 2009). In 3D, the spatial firing regions of 3D border cells could potentially take the form of 2D sheets or 1D lines embedded in 3D. In our study, we found both types of 3D border cells (Ginosar et al. 2021).

3D head-direction cells The survival of many animals depends on successful directional orientation in 3D space, and yet for many years, it remained unclear how 3D head direction is represented in the mammalian brain. We conducted a study in dorsal presubiculum, the first brain area in which 2D head-direction cells were discovered in rodents (Ranck 1985; Taube et al. 1990). We measured the bat's 3D head direction while recording neural activity in presubiculum neurons (Finkelstein et al. 2015). We found a large fraction of 3D head-direction cells (figure 4.1c, right). When bats were upside-down, most neurons remained directionally tuned; surprisingly, however, their azimuth-tuning was shifted by 180 degrees relative to the upright position. This 180-degree shift is inconsistent with the standard system for representing 3D directions—spherical coordinates—because in spherical coordinates, the azimuth tuning should have been stable. Instead, it suggested that 3D head direction is represented on a torus consisting of 360 degrees azimuth × 360 degrees pitch (Finkelstein et al. 2015). These results demonstrated, for the first time, a 3D head-direction mechanism in mammals, which could support navigation in 3D space.

In the original bat studies, we hypothesized that 3D head direction should be represented in other mammals as well, including rats, mice, squirrels, cats, dolphins, and monkeys, as these mammals move in 3D to various degrees (Finkelstein et al. 2015, 2016). Indeed, recent recordings in mice and monkeys found 3D head-direction cells similar to bats (Laurens et al. 2016; Laurens & Angelaki 2019; Angelaki et al. 2020; Mallory et al. 2021).

* * *

The third aspect in which real-world navigation differs from typical laboratory setups is the *spatial scale*. Animals and humans navigate over distances spanning from hundreds of meters to many kilometers. Rats, for example, navigate outdoors up to 1 km per night (Taylor 1978; Russell et al. 2010), and bats navigate from several kilometers to several tens of kilometers every night (Tsoar et al. 2011; Egert-Berg et al. 2018; Harten et al. 2020; Toledo et al. 2020). Migratory species navigate even longer distances during their annual migrations—sometimes thousands of kilometers (Ulanovsky & Moss 2008; Kays et al. 2015; Yovel & Ulanovsky 2017). However, over the last fifty years, hippocampal research focused on spatial representations in small laboratory environments (with two exceptions of somewhat larger indoors environments[4]). Until very recently, nothing was known about hippocampal neural codes for large spatial scales of kilometers—the scales of natural navigation in rodents, bats, and many other mammals. In particular, it was unknown whether place cells and grid cells, as recorded in small, laboratory-sized environments, are relevant at all to large-scale navigation in the real world, over kilometers. Notably, even if place cells do exist on such spatial scales, they *must* differ from place cells as recorded in laboratory environments, for the following reason: Egyptian fruit bats fly every night a distance of up to about 30 km to their favorite fruit trees, with flyways spanning about 2 km width and 0.5 km height (Tsoar et al. 2011; Geva-Sagiv et al. 2015)—namely, the volume of their flyways is about 30 km^3. In other species with longer navigation distances, the volumes of their flyways may be much larger—hundreds or even thousands of cubic kilometers. A simple "back of the envelope" calculation shows that tiling such a huge space of 100 km^3, using place fields of the size that is typically measured in small laboratory boxes, namely about

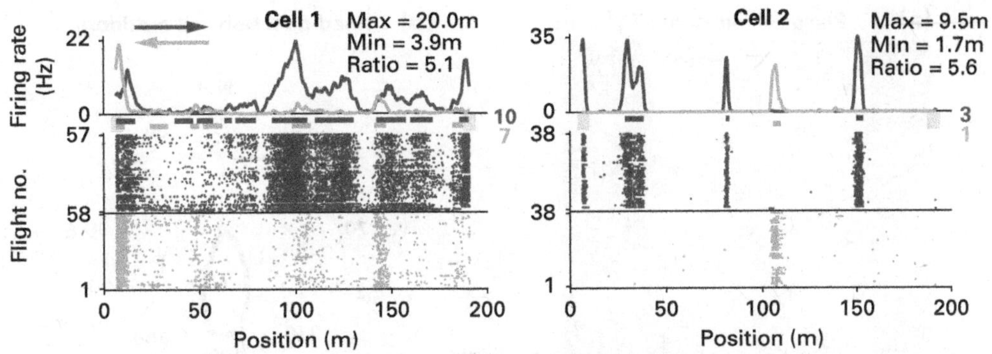

Figure 4.2
Examples of two place cells recorded in dorsal hippocampus CA1 of bats flying in a 200-m-long tunnel. Shown are spike rasters (bottom) and firing-rate histograms as a function of position (top), separated to the two flight directions (dark and light gray; see arrows). Raster plots show spikes (dots) as a function of position (*x*-axis) for different flights (*y*-axis); firing histograms (also known as "firing-rate maps") show the firing rate as a function of position. The detected place fields are marked by thick dark- or light-gray horizontal lines above the raster plots; the number of fields in each direction is indicated on the right. In each example, the smallest and largest field sizes are indicated at the top-right (min, max), together with the ratio between them. Note that in this very large environment, place cells in dorsal CA1 typically have multiple fields with very different sizes for the different place fields of the same neuron: a multifield, multiscale neural code for space. Adapted from Eliav et al. (2021).

10 cm diameter, would require about 10^{15} neurons (here we assume a single place field per neuron and require about 10 overlapping neurons per location, for robustness). This enormous number of neurons is about 10^{10} times larger than the number of neurons in the entire dorsal hippocampal area CA1 (Andersen et al. 2007), suggesting that it is simply not feasible to represent such large spatial scales with laboratory-sized place fields. Thus, there was a major gap between the neurobiology of navigation as studied in the lab and kilometer-scale natural navigation outdoors—both a conceptual gap and a gap in spatial scale.

In the first-ever study that addressed this gap (Eliav et al. 2021), we tackled this problem by developing a unique setup, including a very large environment with an ethologically relevant spatial scale. We built a 200-m-long tunnel where bats could fly freely (by now this tunnel has been extended to 700 m). The bat's position was tracked using a radio-frequency (RF)–based localization device that measures distances from the bat to a ground-based antenna array—yielding a high localization accuracy of 5 cm. For acquiring neuronal data, we used a wireless neural logger, which is carried by the bat and stores the data on-board, which allows for recording neurons over unlimited distances (chapter 3) (Eliav et al. 2021). Recordings from dorsal hippocampus area CA1 revealed a very surprising spatial firing pattern, which was very different from previous findings in small-scale laboratory environments. Single neurons exhibited multiple fields, and a multiscale representation of space, whereby place fields of the same neuron differed up to twentyfold in size (figure 4.2). Surprisingly, this coding scheme was observed from the first exposure to the environment and was similar between lab-born and wild-born bats—suggesting that the multifield multiscale code is a very robust phenomenon, which does not require experience with large environments. Theoretical decoding analysis showed that this multifield, multiscale code enables representing very large environments with extremely low decoding errors—several orders of magnitude more precise than the classical hippocampal place code (Eliav et al. 2021). Overall, these naturalistic experiments, which featured real-world spatial scales, have suggested that in large environments, the hippocampus exhibits a novel coding scheme for

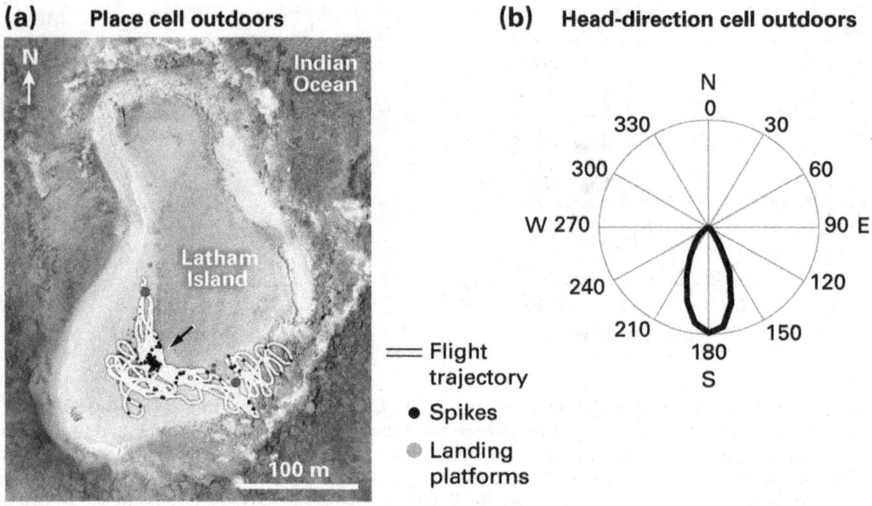

Figure 4.3
Electrophysiological recordings of a place cell and a head-direction cell that we conducted in bats flying outdoors on a remote uninhabited oceanic island: Latham Island, in Tanzania. (a) Example of a place cell. Shown is an aerial photo of the island that we took with a drone, together with the trajectory of a bat during one recording session (white curve), and the spikes recorded from one hippocampal neuron during that session (black dots). Black arrow points to the prominent place field of this neuron (the large concentration of spikes). Note that this place cell, as well as some others, had one place field, while many other place cells had multiple place fields. (b) Example of a head-direction cell; shown is the directional tuning curve (firing rate versus direction) as a polar plot. Analyses courtesy of Shaked Palgi.

space that is radically different from small environments. Moreover, the theoretical analysis elucidated *why* we found such different spatial coding in very large environments: a possible answer is that this type of multifield, multiscale code is much more efficient and precise. In other words, by taking the natural neuroscience approach—in this case by using a real-world navigational scale—we discovered a completely unexpected and superior neuronal representation.

Finally, let us come back to the major gap mentioned previously: namely, it is unknown whether the various spatial cell types in the brain's navigation circuit, as studied for fifty years in small laboratory-sized environments, are at all relevant to large-scale navigation in the real world (i.e., outdoors). One can address this gap by taking experiments one step further—to the great outdoors. To this end, we recently conducted the first-ever neural recordings from the brain's navigation circuit in animals navigating outdoors. We implanted bats with a tetrode microdrive, fitted them with a wireless neural logger and GPS logger, and released them to fly on a remote, uninhabited oceanic island. We already had recorded place cells and head-direction cells in bats navigating outdoors on the island (figure 4.3). We aim to further explore a whole host of questions on the neural correlates of real-world navigation. These experiments will allow us to elucidate, for the first time, the neural basis of navigation in a real-life, large-scale, outdoor environment, and will thus close the major gap between ethological field studies on large-scale navigation and neurobiological laboratory studies on small-scale navigation. More broadly, this research will constitute the first single-cell neurophysiological study in any mammal of neuroscience in the real world.

4.2 Hippocampal Place Cells Are Sensitive to the Details of Animals' Natural Behaviors

In the previous section, I discussed how new and surprising discoveries were made by studying the brains of animals performing more naturalistic navigational behaviors. However, as it turns out, neurons in the hippocampal formation are sensitive to a large variety of natural behaviors, including nuanced details of behavior, even when these behaviors are not navigational per se. In this section, I will survey a few such examples, which will serve to illustrate the importance of conducting more naturalistic experiments and of measuring the detailed characteristics of behavior when studying the hippocampus. Moreover, as will be discussed in subsequent chapters, this does not apply only to the hippocampus; in fact, it is equally important to focus on natural behaviors when studying other brain areas.

One such example was already discussed in chapter 3: Lisa Giocomo and colleagues (Mallory et al. 2021) have found that neurons in the mouse medial entorhinal cortex, which are classically thought to represent the animal's position in space or azimuthal head direction (e.g., grid cells, head-direction cells), often also encoded the eye gaze in 3D. This was a surprising result, as eye gaze was never thought of by researchers studying these brain areas (at least not in mice). This study indicated that perhaps eye gaze should be considered as a relevant navigational variable (e.g., because animals may use eye movements to scan space), even in mice, not just in primates as typically thought. This finding highlights the importance of measuring unconventional details of behavior: in this case, eye-gaze behavior.

A good example for the strong influence that the nuances of behavior can have on hippocampal CA1 place cells comes from a study by James Knierim and colleagues (Monaco et al. 2014). In this study, they examined the effects of attentive head scanning in rats during exploration—investigatory episodes that are manifested by bouts of lateral head movements during pauses in forward locomotion. They found that, remarkably, new place fields tended to form at the location of these attentive head scans the very next time that the rat visited that location. One interpretation of this result is that the heightened attention during these bouts of sensory acquisition is causing plastic changes and potentiation of the neuron's excitability at that location, perhaps through mechanisms of behavioral timescale synaptic plasticity—a powerful type of synaptic plasticity that was recently discovered in the hippocampus CA1 (Bittner et al. 2017; Grienberger & Magee 2022). This result thus sheds interesting light on the mechanisms of place-field formation. Further, this study again demonstrated the importance of considering the detailed nuances of animal behavior—in this case, the rat's investigatory head-scanning movements—as these subtle behaviors may be crucial for understanding important brain mechanisms.

Another example for the importance of behavior for hippocampal activity comes from a study by Anton Sirota and colleagues (A. Sirota, personal communication). They measured the detailed pose and 3D kinematics of head and body movements in rats and investigated their effect on the activity of hippocampal CA1 neurons. They examined both well-known behavioral states, such as locomotion, rearing, turning, and pausing, and two new locomotion states that these researchers have uncovered: "low walking" and "high walking." These two walking states are characterized, respectively, by low head pitch and extensive active-sensing behavior in the form of fast sniffing, whisking, and lateral head-scanning motions; and high head pitch associated with reduced active sensing. The researchers found that the firing rates of hippocampal CA1 place cells were modulated by these behavioral states, while leaving the cells' spatial tuning largely unchanged. Interestingly, for neurons with multiple place fields, the firing within different fields of the same neuron was often tuned to different behavioral states. The researchers then defined joint spatial-behavioral tuning for the neurons—a new way to think about hippocampal coding. Population decoding analysis was able to decode both the animal's position and its behavioral state jointly at any given moment. This study illustrates again the importance of tracking the details of animal behavior, as it shows that these details can have surprising effects on the coding properties of hippocampal neurons.

Other behaviors can influence hippocampal activity even more dramatically. György Buzsáki and colleagues investigated the effects of *jumping*—a highly natural, evolutionarily conserved, and relevant behavior for most rodent species (Green et al. 2022). They constructed a modular apparatus that allowed them to introduce a large gap in the middle of a 1D track, on which rats were running back-and-forth, in order to elicit jumping behavior. The authors found strong remapping of hippocampal place cells during jumping trials compared to prejump trials. "Remapping" is a phenomenon whereby upon a certain change in the environment, hippocampal place cells change their spatial tuning; this change can be manifested either via place fields changing their positions, turning on and off, or changing their firing rates (Colgin et al. 2008; Kubie et al. 2020). In this study, remapping occurred throughout the track, not only near the gap (likely because the rats could see the gap from the beginning of the track, and hence knew that they were facing a jumping trial). Specifically, during jumping trials, most hippocampal neurons modified their firing, either via appearance or disappearance of place fields or via changes in firing rates. This dramatic change in hippocampal network activity suggests that the planning and execution of distinct behavioral actions can strongly affect hippocampal representations. Interestingly, such dramatic action-driven remapping was not reported before for other types of actions, but it did occur for jumping—perhaps because jumping is a particularly demanding action, or perhaps because it is a highly natural behavior. This suggests that when studying the hippocampus, as well as other regions of the brain's navigation circuit, animals should be allowed to behave naturally, in rich settings, and their behaviors must be well documented.

The details of behavior influence hippocampal activity not only in rodents, but also in bats. In an early study of bat hippocampus (Ulanovsky & Moss 2011), we asked whether 2D place cells exhibit rapid dynamics linked to their echolocation (biosonar) behavior. We found that place-field size was modulated by the availability of sensory information on a timescale of a few hundred milliseconds. Bat hippocampal place fields were smallest immediately after an echolocation call, but place fields diffused and became larger with the passage of time after the call, when no new echo information was arriving anymore (figure 4.4a). These results suggested that spatial representation in mammalian hippocampus can be very rapidly modulated by external sensory and behavioral events. We hypothesized that these changes in place fields after each echolocation call occur because during the periods of no-sensory-information between echolocation calls, the bat relies on path-integration; and in turn, the locational information provided by path integration is rapidly accumulating errors over time, and is then being reset by the echo input from the next sonar call, which brings new sensory information. An alternative possibility is that it is not the dynamics of *sensory* information that has modulated hippocampal place coding in that study, but rather the dynamics of *motor* movements of echolocation call production. Both the sensory and motor hypotheses can in principle account for the bat results (figure 4.4a) because bat echolocation is fundamentally a sensorimotor behavior, where motor production of sound pulses is coupled to sensory perception of echoes (Ulanovsky & Moss 2008).

Interestingly, a recent study in rats provided evidence that potentially leans toward the motor interpretation. In this study, Loren Frank and colleagues examined the effects of *foot-stepping* on hippocampal CA1 representations in rats running in a 1D maze (Joshi et al. 2023). They found that hippocampal representations were coupled to the stepping cycle of rats. When the rat's forelimbs touched the ground, the decoded hippocampal representation was closest to the actual position of the rat, whereas between these ground touches, the hippocampal representation swept forward of the rat (figure 4.4b). It is possible that the widening of place fields in bats following echolocation calls (Ulanovsky & Moss 2011) has a similar motor-related source as the forward-sweeping following stepping in rats (Joshi et al. 2023) because the bats moved in 2D, and therefore if the firing was widened along the 1D direction of movement, as in the rat study, this could explain the diffusion of place fields in all directions due to the bat's 2D locomotion in all directions (figure 4.4a).

The sensory and motor hypotheses can be further tested by an experiment on rodents. Specifically, this experiment would allow a rat or a mouse to forage in the dark in an open-field arena surrounded by visual cues while a strobe light repetitively flashes (with olfactory cues being thoroughly eliminated, such as by a layer of water, so the strobe lights provide the only sensory information about position). The prediction of the sensory hypothesis is that rat place fields would diffuse with the passage of time after each strobe flash, similar to our findings for the stroboscopic sonar system of bats. The prediction of the motor hypothesis is that rat place fields would change synchronized to foot-stepping—and *not* synchronized to the strobe lights. This experiment with stroboscopic lights is somewhat artificial, so perhaps a more naturalistic version would be to use arenas that are rich with somatosensory and olfactory cues, and test the dynamics of place fields as a function of the passage of time after each whisk or olfactory sniff. These experiments are still to be done. In any case, these two studies (Ulanovsky & Moss 2011; Joshi et al. 2023) suggest that it is crucial to measure the details of animals' sensorimotor behaviors that are *not* measured routinely in hippocampal studies, such as stepping, wingbeats, echolocation, sniffing, whisking, and eye saccades. Such measurements are important for elucidating neural dynamics in the hippocampal formation, and likely also in other brain regions, across species.

Measuring behaviors in a detailed manner not only provides information on the precise *timing* of sensorimotor behaviors, but also opens a window to cognitive processes, such as the animal's *attention.* We capitalized on this ability in a study that examined the effects of switching between different types of behavior on hippocampal spatial representations (Sarel et al. 2022). In their everyday lives, both animals and humans constantly switch between different behaviors. However, neuroscience research typically studies the brain while the animal is performing one behavioral task at a time. Thus, little is known about how brain circuits switch between natural behaviors. Navigation is a complex dynamic behavior that enables testing these behavioral switches: It requires the animal to know its own location within the environment, while also paying attention to incoming sudden events such as obstacles, predators, or conspecifics. In this study (Sarel et al. 2022), we investigated how brief attentional switches to "things out there" affect the representation of space in the

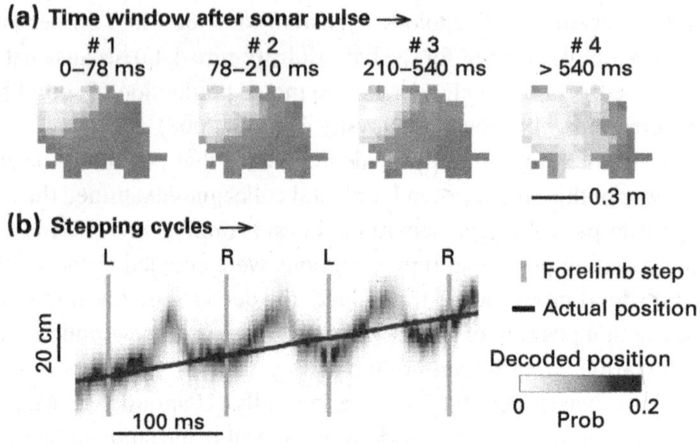

Figure 4.4
The spatial tuning of place cells is dynamically modulated by the animal's sensorimotor behaviors. (a) Example of a place cell in bat hippocampus CA1, plotted as a function of time from the echolocation call (sonar pulse). The bat was crawling here in a small 2D arena. Note the broadening of the place field with the passage of time after the echolocation call. (b) Decoded position of a rat in a 1D maze (a W-shaped maze): shading denotes the probability of the decoded position. The decoding was based on hippocampal CA1 recordings and was modulated by the stepping behavior: note that after each step, the hippocampal representation (decoded position) swept forward compared to the actual position of the rat. (a) Adapted from Ulanovsky & Moss (2011); (b) adapted from Joshi et al. (2023).

hippocampus. We introduced pairs of bats to a 135-m flight tunnel (a shortened version of the 200-m tunnel discussed previously), where the bats alternated between flying alone and flying toward each other ("crossover"; see figure 4.5a). During crossovers, bats had to be attentive to their partner to avoid collisions. Thus, during their flights, the bats switched many times between two types of behavior: navigation, and collision avoidance. They increased their echolocation click rate before the crossover moment, indicating high attention in these collision-avoidance events. We recorded the neural activity in hippocampal area CA1 using wireless-electrophysiology devices (neural loggers) and tracked their position in the tunnel at high resolution using RF tags, as described in section 4.1. We found that during crossovers, many CA1 neurons encoded the distance to the other bat: the neurons switched from coding the bat's own position (place tuning) to a conjunctive coding of distance by position. These neuronal switches were very rapid (up to 100 ms). Next, we utilized the fact that the amount of increase in the bat's echolocation click rate provides a direct behavioral measure of the bat's attention (Sarel et al. 2022). Surprisingly, we found that the neuronal switches were correlated with the bat's spatial attention, as indexed by its echolocation: Specifically, the tuning for the distance to the other bat was stronger on high-attention trials. Thus, spatial attention modulated spatial representation. Interestingly, the different place fields of the same neuron often exhibited very different tunings to interbat distance, creating a nonseparable code of position by distance (figure 4.5b): "nonseparability" means that the 2D firing-rate map for position by distance *cannot* be explained as a multiplication of the 1D marginals (i.e., the 1D tuning to position × 1D tuning to distance). (Of note, the results of Anton Sirota, described earlier, whereby different behavioral modulations were found in different place fields of the same neuron in the rat hippocampus, also indicate nonseparability and are thus conceptually quite reminiscent.) Overall, these results

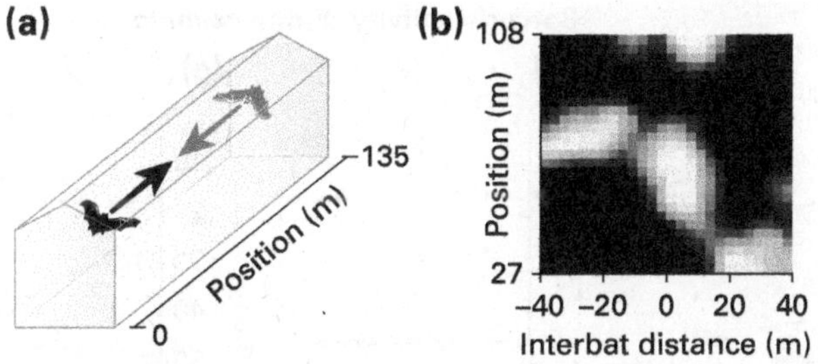

Figure 4.5
In a two-bat collision-avoidance experiment, hippocampal place cells exhibit complex tuning to the bat's self-position and its distance to the other bat. (a) Schematic of the two-bat experiment in a 135-m-long flight tunnel: two bats were flying opposite each other and were passing (crossing over) each other. (b) Example of one neuron from dorsal hippocampus CA1 that was recorded in this task. Shown is the firing rate of the neuron as a function of the animal's position (y-axis) and the distance from the other bat (x-axis), color-coded from zero (black) to maximal firing rate (white). Note the complex tuning of this neuron as a function of position and distance. Adapted from Sarel et al. (2022).

suggested that during natural behaviors, hippocampal circuits can rapidly switch between several computations, representing the relevant behavioral variables.

These latter findings (Sarel et al. 2022) have two important implications. First, they underscore again the importance of conducting naturalistic experiments and of measuring the details of animal behavior. The second implication relates to reductionism versus natural neuroscience. The fact that this coding is nonseparable (figure 4.5b) argues against the classical reductionist approach to scientific research, where you measure one variable at a time in highly controlled 1D experiments and then *assume* that the multidimensional tuning is simply a multiplication of the 1D marginals. This is simply not true! The nonseparability of the coding observed here means that you cannot reconstruct the 2D tuning by multiplying the 1D marginals. This implies that, at least in this case, the reductionist approach is fundamentally wrong. We would *not* be able to elucidate the true complex 2D tuning using simple reductionist 1D experiments. Instead, we *must* let the animal perform a multidimensional complex behavior to truly understand the complex, multidimensional and nonseparable coding of the neurons.

Finally, measuring unconventional animal behaviors can shed light on the interplay between *spatial* signals and *memory* signals in the hippocampus, as demonstrated beautifully in a study by Dmitriy Aronov and colleagues (Chettih et al. 2024). This issue is important because the two most dominant theories of hippocampal function posit that the hippocampus is involved in spatial cognition and navigation, that it is involved in memory, or both (Scoville & Milner 1957; O'Keefe & Nadel 1978; Squire 1992; Buzsáki & Moser 2013; Eichenbaum & Cohen 2014). However, the relation between the navigation perspective and memory perspective on hippocampal function remains unclear. Aronov and colleagues approached this problem by studying a unique animal model: food-caching birds. These birds hide food (seeds) at thousands of sites scattered throughout their home range, and they can remember their locations and retrieve them later (Sherry 1984; Pravosudov 1985). Thus, these birds are spatial-memory specialists. Moreover, they can remember the timing when they hid the food items and subsequently retrieve them in the correct temporal order (Clayton & Dickinson 1998; Salwiczek et al. 2010). Thus, they remember *what* they've

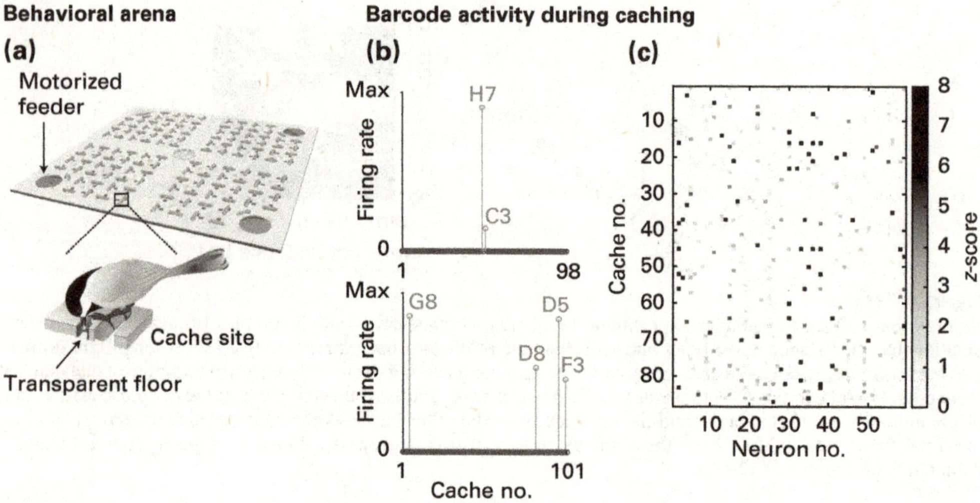

Figure 4.6
Episodic coding of caching events in the hippocampus of a food-caching bird. (a) Schematic of the behavioral arena where the birds were caching seeds. (b) Examples of two neurons that responded very specifically to only a handful of caching events: the top neuron responded to two caching events, and the bottom neuron responded to four caching events (the labels H7, G8, etc. refer to the locations of the caches, using a chesslike notation of columns [letters] and rows [numbers] for positions within the arena). (c) Activity of neurons for different caching events (caching episodes). Across the neuronal population, each caching event created a barcode of activity that was unique to that particular episode. Adapted from Chettih et al. (2024).

placed *where* and *when*—and are thus also episodic-memory specialists. These memories were shown to depend on the bird's hippocampal formation (Sherry & Vaccarino 1989). The study by Aronov and colleagues (Chettih et al. 2024) capitalized on these previous findings, asking if and how episodic memories for particular food caches are represented in the hippocampus in addition to the classical place code. They constructed a behavioral arena in which chickadees, a species of food-caching birds, stored and retrieved seeds (figure 4.6a). Further, they developed machine-learning-based methods (similar to those described in section 3.3) to track the behaviors of the animal in great detail, including its 3D pose, seed-caching and seed-retrieval events, and the presence or absence of seeds in each of the cache sites. First, as expected, they found classical place cells in the hippocampal formation of birds moving around this arena. However, on top of the classical place representation, they found also an episodic-like representation: each caching event was represented by a burst of firing in a unique set of hippocampal neurons, creating a sparse, barcode-like pattern of activity across the hippocampal neuronal population (see figure 4.6b for an example of two neurons that responded very specifically to only a handful of caches; and see figure 4.6c for the population barcode-like responses [rows] that were unique for each of the caching events). These barcodes were different for spatially adjacent caches, and different for separate caching-events at the same location. Importantly, the barcode representing a specific caching event was reactivated whenever the bird later retrieved food from the same cache. Thus, these barcodes seem to comprise a signature of each particular memory episode. This fascinating study opens the door to many future questions on mechanisms of episodic memory. Importantly, these findings were enabled by using an animal species that exhibits unique memory-guided natural behaviors and by tracking those behaviors in great detail.

This highlights again the importance of taking the natural neuroscience approach when studying the hippocampus.

4.3 The Social Hippocampus

In recent years, another perspective has emerged on hippocampal function, the *social* perspective. This new direction was driven initially by findings that certain parts of the hippocampus, like dorsal CA2 and ventral CA1, are important for social memory—specifically, for the memory of individual identity (Hitti & Siegelbaum 2014; Okuyama et al. 2016; Oliva et al. 2020). Importantly, the social perspective is consistent with the classical memory perspective on the hippocampus because social memories are among the most important memories for any social animal.

A set of studies has subsequently examined the relation between the social perspective and the spatial perspective on the hippocampus (Danjo et al. 2018; Omer et al. 2018). The rationale was that social animals need to know the spatial positions of conspecifics—mates, competitors, and offspring. While the classical view on hippocampal place cells is that they represent self-position, it was unknown whether and how the position of *other* individuals is represented in the hippocampus, or anywhere else in the brain. To address this question, we recorded the activity of neurons in the dorsal hippocampus CA1 of bats that performed a mimicry task, wherein one bat had to observe the other bat's flight while being stationary, and then had to fly to the same location. We discovered a subpopulation of hippocampal neurons that represented the position of the other bat, when the recorded bat was completely stationary (Omer et al. 2018). We called these neurons "social place-cells." About half of these neurons also represented the bat's own position, when it was flying (i.e., they were classical place cells). Some CA1 neurons also represented the position of inanimate moving objects; however, their representations were different from the representations of the conspecific bat. A parallel study by Shigeyoshi Fujisawa and colleagues discovered similar neurons in dorsal CA1 of rats: here again, these neurons represented the position of another individual (Danjo et al. 2018). Thus, social place-cells seem to exist across species. The discovery of social place-cells suggested a role for dorsal hippocampal CA1 neurons in social-spatial cognition—namely, representing the positions of other individuals. These neurons thus tie together the spatial and social perspectives on hippocampal function.

These previous studies left several obvious questions open. How are *multiple* individuals represented in the hippocampus? For example, are there neurons that represent the identity (and not just the position) of individual animals? Also, how does the hippocampus respond in a more natural social situation instead of a highly structured task (as used in the previous two studies)? To address these questions, we studied hippocampal coding in a social *group*, where animals did not have any explicit task and were free to interact with whomever they wanted, whenever they wanted (Ray et al. 2022, 2024). Social animals, like bats, rodents, and primates, live in groups and interact volitionally in complex ways. However, prior to this study, almost nothing was known about neural responses—in any brain area—under such natural group conditions. In this study, we investigated hippocampal CA1 neurons in a mixed-sex group of 5–10 freely behaving wild-born bats, which lived 24/7 for many months in a laboratory-based "cave" and formed a stable social network. We developed experimental methods to track the positions, head directions, and identities of all the bats in the room (cave). The bats exhibited three main behavioral states in this highly social group situation: flights, explicit social interactions, and active observation. In the flight state, most hippocampal place cells were socially modulated by the presence or absence of bats in key locations. Interestingly, many neurons represented the identity of conspecifics. A similar finding on identity coding in the bat hippocampus was found by Forli and Yartsev (2023). Remarkably, we found that among *non*-place cells, the large majority of the neurons exhibited significant social modulation (Ray et al. 2022, 2024).

In the social-interaction state, hippocampal neurons responded strongly to social interactions. Moreover, the neurons selectively represented *specific* interaction types, with most neurons responding to only *one* type of social interaction (e.g., responding only to affiliative allogrooming or only to aggressive interactions). Finally, during active observation, neurons encoded the bat's own position and head direction, together with the position, direction, and identity of multiple conspecifics. Thus, here again the representation was fundamentally sociospatial, combining social variables with spatial variables. Interestingly, we found a surprising *invariance* property in these neurons: Identity-coding neurons encoded the *same* bat across different contexts—in-flight and during active observation. Further, we found that the strength of identity coding was modulated by three key social factors: sex, dominance hierarchy, and social affiliation ("Are you my friend?"). Overall, this study provided the first evidence—in *any* brain region of *any* animal species—on single neurons that jointly encode the identity, sex, social affiliation, and dominance hierarchy of multiple other individuals, as well as the locations of all individuals. Thus, in a highly social group situation, hippocampal neurons form a complex, multidimensional, sociospatial representation of the natural world.

Our study in bats (Ray et al. 2022, 2024) tied together all three perspectives on the hippocampus: the social perspective, memory perspective, and spatial perspective. The social perspective was supported by finding representations of sex, hierarchy, and social affiliation, as well as responses to social interactions. The memory perspective comes from the identity coding, as the hippocampal neurons represented the bat identity (social memory), and they did it in a context-invariant manner. And the spatial perspective was supported by all these signals being fundamentally spatial, encoding which bat was at which location. Another interesting aspect of these results is our finding that hippocampal neurons represent simultaneously both self and others. This result goes along with findings in the mouse prefrontal cortex, of neurons that represent positions of self and other (Murugan et al. 2017), as well as classical findings on "mirror neurons" in the parietal cortex of primates: namely, neurons that represent the actions of self and other (di Pellegrino et al. 1992; Rizzolatti & Sinigaglia 2010). The corepresentation of self and other might be a general property of many higher brain regions. Specifically, I propose that for any high brain area, if we consider the functions that are thought to be subserved by that brain area—whether memory, spatial cognition, planning, action execution, or any other behavioral or cognitive function—we would find that in highly social species, this brain area represents these functions for both self and other. Such self-and-other neurons may also explain why several studies have found interbrain synchronization between pairs of brains of interacting animals (Kingsbury et al. 2019; Zhang & Yartsev 2019), because if my brain contains neurons that represent my behavior (self) and your behavior (other), and your brain contains neurons that represent my behavior (other) and your behavior (self), then naturally these two brains will exhibit synchronized activity.

Finally, it is noteworthy that the percentage of socially responsive hippocampal neurons in our study (Ray et al. 2024) was extremely high: over 50 percent. This percentage was *much* higher than the percentage of socially responsive neurons in previous studies of rodents during pairwise interactions (von Heimendahl et al. 2012; Zynyuk et al. 2012), and also much higher than the percentage of identity-coding neurons found in passively viewing humans or monkeys ("Jennifer Aniston neurons") (Quiroga et al. 2005; Tyree et al. 2023). We speculate that our finding of a very high percentage of socially coding cells in the bat hippocampus may be a consequence of two factors. First, the fact that these were wild-caught bats that have experienced the full natural repertoire of social interactions outdoors. Second, likely an even more important factor was our highly naturalistic setup and our use of a mixed-sex group, which promoted rich and natural social interactions—which in turn exposed the full richness of hippocampal neural coding. The conclusion from this is that it is crucial in the future to conduct brain studies on animals in social groups, preferably in mixed-sex groups, where animals can experience and perform the full repertoire of natural social behaviors.

A more general conclusion is that, as discussed in chapter 1, we see impoverished neural representations when we use impoverished environments—and therefore, to expose the true richness of neuronal representations, it is crucial to use richer environments and richer stimuli (in this case, this enrichment was done in the social domain, but it can be done in other domains as well). In other words, it is crucial to take the natural neuroscience approach.

4.4 Summary and Additional Thoughts

This chapter surveyed numerous examples where surprising findings about hippocampal coding were discovered by investigating animals that performed more naturalistic behaviors. For example, when we allowed bats to move in 3D space, we discovered 3D place cells, 3D head-direction cells, and 3D grid cells (figure 4.1c) (Yartsev & Ulanovsky 2013; Finkelstein et al. 2015; Ginosar et al. 2021). The 3D grid cells lacked a global hexagonal lattice, but instead showed a local distance scale—and this argued against classical models of grid cells. When we introduced navigational goals, we found egocentric vectorial coding of these goals (Sarel et al. 2017). When we increased the size of the environment to hundreds of meters, we discovered a radically different type of neural code—a multifield, multiscale code (figure 4.2)—and we showed that this code is much better than the classical place code in terms of coding performance (Eliav et al. 2021). When we introduced other animals, we found a representation of the position of other individuals by social place-cells in the hippocampus (Omer et al. 2018). When we allowed bats to behave completely naturally in a social group, we discovered that bat hippocampal neurons encoded a whole host of social signals, including identity coding, and representations of sex, social affiliation, and dominance hierarchy for multiple other individuals, as well as the locations of all individuals (Ray et al. 2024). Finally, when we allowed bats to dynamically change their behavior, akin to the behavioral dynamism that normally occurs during natural everyday behaviors, we discovered extremely rapid switching in the neural code (Sarel et al. 2022). In other words, every time we made a step toward more naturalistic behaviors, we discovered surprising new findings about the brain.

The same was true in rodents. First and foremost, the initial discovery of place cells was enabled by allowing rats to explore space naturally (O'Keefe & Dostrovsky 1971; O'Keefe

& Nadel 1978). When allowing rodents to move in 3D, multiple recent studies discovered 3D place cells, 3D head-direction cells, and 3D grid cells (Laurens et al. 2016; Laurens & Angelaki 2019; Angelaki et al. 2020; Grieves et al. 2020, 2021; Mallory et al. 2021), which were very similar to those we found in bats. Likewise, when introducing navigational goals, an egocentric vectorial coding of these goals was found (Ormond & O'Keefe 2022), very similar to the vectorial coding that we found in bats. Social place cells were found in rats (Danjo et al. 2018)—again, very similar to our finding in bats. When recording the details of rats' head-scanning behaviors, it was found that place cells tended to form at those locations where active scanning of the environment previously occurred (Monaco et al. 2014). When measuring the body pose of rats (their posture), it was discovered that hippocampal place cells were modulated by the behavioral state of the animal (as found in an unpublished study by Anton Sirota), and exhibited dramatic remapping that was triggered by jumping behaviors (Green et al. 2022). Finally, when allowing food-caching birds to display their natural food-caching behaviors, a cache-unique barcode activity pattern was found in the hippocampal population. These barcodes rode on top of the classical place code (i.e., were additive). Importantly, the cache-unique barcodes were reactivated whenever the birds later retrieved food from the same cache. This suggests that these neuronal barcodes comprise a signature of episodic memory. Thus, across species, studying naturalistic behaviors has uncovered surprising findings about the hippocampal formation.

It is important not only to study more naturalistic behaviors, but also to widen the array of species investigated in the neurobiology of navigation, in order to capitalize on the unique advantages of each species, in the spirit of Krogh's principle (as discussed in chapter 2). For example, the advantage of using food-caching birds is their incredible episodic memory and spatial memory abilities. The advantage of using rats is that they are smart and inquisitive animals, and thus wonderful models for cognitive and behavioral neuroscience. The advantages of using monkeys are that they are even smarter than rats; and in addition, they are of course similar to humans—and therefore findings from the monkey brain are very likely to apply closely to the human brain. The advantage of using mice is the amazing array of molecular tools available for them. An even larger array of molecular tools exists for *Drosophila melanogaster* flies. This elaborate toolkit enabled, for example, the discovery in the fly brain of an incredible ring attractor network of head-direction cells (Seelig & Jayaraman 2015; Green et al. 2017; Kim et al. 2017).

And what is the advantage of using bats? *Why bats*? As demonstrated in this chapter, there are several answers to this question—primarily focusing on the interesting behaviors of bats—and over the years, we have tried to utilize these interesting bat behaviors to shed new light on the brain. See Las and Ulanovsky (2024) for a detailed summary. In particular, note the following points:

1. *Bats navigate in 3D space,* which allowed for studying 3D spatial coding (figure 4.1c).
2. *Bats move very fast,* which allowed for studying the representation of large-scale space (figures 4.2 and 4.3) since these flying mammals can cover large spaces many times within a limited time period. In addition, the bats' speed allowed for studying rapid switches in neural coding (Sarel et al. 2022) since the bats' fast flight created an ultrafast behavioral switch during crossovers, which in turn created an ultrafast switch in neural coding.

3. Bats emit temporally discrete echolocation calls, so we can *measure the timing of sensory inputs with millisecond-level precision.* This allowed us to quantify the dynamics of place tuning around sensory inputs (figure 4.4a). In addition, the bat's echolocation provides *a proxy of attention,* which allowed us to examine the effects of spatial attention on spatial coding in the hippocampus (Sarel et al. 2022).
4. *Bats are highly social,* which allowed us to discover the strong social representations in the hippocampus (Omer et al. 2018; Ray et al. 2022, 2024).

That said, even though a number of species are already being studied in the field of the neurobiology of navigation, I strongly believe that we must further broaden the array of species being researched. For example, we need a species that performs *annual migrations* in order to study the neural basis of navigation in migratory species. This will allow for answering many interesting questions, including: How do migrating animals choose the navigational route? How is the timing of migration controlled? And how can they pinpoint their goal with extreme precision after having migrated for thousands of kilometers?

One conclusion from the studies surveyed in this chapter is the importance of increasing the environment size. The brain's navigation system did not evolve to navigate in 1-m boxes, so we must be missing something by doing experiments in such small environments. And indeed, when we increased the environment size to hundreds of meters, we discovered a completely different type of neural code for space in the hippocampus—the highly surprising multifield, multiscale code (Eliav et al. 2021). Of note, Edvard and May-Britt Moser have discovered grid cells primarily via increasing the environment size to 2 m (Hafting et al. 2005; Moser et al. 2017). The issue of environment size also relates to the findings discussed in chapter 2, wherein completely new facets of mouse spatial-exploration behaviors were discovered following an increase in environment size (section 2.2.1). There are a number of other arguments about why it is crucial to study the neurobiology of navigation in larger and more complex environments, as reviewed by Wolbers and Wiener (2014). One of the key ones is that when navigating in large environments, the animal needs to *integrate* navigational information over time and space. This is unlike navigation in small boxes or small rooms, where the entire environment is visible to the animal at once from one location or one vista. Such "vista-scale navigation" was shown (at least in humans) to engage completely different navigational mechanisms than large-scale navigation, in which integration over time and space is required (Wolbers & Wiener 2014; see also further discussion in section 7.3). Therefore, although our outdoors studies (figure 4.3) are in their early stages, we expect to uncover surprising findings about the brain's navigation circuit when animals navigate in the real world (i.e., outdoors). But also with respect to indoors laboratory experiments, all of these points suggest that it is crucial to dramatically increase the environment sizes for laboratory studies of the neurobiology of navigation.

Besides the issue of spatial scale, there are numerous other aspects of real-world navigation that are poorly understood and remain to be addressed in future research. In particular, successful navigation in real-world environments requires (1) assessing our own *orientation* and (2) *location* with respect to the surrounding environment, and with respect to the goal; (3) *route planning* beyond our current field of view; (4) constantly *updating* the navigational information as we move through the environment toward the goal; (5) *decision-making* at crucial junctions; (6) *reorienting* if the route is blocked or if the way is lost; and (7) *coordinate transformation* from allocentric absolute spatial coordinates to egocentric body-referenced coordinates, which is necessary to enable the motor action of moving toward the navigational goal (Wolbers & Wiener 2014; Park et al. 2018).

Currently, we lack solid understanding of the neuroscientific basis of *any* of these aspects of real-world navigation. For example, with respect to orientation cues during navigation outdoors, animals are known to use a variety of compass cues: the sun, moon, stars, polarization pattern of the sky, Earth's magnetic field, wind direction, odor gradients, wave direction, and major distal geographical landmarks such as coastlines and mountain ranges (Geva-Sagiv et al. 2015; Yovel & Ulanovsky 2017; Mouritsen 2018). It is currently unknown how all this rich compass information is extracted; how it is weighed (i.e., how does the animal decide to use one compass cue at a certain time point, but a different compass cue at a later time point); and how all this information is integrated into the brain's navigation circuit. Also, very little is known about real-world locational cues (i.e., which sensory cues are used for self-localization) and about their coding in the brain. Likewise, very little is known about the neural basis of route planning for real-world navigation; how navigational information is integrated over extended trajectories; or how allocentric-to-egocentric coordinate transformations are performed. These aspects have been investigated very little in the brain's navigation circuit, and thus they form a list of major challenging areas for future research on the neurobiology of real-life navigation.

Another open question relates to border cells, also known as "boundary cells": neurons that fire along the geometric borders of the environment (Savelli et al. 2008; Solstad et al. 2008; Lever et al. 2009). Currently, it is unknown whether border cells are at all relevant to the real world outdoors—and if so, to which real-life physical borders they respond. A related, and perhaps even more intriguing, question is how *territorial borders* are encoded in the brain (Wirth et al. 2021). Many territorial species of animals have territories whose borders are not marked by physical boundaries such as streams, cliffs, or hedges, but comprise mostly imaginary borders. These borders often form straight-line segments connecting landmarks, which mark the corners of the territory (where the landmarks can be either visual landmarks or scent-marked locations); thus, the animal's territory roughly forms a polygon (Dugatkin 2020). How are such imaginary borders being recognized by the animal, and how are they encoded in the brain? One possibility is that these borders are based on olfactory matching or visual matching (Zeil 2023). Another possibility is that territorial borders are encoded by border cells. Although these neurons have been traditionally studied in the lab using physical borders, perhaps they also can encode imaginary borders of a territory—at least in territorial species for which these borders are behaviorally important. Finding such territory-border cells would create an interesting connection between the neurobiology of navigation and the ethology of animal territoriality.

And, given the wide array of hippocampal responses to many behavioral features, as described in this chapter, one may wonder: Would one find in the hippocampus a representation for *any* behavioral feature that one would look for? In other words, perhaps you will find *anything* that you look for in the hippocampus? However, this is *not* necessarily the case—instead, it seems that the hippocampus represents behavioral variables only if they are behaviorally *relevant*. Thus, while we found strong identity coding in the social-group study, with its rich social interactions (Ray et al. 2024), this was *not* the case in the two-bat study in the tunnel (Sarel et al. 2022). In that study, in some experimental sessions, we introduced *two* possible other bats, but we found that hippocampal CA1 neurons did not distinguish between these two other bats (i.e., did *not* exhibit identity coding), presumably because in this high-speed collision-avoidance behavior, it is less relevant to encode what are you colliding with. Similarly, in the social-group study, we found that at very short distances, hippocampal neurons encoded the directions toward other bats much more strongly than the distances between them (Ray et al. 2024)—possibly because at short distances, direction is more relevant than distance. Likewise, in the study of 3D head-direction cells (Finkelstein et al. 2015), we found much weaker representation of head roll as compared to head azimuth and pitch, presumably because the roll angle is much less relevant as a navigational variable, as compared to azimuth and pitch (Finkelstein et al. 2016). Thus, hippocampal neurons do not encode everything—instead, they seem to represent only *relevant* information. If this is true, then perhaps the reason that spatial coding is so prominent in hippocampal and parahippocampal neurons is that movement through space is so relevant for animals—including in the testing paradigms that we normally use in laboratory studies of animals. However, the underlying neural representation in the hippocampus might not be for representing space per se; rather, it may embody a combinatorial, **mixed-selectivity** representation of whatever is most relevant for the animal.

Finally (speaking of relevance), the most relevant things to an animal are (1) the other animals that it encounters in its surrounding world, including conspecifics, but also predators and prey; and (2) the natural stimuli that it encounters in the wild. In this chapter, we discussed some issues related to point (1), on social interactions—and we shall return to social neuroscience later in the book. In the next two chapters, however, I will shift to point (2) and discuss sensory systems in the brain. I will highlight the importance of natural active-sensing behaviors and natural sensory stimuli. In particular, I will show that the brain responds completely differently to natural sensory stimuli and natural sensory-acquisition behaviors, as compared to the passive, artificial, and oversimplified sensory stimuli that have been studied for decades in sensory neuroscience.

5
Active Sensing

Sensory neuroscience has traditionally focused on experiments with head-fixed, often anesthetized animals, which are *passively* presented with sensory stimuli. It is possible that this mode of research will eventually lead to full understanding of the neural basis of sensation—and that the natural-neuroscience type of experiments promulgated in this book will provide only a slight modulation of our knowledge. However, I consider this to be quite unlikely. In fact, I believe that conducting more naturalistic experiments is crucial and will fundamentally change our understanding of sensory neuroscience. The evidence leading to this belief will be discussed in this chapter.

If one observes how an animal senses objects under normal natural conditions, it immediately becomes evident that real-life sensing is a very *active* process. When a rat explores an environment and touches an object with its long whiskers, it will typically turn toward the detected object and start actively palpating it with the short whiskers (microvibrissae) located under the nose, and will also palpate the object using the nose itself and the lips. We humans also actively explore objects using our fingers. When an animal or human detects a smell, they will start actively sniffing it, inhaling the odor into the nose (Porter et al. 2007). When we want to inspect an object or a scene with our vision, we move our eyes through multiple saccades, visually palpating the object with our fovea. Many studies have shown that these unitary movements—the whisk, the sniff, and the saccade—constitute a fundamental unit of sensory processing and may even be an integral part of the sensory percept itself (Yarbus 1967; Kepecs et al. 2006; Kleinfeld et al. 2006; Mainland & Sobel 2006). Thus, it is clear that if we wish to understand sensory perception, we have to start thinking of sensation as an active sensory acquisition process—or *active sensing*. This chapter will survey the evidence for the active nature of sensation, as well as the dramatic consequences that active sensing entails for understanding the neural basis of sensory perception.

Further, classical two-alternative forced choice (2-AFC) experiments in sensory neuroscience—in both monkeys and rodents—artificially create a *temporal separation* between perception and action. This separation in time assumes that sensory systems work in a feedforward, open-loop manner. But in real life, there is no such separation: perception and action constitute a closed-loop process that keeps going continuously. Perception and action cooccur simultaneously all the time in the real world—they are *not* separated in time. This is another key distinction between thinking about the senses as passive feedforward systems and the active sensing approach, which emphasizes the massive, closed-loop feedback between perception and action.

Active sensing is important in almost all sensory modalities. Here, we will limit ourselves to discussing mostly three senses: vision, somatosensory whisking, and olfactory sniffing—the sensory systems for which the most extensive evidence exists for the dramatic effects of active sensing on the neural mechanisms and algorithms of sensory processing. We will also briefly discuss echolocation. But we will start with the classical, reductionist approach to sensory neuroscience—in particular, vision.

5.1 The Reductionist Approach to Visual Neuroscience

The classical approach to visual neurophysiology was to study the visual system—from retina to cortex—in anesthetized, paralyzed, head-fixed animals, and to use well-controlled, simple visual stimuli to decipher the response properties of visual neurons. This approach was pioneered in the first half of the twentieth century by Adrian, Hartline, Granit, and Kuffler, who used primarily black-and-white spots as visual stimuli. One of Kuffler's notable discoveries in the early 1950s was that cat retinal neurons have center-surround receptive fields—namely, they are excited by a spot of light illuminating the center of their receptive field but are inhibited by shining light on the surround, or vice versa (Kuffler 1953). Subsequently, Hubel and Wiesel experimented on anesthetized and paralyzed cats and found that many V1 neurons are tuned to the orientation of bars that are moving in specific directions (Hubel & Wiesel 1959). The careful analysis that Hubel and Wiesel conducted for the properties of V1 neurons was absolutely incredible—and included the elucidation of the neurons' receptive fields, tuning curves, and the structure of cortical columns and cortical maps (Hubel & Wiesel 1959, 1962, 1963b). The overwhelming success of Hubel and Wiesel's early discoveries and subsequent research program has influenced visual neuroscientists so much that for several decades, it seemed impossible to study the visual system in any other way. This approach also greatly influenced investigators of other sensory systems. However, this amazing success also had a downside: it propelled sensory neuroscience as a field into several decades of experimentation in head-fixed, anesthetized, passive conditions—very different from natural active-sensing behavior. Hubel explained that he chose the anesthetized, paralyzed cat for two main reasons: the technical reason of greater recording stability and the more fundamental reason of having a precise control over the sensory stimulus (Hubel 1959).[1]

To this day, a key emphasis of reductionist experiments in visual neuroscience is to *control behavior*. In practice, experimenters strive to *eliminate behavior*. This behavioral control is typically expressed by conducting experiments in head-fixed animals, controlling for potential confounds that may be introduced by head movements; as well as eliminating eye movements, controlling for potential confounds from eye saccades (figure 5.1). The elimination of eye movements is achieved in one of two ways: either by anesthetizing and paralyzing the animal, as was done in Hubel and Wiesel's original experiments in V1 of cats and monkeys (Hubel & Wiesel 1959, 1968), or by training an awake subject to fixate its eyes on a stable fixation point in the middle of the screen (Wurtz 1969). Figure 5.1 shows schematically these experimental setups, which enabled a nearly perfect behavioral control, thus achieving highly repeatable sensory stimuli.

However, let us pause and consider the reasons for wishing to control stimuli by eliminating motor behavior. There are two separate reasons for this, as pointed out by Hubel. First,

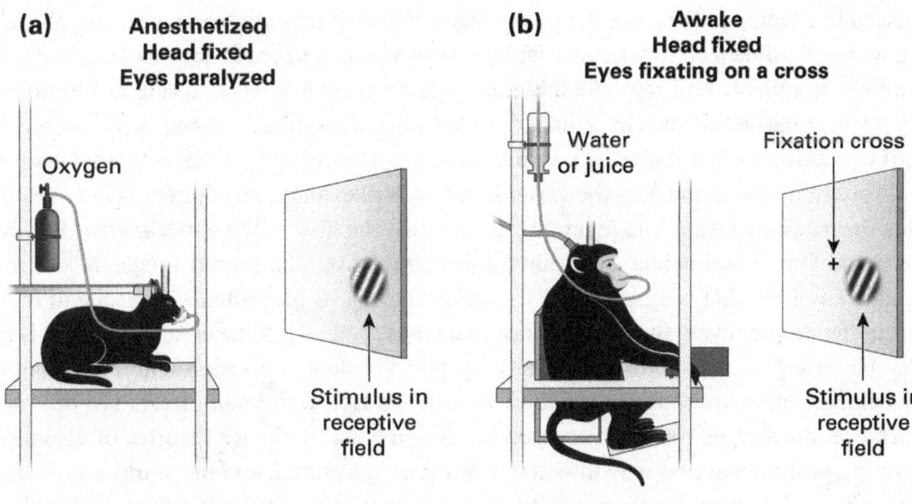

Figure 5.1
Classical controlled setups for electrophysiological studies of the visual system. (a) Head-fixed, anesthetized, and paralyzed cat, as studied by Kuffler and by Hubel and Wiesel. Paralysis prevents eye movements. (b) Head-fixed, awake monkey that was trained to fixate its eyes on a "fixation cross," to minimize eye movements. The drawing in (b) was inspired by Wurtz (1969).

the technical reason: Over half a century ago, when neural recordings in the visual system (and other sensory systems) were pioneered, it was not easy to obtain stable and artifact-free neural recordings. To achieve this, it was necessary to fix the head of the animal and train it to fixate its eyes to reduce head and eye movements, which cause recording artifacts. Conducting recordings in freely moving animals, although possible, was very difficult at that time (Hubel 1959). This technical reason can no longer be considered valid today because, as described in chapter 3, modern electrophysiological and optical methods allow reliable, stable, and nearly artifact-free neuronal recordings from freely moving animals.

But there is a second, more fundamental reason for the wish to eliminate behavior in sensory-neuroscience experiments. Namely, uncontrolled behavior and movement are considered sources of "noise," which may prevent the experimenter's ability to present precisely repeatable sensory stimuli at the exact same retinal location. Consequently, this noise is thought to hinder our understanding of the key phenomena of sensory perception. This way of thinking is reflected even in the term "sensory processing," which implicitly assumes a feedforward flow from sensory input (perception) through some neural processing, eventually leading to motor output (action). This feedforward way of thinking has deep roots in physiology, in the concept of the reflex arc, and in psychology, in the behaviorist approach (Wallach et al. 2016). If the feedforward notion is true, then it seems justified to eliminate movement altogether. However, as will be argued next, sensation and movement are tightly interconnected via multiple feedback loops, and therefore they cannot be thought of as "input" and "output." Thus, even in principle, we cannot think of the brain in feedforward terms. Consequently, movement of the sense organs (such as eyes, fingers, nose, or ears) is not noise, but rather an integral part of active perception.

The classical line of thinking about feedforward sensory processing can be traced to two sources: a biological rationale and a computational rationale. The biological rationale for eliminating behavior comes from the heart of the reductionist method in biology, as discussed in chapter 1. This method posits that if we wish to understand a certain phenomenon, which is influenced by a key variable X, then we need to create an experimental setup that allows to reliably and reproducibly manipulate variable X while fixing or eliminating any irrelevant variables, such as Y and Z. In the case of the phenomenon of visual perception, if one assumes that the brain's visual system is strictly concerned with analyzing the visual pattern on the retina (i.e., the variable X that is relevant to this system is the stream of photos that create a visual image on the retina), then the task of the experimenter becomes quite clear. The experimenter must have full control over the retinal image X and must eliminate anything that might weaken this control, such as movements of X due to movements of the head or eyes. But this idea has two flaws (table 5.1). One conceptual flaw is that it may be wrong to assume that the visual system is solely concerned with analyzing the retinal image and extracting information from it. In fact, top-down effects are of crucial importance. Further, as will be discussed in section 5.2.2, there are theories of visual perception suggesting that two-way interaction between the animal and the world is crucial for high-acuity vision, and therefore, vision research should be centrally concerned with the behavioral repertoire and movements that animals perform. The active-sensing approach, in particular, disagrees with the notion that the visual system is concerned only with analyzing the retinal image; instead, it suggests that action is an inherent part of perception, and therefore motor-sensory interactions should always be studied together. During active vision, in real-life situations, we move our eyes and head all the time and actively look at things. The crucial importance of active vision has led people like James Gibson to propose that one *cannot* study feedforward visual processing in isolation from the normal eye movements and head movements that occur during active vision (Gibson 1979). In chapter 6, I will further describe Gibson's approach. But crucially, in this view, eliminating behavior is eliminating the very thing that we ought to be studying if we truly wish to understand visual perception.[2]

Another conceptual flaw in the biological-reductionist approach is that even if its basic assumption is correct—namely, if the visual system *does* care only about the retinal image—then in real-world situations, this retinal image itself is highly dependent on the animal's behavior. For example, when an animal or human walks in the woods, the main source of changes in its retinal image is coming from its own movements—whole-body locomotion, head turns, and eye movements. Therefore, if one wishes to analyze the responses of neurons to moving bars, as Hubel and Wiesel did, then one must not forget that during a walk through the woods, the main source of moving bars on the retina will be the animal's *own motion*, which will cause trees to move across the retina. Trees do not walk—the animal does. Therefore, even if we embrace the reductionist approach that the visual system strictly analyzes the retinal image X, then I would argue that the animal's behavior is a key determinant of X: behavior is *not* an irrelevant phenomenon that needs to be eliminated; rather, behavior is part of the crucial variable X that needs to be studied explicitly. In other words, instead of defining X as: X = [retinal image], we should redefine it as: X = [retinal image as determined by the interaction between the external world and the animal's behavior]. To this we may also add some top-down factors that strongly influence vision, such as context, experience, and memory. As I will discuss in detail in section 5.2, the need to redefine X has been strongly supported by a recent series of studies from a number of laboratories, all of which demonstrated directly that V1 neurons are indeed centrally concerned with the interaction between the external world and the

Table 5.1
Rationales for conducting visual-neuroscience experiments in nonbehaving animals, and critiques of these rationales

Rationales for Highly Controlled Experiments in Visual Neuroscience	Critiques of These Rationales
Technical rationale: Fixing the head of the animal allows stable neural recordings. In addition, to reduce artifacts from eye movements, animals should be anesthetized or trained to fixate their eyes.	Modern electrophysiological and optical recording methods allow stable neural recordings in freely moving animals—even flying bats—with minimal artifacts from head, eye, or body movements (see chapters 3 and 4).
Biological rationale: Vision is about analyzing the retinal image. Therefore, the reductionist approach to biology posits that we must control for any variables besides the retinal image; in particular, we should eliminate behavior.	First, the assumption that vision is strictly about analyzing the retinal image may be wrong: Top-down effects are of vital importance. Further, the active sensing approach proposes that vision—and in fact almost all sensory systems—is fundamentally about the interactions between the animal's motion and the sensory input (motor-sensory loops). Indeed, recent recordings in V1 of behaving animals revealed very strong motor signals in the visual system: in fact, only a minority of V1 cells are strictly "visual" neurons (see section 5.2.1). Second, even if vision is strictly about analyzing the retinal image, the main real-world source of image motion is the animal's own movements; hence, we cannot ignore behavior.
Computational rationale: Marr (1982) proposed that the "job of vision" is to derive a representation of shapes and positions of objects from images, and therefore, all things besides the retinal image are irrelevant to the main job of vision. The extraction of shape was proposed to proceed through a series of feedforward computations and representations of increasing complexity.	First, Marr's assumption about the job of vision may be incorrect: While the ventral stream of the visual cortex may indeed extract *shape*, the dorsal stream is concerned with visual *motion* and *actions* (see chapters 6 and 7). Second, even if the job of vision is to extract the shapes of objects, it is strongly aided by movements and looking at 3D objects from various angles; hence, movement is tightly linked to visual perception. Third, visual processing is not feedforward: there are massive feedback projections from high visual areas to V1, which carry information relevant to vision, regarding attention, memory, and behavior, and this information is missing if one eliminates behavior in laboratory experiments. In addition, there are strong projections into V1 from the motor cortex and other brain regions (figure 5.2).

animal's behavior and movement. In fact, as we will see, some V1 neurons don't even care about retinal input—they care only about the animal's behavior and movement.

The computational rationale for eliminating behavior in visual-neuroscience experiments is equally illuminating. It can be clearly appreciated from David Marr's influential book *Vision* (Marr 1982), which laid much of the foundation for both computer vision and computational neuroscience, and has also greatly influenced many of the subsequent experiments in visual neuroscience. Marr wrote:

> My approach to these problems was very much influenced by the fascinating accounts of clinical neurology. . . . Particularly important was a lecture that Elizabeth Warrington gave at MIT in October 1973. . . . Warrington had put her finger on what was somehow the quintessential fact of human vision—that it tells about shape and space and spatial arrangement. Here lay a way to formulate its purpose—*building a description of the shapes and positions of things from images*. Of course, that is by no means

all that vision can do; it also tells about the illumination and about the reflectances of the surfaces that make the shapes—their brightnesses and colors and visual textures—and about their motion. But these things seemed *secondary*; they could be hung off *a theory in which the main job of vision was to derive a representation of shape.* [pp. 35–36; italics added]

Marr (1982) then went on to describe his theory of the "job of vision," which relies on a series of feedforward transformations, as follows:

Visual image (retinal image)

↓

"Primal sketch" representing intensity changes and their geometric organization

↓

"2½-D sketch," representing rough depth of surfaces and contours of discontinuities

↓

"3D model representation" of three-dimensional (3D) objects in the environment

Marr's theory provided an important first step and was extremely influential in launching a fruitful approach to understand the visual representation of object shape. But we may ask: What if it is not entirely correct that the "main job of vision was to derive a representation of shape"? Moreover, what if the visual system does not work via a series of strictly feed-forward transformation, as Marr envisioned?

Indeed, we know today that Marr missed part of the picture of the function of vision (see table 5.1). What Marr did not know in the late 1970s is that the primate visual system has *two* major cortical processing streams, or pathways (Ungerleider & Mishkin 1982; Milner & Goodale 1995), as depicted in figure 5.2a. The ventral stream, which is concerned with *object representation*, indeed seems to perform a series of analyses of increasing complexity, close to what Marr's theory described. However, there is also another pathway—the dorsal visual stream, which is concerned with analyzing *visual motion* and enabling vision-based *actions*; the dorsal stream was not accounted for by Marr's theory. I shall come back to visual motion analysis in chapter 6.

Furthermore, we know today that in primates, there are strong top-down feedback projections from high visual areas to V1 (see figure 5.2a, descending arrows), as well as to the thalamic dorsolateral geniculate nucleus (dLGN) (see Clavagnier et al. 2004, Gilbert & Li 2013). In the mouse, too, the top-down projections to V1 are extremely strong: figure 5.2b shows the quantitative distribution of input strength into V1, as measured by injecting a retrograde monosynaptic rabies tracer into V1. This experiment, by Georg Keller and colleagues (Leinweber et al. 2017), has shown that the classical feedforward visual inputs from thalamus dLGN into V1 are dwarfed by massive top-down feedback coming into V1: There are very strong top-down monosynaptic inputs from a variety of brain regions, including various subregions of V2, but also from many nonvisual areas, such as the retrosplenial cortex (figure 5.2b, RSP), auditory cortex (AuD), anterior cingulate cortex (ACC), and motor areas (M2).

The descending input from the auditory cortex and motor cortex into V1 are each about threefold stronger than the ascending visual input from dLGN, and the descending input from retrosplenial cortex into V1 is about sixfold stronger.[3] Even the input from the hippocampal formation (HPF) into V1 is as strong as the visual input from dLGN into V1. In

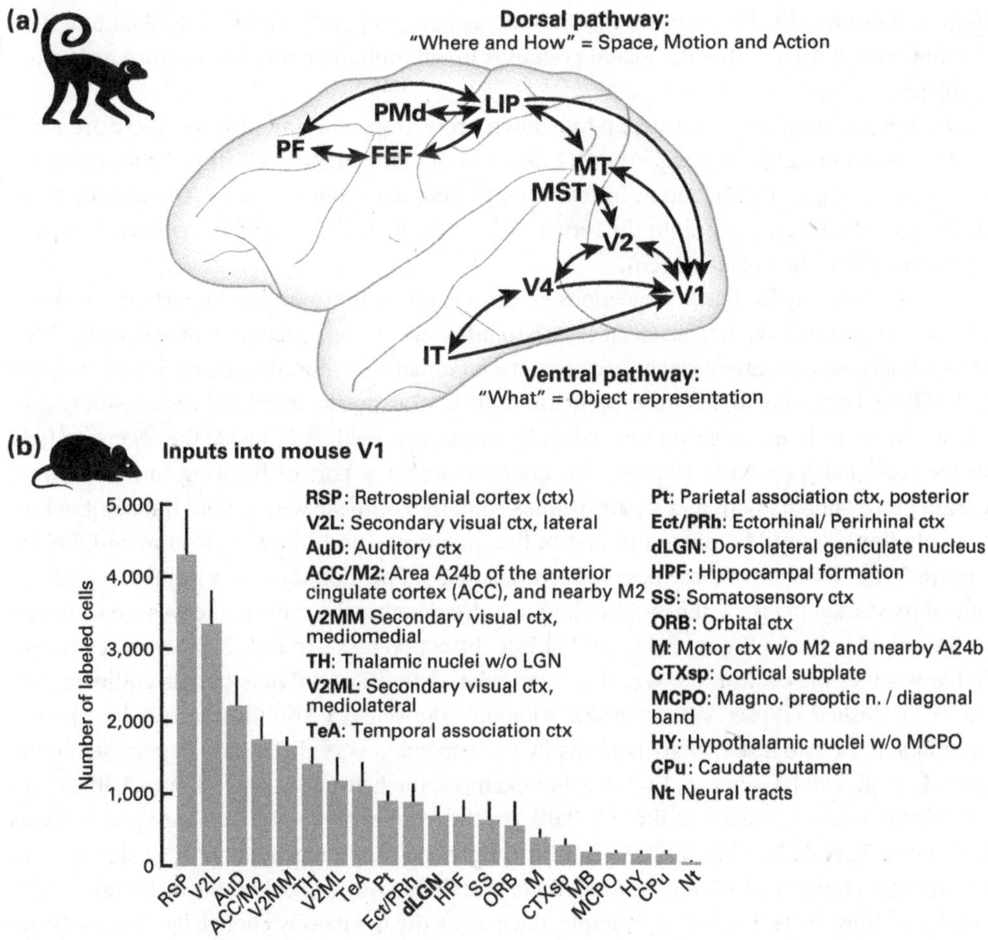

Figure 5.2
Strong top-down descending inputs into the primary visual cortex, V1. (a) Simplified schematic of ascending and descending projections between some of the major cortical visual areas in primates, as well as beyond the visual cortex. Note the multiple top-down inputs coming into V1 directly (monosynaptically) from visual areas like V2, V4, inferior temporal (IT) and medial temporal (MT) areas, as well as from areas beyond the visual cortex such as the lateral intraparietal area (LIP). When considering two synapses (disynaptic inputs), primate V1 receives wide-ranging inputs from areas in the motor system (e.g., dorsal premotor area, PMd), prefrontal cortex (PF), and more. These top-down inputs send a variety of complex behavioral signals into V1. Adapted from Lehky and Sereno (2007); Gilbert and Li (2013). (b) Inputs into V1 of the mouse: Shown are the numbers of cells labeled in different brain regions by monosynaptic rabies tracing from V1. This virus jumps only one synapse retrogradely, labeling the direct monosynaptic inputs into V1 cells. Note the major direct inputs into V1 from a variety of non-visual areas, such as the retrosplenial cortex (RSP), auditory cortex (AuD), anterior cingulate cortex (ACC), motor cortical areas (M2), temporal and parietal association cortices (TeA and Pt), and the hippocampal formation (HPF). All these areas provide very strong inputs into V1, stronger or similar in strength to the input from the thalamic dorsolateral geniculate nucleus (dLGN), which is the main source of visual input coming into V1. Error bars indicate SEM. Adapted from Leinweber et al. (2017).

light of these results, how can we think of V1 as being a purely visual area? Just based on the anatomy, it is clear that the visual system is much more than a strict feedforward visual analyzer.

Notably, in addition to the strong top-down projections coming into V1, there are massive feedback projections going from V1 down to dLGN (Cudeiro & Sillito 2006; Briggs & Usrey 2011; Briggs 2020), and even feedback projections from the brain (specifically from the hypothalamus) directly into the retina (Warwick et al. 2022). Thus, feedback is pervasive throughout the visual system.

The very strong feedback projections in the visual system provide important top-down behavioral information that has been shown to influence visual processing dramatically. This information includes attention, expectation, task demands, and multisensory inputs (Gilbert & Li 2013). For example, in one experiment (figure 5.3a), awake monkeys were visually presented with five lines: a central line, whose position was held fixed inside the receptive field of the recorded V1 neuron (figure 5.3a, gray rectangle); a pair of flanking lines that were parallel to it (side flanks); and a pair of lines roughly collinear with it (end flanks). Each of the side flanks could be placed in one of five positions, and likewise for the end flanks. Figure 5.3a shows one of these possible arrangements. The monkeys were trained to perform one of two tasks. In task 1, the monkey had to judge whether the central line was closer to one of the two side flanks (figure 5.3a, solid ellipse: bisection task). In task 2, the monkey had to judge whether the central line was displaced relative to the end flanks or was collinear (figure 5.3a, dashed ellipse: Vernier task). Although the sensory stimuli (the five lines) were *identical* in the two tasks, many neurons in V1 responded very differently depending on the task (Li et al. 2004; Gilbert & Li 2013). For example, the neuron shown in figure 5.3b did not encode the relative position of the side flanks in the Vernier task—the tuning to position was flat (figure 5.3b, dashed black line)—but it strongly encoded the position of the side flanks in the bisection task, in which the position of the side flanks was critical for the task (figure 5.3b, solid gray line). Note that in this example, the neuron did not merely change its *average* firing rate via a multiplication by some scalar gain factor; instead, the neuron totally changed the *shape* of its tuning curve, even though the visual stimulus was exactly the same. In other words, a top-down signal (in this case, a signal conveying task demands) caused this V1 neuron to qualitatively change its response and the kind of information that it conveys.

In another experiment, rats learned the association of two visual stimuli with two different reward timings (Shuler & Bear 2006); surprisingly, a large fraction of neurons in the deep layers of rat V1 (43 percent of the recorded neurons) developed very strong responses whose timing predicted the timing of rewards. Such temporally precise, reward-predicting responses are certainly not expected from the classical view of V1 neurons as visual feature detectors. Gilbert and Li (2013) concluded from many such experiments that a sensory neuron—even in V1—is "an adaptive processor, changing the calculation it performs in accordance to the top-down instruction received from higher-order cortical areas." Therefore, I would argue that sensory systems must be studied very extensively in the context of behavior because essential information is missing if one ignores these "top-down instructions." Using Marr's own terminology, it is now becoming clear that the "primal sketch" in V1 and V2 receives a plethora of top-down information, including from the higher-level 3D model representation, which is derived in part from the animal's own visually guided behavior—for instance, as the animal walks around the 3D object and looks at it from

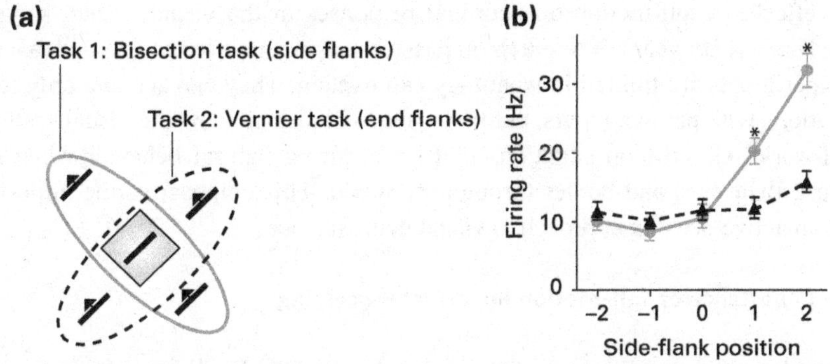

Figure 5.3
Example of top-down influences on neuronal tuning in V1. (a) The exact same visual stimuli were used for two behavioral tasks: Either a bisection task, where the monkey had to judge whether the central line is in the middle between the two flanking lines (side flanks) highlighted here by the gray solid ellipse 1; or the Vernier task, where the monkey had to judge whether the central line is collinear with the other two flanking lines (end flanks) highlighted here by the black dashed ellipse 2. The position of the four flanking lines was varied from trial to trial. (b) Responses of one V1 neuron in these two tasks, as a function of the position of the side flanks. This neuron exhibited flat tuning to the side flanks during the Vernier task (black dashed line), when the side flanks were irrelevant, but it showed selective tuning to the position of the side flanks during the bisection task (gray solid line), when the side flanks were relevant. Note that in both tasks, the visual stimulus was identical. This example demonstrates that V1 neurons respond to top-down information such as task demands and relevance, and not just to the visual stimulus. In addition, it shows that top-down information may go beyond a simple global gain mechanism and can instead selectively change the tuning shape of V1 neurons. Adapted from Gilbert and Li (2013).

various directions (see more on this in chapter 7). This implies that visual experiments should not aim to control for behavior, but instead should incorporate active behavior into their design. Behavior is not merely noise, or an experimental nuisance that should be eliminated—rather, guiding behavior is why we need perception in the first place.

More specifically related to the theme of this chapter on active sensing, it is noteworthy that, at least in rodents, there are strong direct anatomical projections from the motor cortex to V1 (Leinweber et al. 2017) (figure 5.2b, "M2"). These inputs suggest that the brain does *not* work in a unidirectional fashion, from sensory → motor, but it also goes from motor → sensory: it is a sensorimotor loop. Further, Margrie and colleagues (Vélez-Fort et al. 2018) have shown that the majority of neurons in V1 layer 6 of the mouse are encoding head movements in complete darkness, via vestibular inputs that arrive to V1 from the retrosplenial cortex, which is the most prominent input region to V1 (figure 5.2b, "RSP"). They placed mice on a rotating plate, along with a screen for visual presentation, a setup that allowed presenting visual, vestibular, or visual+vestibular inputs. They found that V1 layer 6 neurons linearly summed the visual and vestibular inputs, suggesting that V1 layer 6 neurons are not purely visual but are instead multisensory, visual+vestibular neurons, which may be involved in processing both external and internal motion stimuli (Vélez-Fort et al. 2018). This experiment was performed in head-fixed animals, providing a nice demonstration that head-fixed preparations *can* in fact be powerfully used to understand more naturalistic situations.

These results highlight again that the visual system is not strictly an analyzer of retinal images. It cares about movement and behavior. Therefore, subjects must be allowed to behave and *move*. In the next section, we will discuss a series of studies showing that the self-motion of the animal strongly affects V1 processing, and moreover, that self-generated motion affects V1 neuronal responses differently than if the motion were passive. The dramatic effects of self-motion on neuronal responses in the visual system suggest that while the last seventy years of research on passive sensing have led to great advances, such classic experiments are limited in what they can explain. They can account only for open-loop situations with passive inputs, but they cannot illuminate how the visual system deals with real-world, closed-loop situations that arise during natural behavior when animals freely move their eyes and bodies through the world. This emphasizes the importance of pursuing an active-sensing approach to visual neuroscience.

5.2 The Importance of Self-Motion for Visual Processing

Although substantial visual processing occurs subcortically, I will focus for brevity mostly on the visual cortex, and in particular on V1. The discussion will center on how the response properties of V1 neurons might be affected by animal movements: both whole-body locomotion and head and eye movements. I will make a distinction here between the specific effect of movement and more general effects, such as those of wakefulness or arousal. Multiple studies over several decades have shown that the visual cortex—and the entire brain, for that matter—behave quite differently when the animal is awake and aroused, as opposed to anesthetized or sleeping animals. For example, Livingstone and Hubel (1981) showed that the signal-to-noise ratio (SNR) of V1 neurons improves in aroused versus sleeping cats, and this is mediated both by an increased signal (enhanced visual responses) and a decreased noise (reduced spontaneous spiking activity). Omer et al. (2019) showed that while the V1 of anesthetized monkeys exhibits spontaneous reactivation of evoked activity maps, which also occurs in anesthetized cats (Kenet et al. 2003), such widespread reactivations are not seen in awake monkeys, so the findings from anesthetized monkeys are likely less relevant for real-life situations. These and many other studies have made it clear that it is important to study the *awake* brain. Here, however, I will consider a more specific question—namely, how visual neurons are affected by *self-motion*.

As mentioned in section 5.1, the main situation in which lines or bars could move across the retina is when the animal itself is moving. For example, when walking in the woods, the animal's locomotion will cause tree trunks and oriented branches to move across the retina. Similarly, head movements and eye saccades provide major sources of visual motion across the retina. Moreover, as we will discuss, animals and humans actively move their eyes over objects as part of their active sensory acquisition process. Therefore, we might expect that the animal's own motions—whole-body locomotion, head motion, and eye movements—could be represented as early as V1 (or perhaps even earlier). Moreover, the animal's own movements may even be an integral part of visual perception itself.

We will begin by discussing the case of whole-body locomotion.

5.2.1 Mouse Vision and Whole-Body Locomotion

The question of the interplay in V1 between whole-body locomotion and visual motion was first tested systematically in an experiment by Niell and Stryker (2010). They conducted electrophysiological recordings of V1 neurons from head-fixed mice that ran on a spherical treadmill while drifting gratings were presented on two screens in front of them (figure 5.4a). The mice spontaneously exhibited periods of running and standing still, and importantly, the visual input in this experiment was independent of whether the animal was

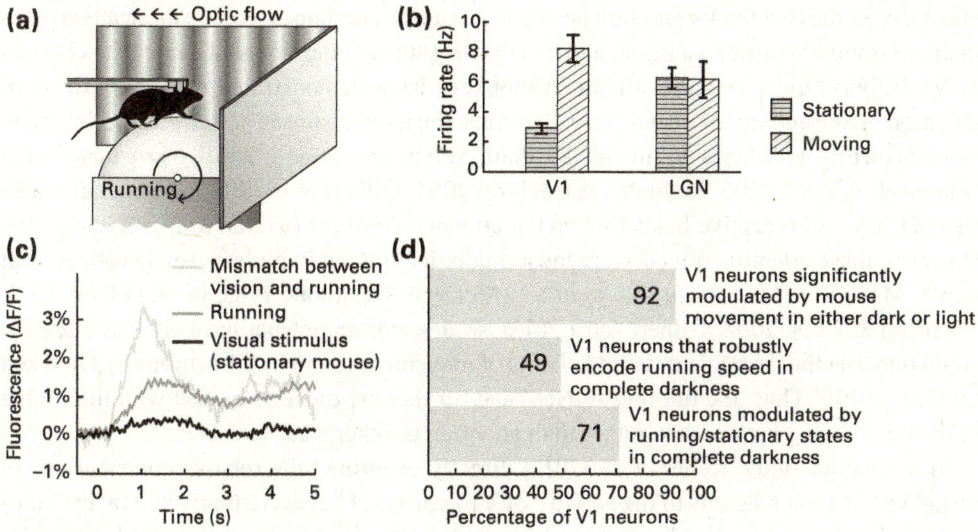

Figure 5.4
Most neurons in mouse V1 are not purely visual; rather, they encode motor signals or the mismatch between locomotion and vision. (a) Experimental setup that allowed dissociating vision from locomotion. (b) The response of V1 neurons to the same visual stimulus increased on average by almost three times when the animal was moving versus when it was stationary (left, V1); no such effect was found in the visual thalamus (right, LGN). (c) In another study, responses of mouse V1 neurons were weakest to visual stimuli, stronger to the onset of running, and strongest for a mismatch between the visual stimulus and running. Shown are population average responses for all the recorded neurons: the same set of neurons was tested under the various conditions. (d) In yet another study, a strikingly high percentage of neurons in mouse V1 responded significantly to running in complete darkness: 49 percent, 71 percent or 92 percent, depending on how responses were assessed. These results indicate that mouse V1 neurons, which receive strong anatomical inputs from motor cortex (figure 5.2b), also exhibit strong functional responses to movement: even stronger than to visual stimuli. (b) Adapted from Niell and Stryker (2010); (c) adapted from Keller et al. (2012). Numbers in (d) are based on Saleem et al. (2013).

running or standing (i.e., this was *not* a VR closed-loop system), but it did allow for comparing the activity of single neurons while running versus standing. The differences were quite striking: Although the shapes of the orientation-tuning curves did not change between standing and running, the firing rate of V1 pyramidal neurons increased on average by almost three times (figure 5.4b, left). By contrast, no difference was found between standing and running in the responses of thalamic neurons in the lateral geniculate nucleus (LGN) (figure 5.4b, right; however, note that another study, which used a different methodology, *did* find motor-related signals in mouse LGN (Roth et al. 2016)). The absence of firing-rate modulation by running speed in the LGN argues against the notion that increases in V1 firing rates are caused by increased body temperature, by increased blood perfusion due to running, or by peripheral effects in the retina, because all these factors are common to the V1 and LGN neurons. Instead, the results of Niell and Stryker suggested that the large gain effects of locomotion on V1 activity have a neural origin. Moreover, the V1–LGN difference suggests that the effects in V1 result from cortical mechanisms.

The dramatic changes in V1 neuronal activity as a function of the animal's behavioral state—running versus not running—could be explained in principle in two ways. First, locomotion itself could have modulated the responses of V1 neurons; for instance, via a motor *efference copy* (an internal copy of the movement signal produced by the motor system). Second, it could be that something about the *state* of running could underlie the effects, regardless of the locomotion per se. For example, perhaps mice are more attentive to their environment during running, and it is the heightened attention that causes the changes in V1 activity. Indeed, many studies in monkeys have demonstrated dramatic effects of attention on the activity of visual cortex neurons: the most common effect was the enhancement of neural responses to attended stimuli versus the same stimuli when unattended (Maunsell & Cook 2002; Reynolds & Chelazzi 2004; Gilbert & Li 2013); some studies also showed shifts of receptive fields toward the attended stimulus (Moran & Desimone 1985). However, these attention effects were most dramatically seen in higher visual cortices such as V4, MT, and IT, and much less so in V1 (Moran & Desimone 1985; Ito & Gilbert 1999; Maunsell & Cook 2002; Gilbert & Li 2013), so it seems somewhat unlikely that attention could underlie the dramatic threefold effects of movement found in V1 neurons by Niell and Stryker (2010).[4] Thus, the findings of Niell and Stryker are most consistent with the source of the effect being locomotion, rather than attention or other state variables.

In a separate study, Keller et al. (2012) directly examined the relative contributions of visual versus motor signals to the activity of V1 neurons. They were motivated by the ethological observation that in the real world, oriented bars move across the retina primarily because of the animal's own motion. Therefore, they decided to perform *closed-loop experiments,* where the sensory stimulus moves according to the animal's movement, and compare them to *open-loop experiments,* where the sensory stimulus moves independent of the animal's movement. They used a spherical-treadmill system similar to that of Niell and Stryker (figure 5.4a), with the important distinction that here, they closed the loop between the animal's locomotion and the visual stimulus—namely, the visual gratings moved according to the animal's movement. In other words, they used a VR system. As a control, the visual gratings were moved in a noncongruent manner. Keller et al. performed two-photon calcium imaging of V1 neurons in head-fixed mice under four conditions: (1) mice running with a visual feedback from a drifting-grating stimulus, such that the visual stimulus moved according to the running velocity of the animal ("visual+motor"); (2) mice running while the visual grating was not moving ("motor-only"); (3) the visual grating being moved without the animal running ("visual-only"); (4) mice running in complete darkness. They found that only 22 percent of the cells exhibited positive correlations between their responses in the visual+motor condition and the visual-only condition—that is, only 22 percent of the neurons could be classified as "visual neurons." Moreover, some of these cells were also modulated by the running, so the fraction of *purely* visual neurons in mouse V1 was in fact much smaller than 22 percent. This dramatic result is at odds with the classical view that V1 is primarily a visual-processing area.

Further, Keller et al. (2012) found that 37 percent of the cells responded strongest either during the visual+motor condition or during the motor-only condition. In the motor-only condition, the mouse was running and anticipating appropriate visual feedback, but unexpectedly this feedback did not occur. Hence, this created a sensorimotor *mismatch* between visual input and running. Figure 5.4c shows the average population response to the onset of visual stimuli in a stationary mouse (black), at the onset of running (dark gray), and at the onset of sensorimotor mismatch (light gray). Clearly, mouse V1 neurons were weakly activated by purely visual input in a stationary animal (black), were activated more strongly by running per se (dark gray), and were activated the strongest by sensorimotor mismatch (light gray). Keller et al. (2012) showed that eye movements could not explain these major differences in V1 responses between the different conditions. So what is the source of the strong motor signals in V1? Most likely, they arise from the strong direct anatomical inputs from the motor system into V1 (Miller & Vogt 1984; Leinweber et al. 2017). The dramatic, about tenfold, enhancement between the purely visual neural activity (black) and the activity driven by sensorimotor mismatch (light gray) suggests that the function of V1 is not just feedforward visual processing of retinal images. Instead, V1 (at least in the mouse) seems to be primarily a region involved in sensorimotor integration and sensorimotor predictive coding, where the mismatch signal could be used as a feedback for learning an internal model of sensorimotor contingencies. Alternatively, this mismatch signal may reflect the detection of nonself-generated visual motion "out there" (as opposed to self-generated visual motion)—for instance, the motion of other animals, such as conspecific or predators.

The study by Keller et al. (2012) underscores the importance of studying sensory neuroscience under active-sensing conditions rather than focusing on passive-sensing conditions, as has been done for decades. Although the authors used a head-fixed preparation, they did ask the right ethological question (namely, what are the natural conditions when oriented bars would be moving across the retina?), and they effectively used VR to dissect the contribution of visual versus motor variables. This is a good example of the power of head-fixed VR setups for behavioral neuroscience.

Finally, a study by Saleem et al. (2013) conducted a VR experiment that bore similarities to that of Keller et al. (2012) but was designed differently. Mice ran on a spherical treadmill along a virtual corridor with several prominent virtual landmarks, while V1 neurons were recorded electrophysiologically. Three experimental conditions were used: (1) mice ran with normal visual input that moved according to the running velocity of the animal; (2) mice were running while the visual stimulus was incongruent with the animal's movement, but instead consisted of visual sequences played back from previous sessions; (3) mice ran in complete darkness.

The results were quite dramatic. First, in complete darkness, 71 percent of the V1 neurons were significantly modulated by whether the mouse was running or stationary. In 49 percent of the V1 neurons, the firing rate robustly encoded the running speed in complete darkness (figure 5.4d). Second, 92 percent of the V1 neurons were significantly modulated by the mouse movement in either dark or light conditions (figure 5.4d). Thus, only 8 percent of the neurons were *not* modulated by the mouse movement—that is, only 8 percent of V1 cells were purely visual neurons. Third, Saleem et al. (2013) found that a large majority of V1 neurons computed a weighted *sum* of visual-motion speed and running speed, and they interpreted it as an estimate of the animal's speed through the world, which could be used for path integration and navigation. This is quite different from the findings of Keller et al. (2012), which emphasized the neurons that compute the *difference* (error) between visual-motion speed and running speed. These substantial differences in results between the two studies could be due to the differences in recording methodology (imaging versus electrophysiology), the recorded cortical layers (imaging targets superficial layers while electrophysiology targets mostly deep layers), the experimental design, and analytical methods. Importantly, however, both studies agree that a very large fraction of V1 neurons are strongly affected by motor variables, regardless of the sensory input.

Overall, the high fraction of V1 neurons in Saleem et al. (2013) that responded to locomotion in darkness (71 percent or 49 percent, depending on definition)—together with the remarkably low percentage of purely visual neurons that they found (only 8 percent !)—speak strongly against the classical notion that V1 is part of a purely feedforward visual-processing stream. On the other hand, we have to remember the caveat that this study, like Niell and Stryker (2010) and Keller et al. (2012), was done in head-fixed mice—and maybe the picture would be different in freely moving mice; maybe then the visual cortex would turn out to be more visual. However, I suspect that the opposite will be true: the visual cortex will likely exhibit even stronger nonvisual responses during free behavior. Visual neuroscientists are starting to address these issues in freely moving mice during a variety of natural behaviors, such as jumping, predator avoidance, and prey capture (Skyberg & Niell 2024).

In summary, I discussed here several experiments that examined mouse V1 responses to combinations of visual and motor inputs (Niell & Stryker 2010; Keller et al. 2012; Saleem et al. 2013). These studies differed in their details; importantly, however, *all* these studies suggested together the following picture: At least under the head-fixed conditions of a VR setup, mouse V1 neurons are *not* primarily visual, but rather seem to be more strongly modulated by motor signals than by purely visual signals; or sometimes they most strongly encode the combinations or differences (mismatch) of motor and visual signals. In fact, it is possible that some of the V1 neurons recorded by Hubel and Wiesel and by Wurtz may have been mismatch neurons rather than visual neurons because their subject animals were presented with moving visual stimuli while being unable to move, which could elicit mismatch signals. Further, as discussed previously, V1 neurons exhibit not only motor or motor-sensory mismatch signals, likely derived from the strong monosynaptic inputs from the motor cortex and anterior cingulate cortex into V1 (figure 5.2b, "ACC/M2"), but they also exhibit vestibular signals that strongly modulate the responses of V1 layer 6 neurons (Vélez-Fort et al. 2018), derived from the massive monosynaptic inputs coming from retrosplenial cortex into V1 (figure 5.2b, "RSP"). Finally, the strong monosynaptic inputs from the hippocampal formation into V1 (figure 5.2b, "HPF") also exert a major functional impact on V1 neurons, as was shown in another study (Saleem et al. 2018) that found that V1 neurons robustly encoded the animal's spatial position, akin to hippocampal CA1 place cells.[5] Further, another study found very strong movement-related signals in the V1 of mice, which were related to the animal's facial movements and were unrelated to visual sensory stimulation (Stringer et al. 2019). Yet another study—this time in freely moving mice—found that over 50 percent of V1 neurons showed significant responses to head movements in complete darkness (Meyer et al. 2018): this demonstrates again the strong motor signals in V1. Together, these results are inconsistent with the view that V1 is primarily part of a feedforward hierarchical system whose job is purely to extract object shapes, à la David Marr. Instead, these data are consistent with several possibilities:

1. V1 is part of a predictive-coding system that computes mismatch (prediction error) between the visual scene and the animal's motion (Keller et al. 2012). In other words, this possibility suggests that, because the visual image on the retina is continuously changing due to self-motion of the animal's body, head, and eyes, it is the role of the visual system—including V1—to identify the genuine visual motion of things in the external world, and to subtract the self-motion of the animal. This subtraction (mismatch computation) is carried out in V1. This idea is part of a broader perspective that a key role of the brain is to perform predictive processing, wherein it compares bottom-up sensory signals to top-down predictions or expectations and updates the internal

representations accordingly (Rao & Ballard 1999; Bastos et al. 2012; Keller & Mrsic-Flogel 2018; Khan & Hofer 2018). This idea also is consistent with a recent study showing that stimulus selectivity of V1 neurons is well suited to generate error signals in response to unexpected changes in the visual stimulus unrelated to movement (Muzzu & Saleem 2022). Predictive coding would have played little role in classic visual neuroscience paradigms (i.e., randomized stimulus presentations to head-fixed monkeys), but it might be very important in the real world, when an animal or a human uses vision under challenging conditions and needs to predict what is going to happen next.

2. V1 is the first cortical stage of a visual-based navigation system (Saleem et al. 2018), which forms a loop between bottom-up visual information coming from V1 toward the hippocampus, and top-down navigational information that is sent back from the hippocampus to V1.
3. V1 is part of the visual sensory system, but the internal representation that it builds is not purely visual; rather, V1 represents "sensorimotor contingencies"—namely, it represents the dependencies (contingencies) between the animal's actions and the subsequent changes in the sensory signals (O'Regan & Noë 2001).
4. V1 is not a pure sensory area but rather part of a sensorimotor loop (Ahissar & Assa 2016).

There can be additional explanations for the mixed selectivity observed in V1, whereby V1 neurons carry mixtures of signal types: visual, motor, vestibular, task relevance, and navigational information. While each of these particular interpretations might be criticized by some, there are two clear consequences that emerge from this overall set of studies. First, if we ever wish to understand the brain's sensory systems, including the visual system, then sensory-neuroscience experiments must be conducted in behaving animals. Second, more experiments must be conducted under active-sensing conditions rather than passive-viewing conditions, so all the massive top-down anatomical inputs descending into V1 (figure 5.2) will have the opportunity to operate normally.

These free-behavior experiments should include full measurement of the animal's position, head direction, body posture, and eye gaze. It is possible nowadays to measure *all* these factors in freely moving animals (chapter 3), and all these factors may dramatically influence neuronal coding in the visual system. Further, as the studies surveyed here have elegantly shown, often the best strategy is to contrast and compare passive versus active conditions to systematically dissect the influences of the various top-down inputs. The importance of active-sensing experiments is yet another manifestation of the general theme emphasized throughout the book—namely, that it is crucial to study the brain as close as possible to the natural, real-life situation.

5.2.2 Primate Vision and Eye Movements

The image on the retina never stands still—it dynamically moves due to the animal's natural motion. There are three main components to this motion: (1) whole-body locomotion, which consists of both translation and rotation; (2) head movements, which mostly comprise rotations; and (3) eye movements, which are rotational (the eyes can also exhibit focusing movements by changing the lens shape—aimed to focus vision at various depths—but I will not consider these movements here). Whole-body locomotion has already been discussed—and it is known to contribute prominently to image motion on the retina. Head movements can also create large image motion on the retina. But the largest and most-studied source of image motion on the retina are eye movements, which will be the focus of this section. Discussing eye movements is of particular importance because most studies in visual neuroscience attempt to entirely abolish eye movements (figure 5.1), which necessarily abolishes much of the natural visual signal.

For most animal species, it is unknown how much they normally move their eyes during natural behavior. Historically, it was thought that, among mammals, eye movements occur mostly in primates, and small mammals move their eyes very little. However, recent studies have shown that this view is wrong. Kerr and colleagues (Wallace et al. 2013) placed miniature eye-tracking cameras on the heads of freely moving rats and used infrared illumination to track the motion of the eyes' pupils. They found that as the rats moved through complex environments, their eyes moved extensively over a wide angular range of up to 80 degrees (±40 degrees), and at fast angular velocities of up to 1,000 degrees/s (similar to the fastest saccades in humans). Surprisingly each eye moved quite independently, but this motion was nevertheless coordinated such that the animal constantly maintained a binocular region (a region of visual space seen by both eyes) that was directed *above* its head, even while the head rotated and pitched quite substantially. The function of this eye-coordination behavior was proposed to allow constant 3D vision in the upward direction—the direction from which birds of prey normally would attack wild rats (Wallace et al. 2013). Rich eye movements were also found in freely moving mice, during a variety of behaviors, including foraging, prey hunting, social interactions, and object tracking (Meyer et al. 2020; Michaiel et al. 2020, Skyberg & Niell 2024). Thus, it is becoming clear that, contrary to dogma, rodents do move their eyes substantially—and they seem to move them in functionally meaningful ways. This finding calls for a major effort to study these eye movements in detail and to elucidate their effects on rodent visual processing. These studies must be done in freely moving rodents because head-fixed rodents do not exhibit such dramatic eye movements (Wallace et al. 2013). In other words, as is the case for many other behaviors, eye movements must be studied under normal, unrestrained, natural conditions.

Let us now turn to eye movements in primates, which have been studied extensively, albeit mostly in head-fixed conditions (which, based on the experiments in rodents, may constitute a major limitation). Primate peripheral vision (outside the fovea) is very low resolution and does not provide much spatial detail—and therefore, to see the details of the visual scene, primates have to move their eyes actively to place the fovea over the regions of interest. Primate eye movements are usually divided into two main categories:[6] saccades and fixational eye movements (Martinez-Conde et al. 2004; Ahissar & Arieli 2012; Rucci & Victor 2015; Snodderly 2016).

The first main category is eye saccades, also known as "macro-saccades"—the large, rapid ballistic eye movements that primates use to scan the visual scene actively (Yarbus 1967). Saccades typically span between 2–20 degrees of angle and take up a relatively fixed duration of 20–40 ms, regardless of their amplitude (angular extent); consequently, saccades exhibit a linear relationship between their amplitude and speed. Figure 5.5a shows an example of an image that was visually scanned by a human: The white curve shows the eye trajectory over 20 seconds after image presentation, with long straight lines corresponding to

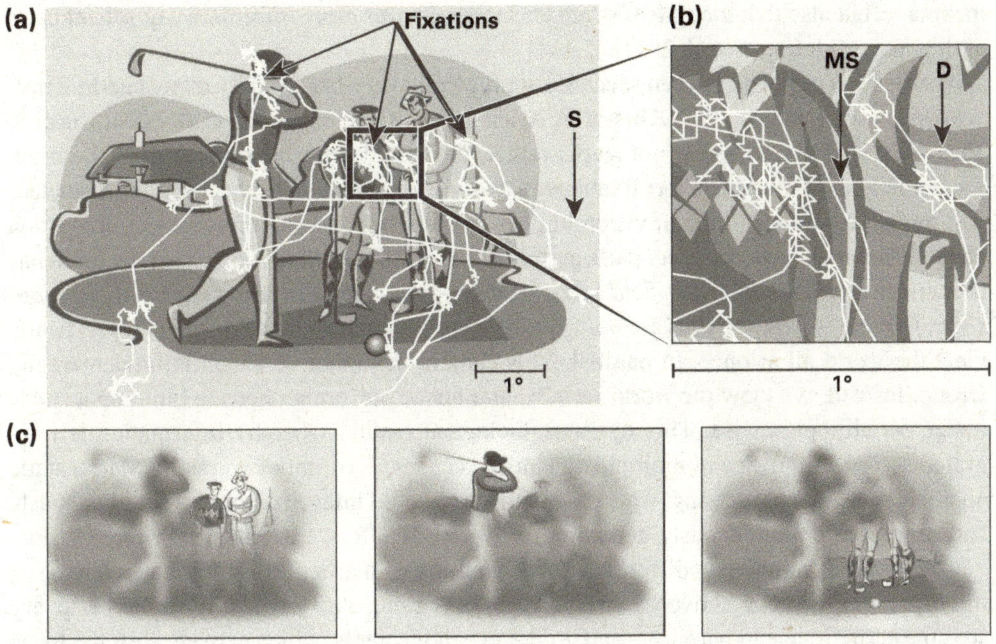

Figure 5.5
Examples of eye movements in a human inspecting an image. (a) Examples of saccades (S)—see the long straight lines; and fixations—see the dense spots (marked with arrows). (b) Zoom in, showing examples of microsaccades (MS) and drift (D) during a fixation. Microsaccades and drift are key classes of fixational eye movements. The third class of fixational eye movements—tremor—is not shown here. (c) Illustration of the sequential nature of human vision: natural human vision proceeds via sequences of saccades and fixations, where during each fixation, we see only a small portion of visual space in high resolution (foveal vision). A key challenge for the visual system is to assemble a stable image of the world from such sequential snapshots of visual information: thus, vision is an active image-assembly process. Another key challenge is that each snapshot (fixation) in itself consists of a highly unstable jittering image, due to the ever-present microsaccades and drift. (a and b) Adapted from Ahissar et al. (2016).

saccades (an example saccade is marked by "S"), and dense, noisy spots corresponding to fixations. During this saccade-and-fixate behavior, humans and monkeys preferentially fixate on the most important regions of the visual scene—those regions that convey the key behaviorally relevant information, such as people, objects, outlines, eyes, mouth, and hands. In the image in figure 5.5a, for example, the subject fixated mostly on the humans and on the houses. Interestingly, the saccadic scanning pattern sometimes depends on context. For example, Yarbus showed in his seminal work (Yarbus 1967) that the scanning pattern changes dramatically depending on the task: a very different pattern of saccades was seen when subjects were asked to estimate the wealth of a family portrayed in a painting, or to estimate the ages of the people in the same painting. The subjects looked at very different things to answer these different questions. On the other hand, the saccadic scanning pattern also sometimes does *not* depend on context. For example, Kanwisher and colleagues showed that the pattern of saccades as measured in the lab is very similar to the pattern of saccades displayed by the same subjects when they were freely walking outdoors with head-mounted eye trackers (Peterson et al. 2016), although another study found also interesting differences in saccade patterns between lab and outdoor conditions (Papinutto et al. 2020). Further studies showed that saccades not only scan preferentially the most important *locations* in the image, but also that the *rate* of saccades increases over more informative or salient parts of the image (McCamy et al. 2014).

The fixation periods between saccades are typically very brief—only a few hundred milliseconds each (Yarbus 1967; Ahissar & Arieli 2001; Martinez-Conde et al. 2004; Rucci & Victor 2015). Thus, vision is not static; rather, the visual information arrives at the brain sequentially over multiple short fixations that are interspersed with saccades. This sequential, dynamic nature of human vision implies that the high-resolution (foveal) information about different objects or object parts arrives at the brain at different times, so the brain has to *actively construct the wide-field scene from this sequential series of narrow-field snapshots*. In other words, the eyes are *not* wide-field static cameras, through which we statically view the world all at once, in parallel. This camera metaphor of vision is fundamentally wrong. Instead, we view the world serially, and hence our brain needs to build up a stable image serially, piecewise. This dynamic biological reality, whereby information is integrated across a sequence of multiple dynamic saccades, is in stark contrast to Marr's static approach to visual processing, where the entire wide-field image is received simultaneously and processed simultaneously according to Marr's feedforward scheme. Primate vision, however, works completely differently: one can think of it as a narrow "pinhole view" (i.e., the fovea) that is being actively moved around over time. The low-acuity visual periphery sees the entire image in parallel—and guides eye movements, which actively shift the fovea toward important scene-details. The brain then has to reconstruct the high-acuity visual scene based on the known times of saccades and known positions of saccades (which are known to the brain, for example from the motor efference copy of the eye), convolved with the visual input that was acquired during each fixation. An illustration of this active image-assembly process is shown in figure 5.5c. This view, which is dictated by the natural behavior of primate active vision, also means that if we want to understand normal vision, it is highly problematic to study the visual system in a monkey that is trained to fixate its eyes, or in a head-fixed mouse that volitionally abolishes its eye movements because of head fixation. It is also problematic to use brief, flashed visual stimuli that are too short to allow a series of saccades because this precludes the natural sequential accumulation of image information.[7] Since saccades are the mechanism by which the visual image is analyzed, then abolishing saccades also abolishes our ability to fully elucidate the neural mechanisms of natural vision.

To date, only two studies have ever recorded single neurons from the V1 of monkeys during free viewing (Livingstone et al. 1996; Gallant et al. 1998). In these studies, the animals were head-fixed and restrained, but eye movements were allowed. These two studies found that neuronal responses to visual stimuli during free eye movements were much weaker and more variable than during eye fixation. Moreover, the neuronal responses during free viewing were only poorly predicted by the classical receptive fields and tuning curves of the neurons as measured during fixation. One of these studies (Livingstone et al. 1996) suggested that bursts of spikes encoded the information about the image more reliably than single spikes, so perhaps bursts carry privileged sensory information to downstream regions—but, notably, the vast majority of spikes were single spikes, and these were very poorly explained by the classical tuning curves. This failure to explain most spikes could stem from complex nonlinear processes in cortical circuits, which may dominate natural vision (Gallant et al. 1998). In addition, this failure could stem from visual-cortex activity being dominated by ongoing intrinsic circuit activity during natural vision: Such intrinsic dynamics is known to be much larger than the sensory-evoked responses (Arieli et al. 1996; Fiser et al. 2004), and this intrinsic activity may grow substantially during free eye movements. Thus, there could be a very large gap between brain activation in classical laboratory studies versus natural free vision. This gap was indeed demonstrated in a recent study by Scanziani's group, who investigated V1 in freely viewing, freely moving mice (Miura & Scanziani 2022). They found that mouse V1 neurons jointly represented information about the direction of image motion and about saccade direction, thus distinguishing between external motion and saccade-induced motion. This finding could have been made only because the mice were allowed to move and behave naturally. Consequently, if we want to understand the neurobiology of natural vision—in either primates or rodents—then it is absolutely crucial to study the visual system under conditions of natural vision. It's a challenging endeavor, but it's a must, and it will surely lead to new discoveries about the neurobiology of sensory perception.

Another problematic issue is that all the studies on visual neurophysiology in primates have focused on thalamic and cortical neurons with parafoveal or peripheral receptive fields, rather than on the foveal neurons, which perform the actual "job of vision" in primates: These are the neurons that analyze the high-resolution information acquired through natural eye movements. While there are good technical reasons why foveal neurons are difficult to record from (primarily the extremely small size of their receptive fields), this does not alter the fact that we know nearly nothing about these neurons—not in the thalamus, nor in the V1, nor in higher-order retinotopic visual areas—even though the foveal neurons are the most important ones for high-acuity natural vision. (This critique does not apply to the highest-order visual areas, such as the IT cortex, which are not retinotopic; in addition, this critique is less relevant to visual neuroscience in rodents, which do not have a fovea). Thus, a rather grim conclusion emerges: despite the large progress made by research on visual neuroscience over the last seventy years, only very limited information has been gleaned about how the job of vision is done. This shortcoming occurred because, by and large, as a field we have neglected the most important visual neurons, and we almost did not consider natural visual behavior.[8]

* * *

The second main category of eye movements is called "fixational eye movements"—the small movements that always occur even when we are fixating. The eyes are never still and are always moving: even when subjects are trying to fixate, their eyes exhibit small movements that span tens of retinal photoreceptors (Riggs et al. 1954; Steinman & Levinson 1990; Martinez-Conde et al. 2004). Thus, even during fixation, the retina does not see a static image, but rather a very dynamic "movie" of sorts. Fixational eye movements are found across vertebrates, including fish, birds, and many mammalian species (Martinez-Conde & Macknik 2008), but here we will focus mostly on primates. Fixational eye movements in primates are subdivided into three types, based on their size and spectral content: drift, tremor, and microsaccades (Ahissar & Arieli 2001; Martinez-Conde et al. 2004). "Drift" are slow fluctuating eye movements (about 1–20 Hz) covering amplitudes of up to 10 or 20 arcminutes. "Tremor" are faster eye movements (about 40–100 Hz) that have smaller amplitudes of up to a few arcminutes. Both drift and tremor occur continuously during fixation and are interrupted several times per second by "microsaccades," which are small saccades that move both eyes together over distances from a few arcminutes to a few tens of arcminutes. Figure 5.5b shows examples of drift and microsaccades that occurred during fixation ("D" and "MS"). For comparison, the smallest photoreceptors have a diameter of about 0.3 arcminutes, so all three types of fixational eye movements can cover more than 10 photoreceptors in a brief fraction of a second. Thus, fixational eye movements may seem to be a bug, which dramatically smears the visual image. However, visual perception can in fact be spatially accurate to within a small fraction of a photoreceptor, being as accurate as 0.05 arcminutes, as shown by "hyperacuity" phenomena such as the ability to tell apart the spatial offset of two noncollinear lines (Westheimer 1981). This constitutes a discrepancy of two orders of magnitude between perceptual accuracy and the retinal "smear" that would be caused if the fixational eye movements were a "bug" (Ahissar & Arieli 2001). It is a true challenge to explain how humans can exhibit spatial visual resolution in the hyperacuity range—two orders of magnitude better than the smearing magnitude of their fixational eye movements.[9]

So, are fixational eye movements a bug, an irrelevant nuisance—or are they a functional feature of visual perception? At least four lines of evidence suggest that they are a feature, and that—just like large saccades—small microsaccades and other fixational eye movements are an integral part of visual perception. The first line of evidence comes from an experiment in which muscle paralytics (curare or succinylcholine) were injected systemically into a researcher who was connected to a breathing machine (Stevens et al. 1976). This achieved a full paralysis of the body, including of the eye muscles, and entirely abolished all eye movements, all while the subject was still fully awake and conscious. Remarkably, this elimination of eye movements led to fading and disappearance of the visual image. This fading away of vision indicates a causal role for fixational eye movements in visual perception.

The second line of evidence comes from stabilization experiments, which used a number of methods to stabilize the image rapidly relative to the eye such that when the eye moved, the projected image also moved in a corresponding way, so the image remained fixed relative to the retina. Using such methods, fixational eye movements no longer created a retinal slip of the image (Jones et al. 1972; Ditchburn 1973; Coppola & Purves 1996; Martinez-Conde et al. 2004; Rucci et al. 2007; Snodderly 2016). In stabilization experiments, similar to the muscle-paralysis experiments, the image faded rapidly, within between 80 milliseconds to several seconds, depending on the method of stabilization. Subjects reported that all the static parts of the image have vanished from perception except for moving objects, which were still perceived normally. This again suggests a causal link between fixational eye movements and visual perception.

A third line of evidence comes from a study of a human ophthalmoplegic patient (Gilchrist et al. 1997), who had extraocular muscle fibrosis and was not able to make any eye movements, but developed instead a suite of saccade-like movements of the head that enabled them to read and scan images via a head-saccade-and-fixate strategy. In addition, this patient displayed a jitter of the head during the head fixations, which could replace the missing fixational eye movements (Gilchrist et al. 1997).

A fourth line of evidence comes from a series of studies by Martinez-Conde and colleagues on the role of microsaccades. In one study (Martinez-Conde et al. 2006), they found that microsaccades counteract visual fading. They asked subjects to fixate on a central spot and report on the strength of a dim peripheral stimulus. The subjects reported that the

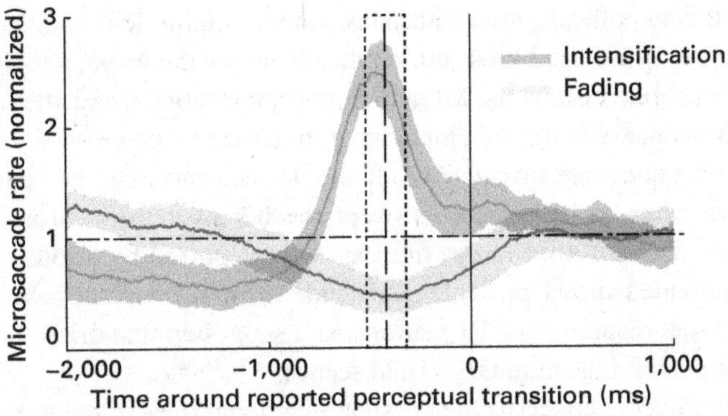

Figure 5.6
Microsaccades prevent visual fading. In this experiment, the microsaccade rate increased almost three times before perceptual intensification occurred (dark gray), while microsaccade rate decreased more than two times before perceptual fading occurred (light gray). The vertical solid line at $t = 0$ denotes the moment of reported perceptual transition; the dashed box denotes the estimated time of the actual perceptual transition, considering the reaction time. Adapted from McCamy et al. (2012).

visibility of the fixed stimulus spontaneously fluctuated. The researchers then found that this waxing and waning was explained by microsaccades. When the rate of microsaccades decreased, their human subjects reported that the visual image has faded; and conversely, just before restoration of visibility, the rate and magnitude of microsaccades increased. In a second study (McCamy et al. 2012), they showed that microsaccades prevented visual fading to a greater degree than the other fixational eye movements, in particular drift—and this rescue from visual fading occurred both in foveal and peripheral regions. The effect was quite dramatic: microsaccade rate decreased more than twofold just before perceptual fading occurred, while microsaccade rate increased almost threefold before perceptual intensification occurred (see figure 5.6). Importantly, this same study also demonstrated a genuine (albeit smaller) contribution of drift to the prevention of visual fading (McCamy et al. 2012).

Together, these studies suggest that all types of fixational eye movements—with the emphasis on microsaccades—are causally necessary for natural vision. These studies further suggest that the visual system is mostly concerned with *detecting transients,* and these transients could be caused by object motion or, for the static parts of the image that are not moving, the transients are introduced into the visual system by fixational eye movements, which serve to refresh the image. Given this new dynamic view, the term "retinal slip," which is often used when discussing fixational eye movements and carries a negative meaning as a nuisance factor, should be replaced with a more positive term, like "retinal sampling," to emphasize that eye movements are not a bug, but a feature; not a nuisance, but an integral part of visual perception (Ahissar & Arieli 2001; Rucci & Victor 2015).

Next, it is useful to consider separately microsaccades as opposed to drift and tremor. Microsaccades are in many ways very similar to large saccades, in that they obey the same linear relationship between their amplitude and speed and are controlled by the same neural circuits as the large saccades (Zuber et al. 1965; Martinez-Conde et al. 2004; Havermann et al. 2014; Krauzlis et al. 2017). It is possible, therefore, that microsaccades have a somewhat similar functional role to large saccades—namely, to actively scan the most important portions of the scene and assemble the visual layout of the scene in a piecewise, sequential manner. Under this hypothesis, microsaccades, whose amplitude is smaller than the foveal diameter (< 2 degrees), are used for scanning details *within* the fovea, while the role of large saccades is to sequentially direct the fovea to important locations, one after the other. In line with this idea, Martinez-Conde and Hubel have shown that when monkeys were presented with stationary bars and were instructed to fixate, V1 neurons exhibited particularly strong bursty responses when a microsaccade swept the bar over the neuron's receptive field (Martinez-Conde et al. 2000). Similar findings were reported by Snodderly et al. (2001). These studies provided direct physiological evidence that microsaccades are crucial for explaining the visual responses of V1 neurons, and suggested that microsaccades might be used actively for perceiving stationary visual scenes.

The bigger controversy concerns the possible functional roles of the slower types of fixational eye movements: drifts and tremor.[10] Several researchers proposed that these fixational eye movements serve to transform the spatial structure of the static image into a temporal code—that is, transforming space into time (reviewed by Rucci & Victor 2015; Rucci et al. 2018). The basic idea of these suggestions is that the small, ongoing eye movements are used by retinal ganglion cells to transform the spatial layout of the image into a temporal-difference code. Thus, a spatial offset Δx between two image parts is transformed into time difference Δt, via $\Delta t = \Delta x / v$, where v is the instantaneous velocity of the eye movement along the *x*-axis. This coding scheme can translate in a direct manner the spatial layout of the image into relative-spike-timing information, and according to this proposal, this relative spike timing provides crucial information for visual image processing. Further, an interesting possibility was proposed by Ahissar and Arieli (2001), who suggested that during natural vision, the *coarse* details of the visual scene are analyzed via classical static *spatial* visual mechanisms, à la Marr, while the *fine* spatial details of the visual scene are analyzed via the *temporal* difference code outlined here, mediated by fixational eye movements. Thus, this proposal posits that low-acuity peripheral vision is based on classical spatial mechanisms, as have been studied since Hubel and Wiesel, and which do not require any behavior or eye movement, while high-acuity foveal vision is based on temporal mechanisms that utilize fixational eye movements, especially drifts and tremor (Ahissar & Arieli 2001, 2012). In other words, the proposal is that high-acuity foveal vision operates not *despite* fixational eye movements, but rather *through* fixational eye movements.[11]

Another way to think about this temporal-coding mechanism is that rapid fixational eye movements may serve as a synchronizing timing signal (time stamp), which precisely synchronizes visual processing across large populations of neurons, allowing reliable encoding of information about spatial layout via the small *timing differences* from the synchronizing signal. An experimental study by Gollisch and Meister (2008) found that such a temporal-difference code may indeed operate in the salamander retina. They measured the precision of first-spike latency of retinal ganglion cells to flashed stimuli and found that this precision was extremely high. For example, when they flashed a natural image at 1,000 locations (figure 5.7a), they found that the temporal latency of individual neurons represented very sensitively the spatial offset of the image (figure 5.7b), and they could use this temporal-latency (temporal-difference) information to accurately reconstruct the image (figure 5.7c). This image reconstruction was more accurate than that based on a firing-rate code (figure 5.7d), suggesting that the proposed temporal code, which operates via the transformation of spatial layout into temporal latencies, can provide a very robust and precise coding for object detail.

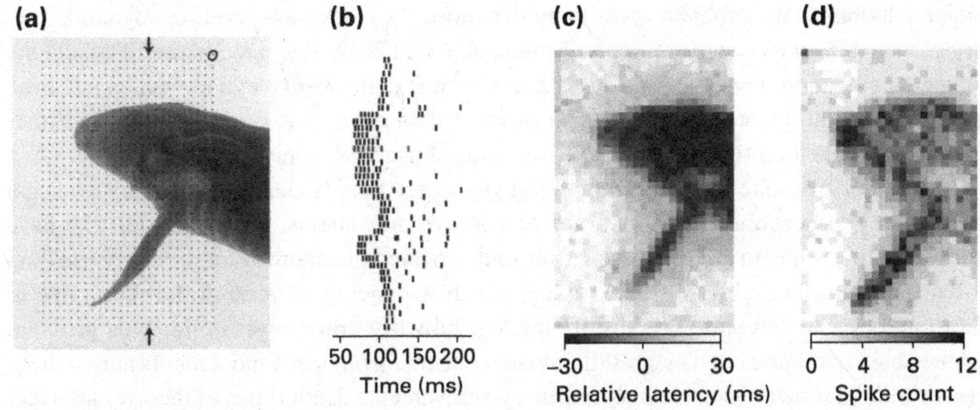

Figure 5.7
Responses of a single retinal ganglion cell to a flashed natural image. (a) The image and the 1,000 locations at which this image was flashed (shown by dots). The same image was presented repeatedly with 1,000 different spatial translations, so that for each presentation, a given retinal ganglion cell responded to a different patch of the photograph. The ellipse at the top right shows the typical receptive-field size of a ganglion cell. (b) Spike trains of one ganglion cell for presentations of the image along a set of spatial locations (y-axis); the receptive-field locations of this ganglion cell during these image presentations were positioned along the column marked by the two arrows in (a). Note the high temporal precision of the visual responses—in particular, the precision of the first-spike latency. (c) Image reconstruction based on spike timing: plot of the differential spike latency on single-trial presentations at the 1,000 locations marked with dots in (a). (d) Image reconstruction based on spike rate: shown is the corresponding plot for the spike counts within a fixed time window. Reproduced from Gollisch and Meister (2008).

A central drawback of this dynamic theory of natural vision is that it requires knowing at all times the full trajectory, velocity, and direction of the fixational eye movements for both eyes in order to transform the temporal code back into retinal coordinates correctly. This is an absolutely essential requirement of the theory. However, to date it remains unclear whether and where in the brain there is encoding of the trajectory of fixational eye movements. One theoretical study has suggested that this crucial information—the full trajectory of retinal motion due to fixational eye movements—may be encoded by a separate network of neurons in the foveal region of V1, which specialize in tracking the position of the image on the retina (Burak et al. 2010). This study has suggested that to see fine spatial detail, the visual cortex must contain two populations of neurons: one that represents the details of the objects in the image, and another that tracks the position of the image, which changes due to fixational eye movements. This latter population of visual-cortex neurons could either *stabilize* the perceived image or contribute to the perception process itself, *through* eye movements, as suggested by dynamic theories of vision. Such a vision-based mechanism should be much more precise in tracking image position than a mechanism based on motor corollary discharge of eye position. However, to date there is no direct experimental data to support the existence of such an image-tracking population of neurons in foveal V1. Therefore, it is crucial to attempt and conduct the difficult recordings in the foveal region of V1, in search of such a population.[12] It is also possible that foveal V1 neurons are not segregated into two distinct populations, but each neuron carries mixed signals about both image position and image details. In any event, it is still unknown where in the brain the detailed trajectory information of fixational eye movements is stored and interpreted. Hence, the two major solutions to the problem seem to be *not* entirely satisfactory—both the dynamic theory of vision *through* eye movements (Ahissar & Arieli 2012; Rucci & Victor 2015) and the image-stabilization theory, which posits that vision is done *despite* eye movements (Burak et al. 2010). Both theories lack a crucial piece of data to support them—namely, whether there is an explicit representation of eye trajectory during fixational eye movements.

In summary, despite the major remaining gaps, one thing is clear—any theory of vision must take into account the dramatic effects of eye movements. Most theories that have addressed this problem are relying on a yet-undiscovered mechanism for precisely tracking the position of the highly unstable image, which is dancing all around the retina due to fixational eye movements. It is totally unclear how this tracking is done. This problem applies both to theories that suggest that vision occurs *through* eye movements and to theories that suggest that vision is done *despite* eye movements. Both types of theories are lacking any direct experimental support for the most critical mechanism that is required to make the theory work. Thus, to date it remains enigmatic how is it possible that monkeys and humans are at all capable of high-acuity vision, which should have been dramatically smeared due to the ever-dancing fixational eye movements. Our ability to exhibit high-acuity vision remains one of the biggest mysteries in sensory neuroscience. To solve this mystery, it will be necessary to conduct electrophysiological recordings in foveal regions of the monkey visual cortex, comparing image-stabilized conditions versus naturalistic free-viewing conditions. This is crucial for understanding the mechanisms of natural vision.

5.3 Rodent Whisking: Active Sensory Probing at Short Ranges

Rats and mice—like many other animal models in neuroscience—are largely nocturnal. Because their senses of vision and hearing are not well developed (at least compared to some other nocturnal animals), they rely primarily on their keen somatosensory and olfactory systems to orient in the environment, forage for food, and find conspecifics. We will discuss olfaction in the next section, while here, we will focus on the rodent somatosensory system—and in particular, on the set of facial hairs, or vibrissae, which are heavily used by rodents for active sensing and are dramatically overrepresented in the rodent brain (Woolsey & Van der Loos 1970). The vibrissae are divided into two classes (figure 5.8a): long *macrovibrissae*, also known as whiskers, approximately thirty thin long hairs on each cheek, which are arranged in five distinct rows and are actively moved by the animal in periodic "whisking" movements for sensing the environment; and short *microvibrissae,* approximately forty to seventy small hairs on each side of the chin and snout, which are arranged less regularly than the macrovibrissae (Brecht et al. 1997). The density of the microvibrissae is 40–100 times higher than that of the macrovibrissae, and hence the microvibrissae are often regarded as a "tactile fovea" (Brecht et al. 1997; Grant et al. 2012). The definition of the microvibrissae as a "tactile fovea" comes from the high receptor density, but also from behavioral studies, which showed the following: First, removal of the microvibrissae impairs object recognition, but not object detection; and conversely, removal of the long macrovibrissae impairs object detection but not object recognition. This is consistent with a specific role for microvibrissae in high-acuity, foveal-like tactile object recognition, while macrovibrissae (whiskers) are important mostly for object detection and localization (Brecht et al. 1997). Second, in their natural whisking behavior, rats usually detect an object

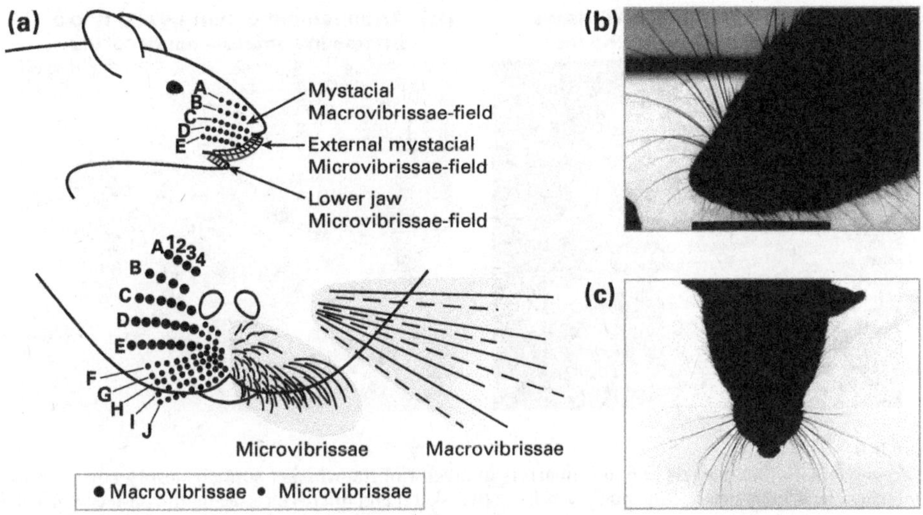

Figure 5.8
Rat vibrissae. (a) View of the rat vibrissae, showing the long macrovibrissae (whiskers) and shorter microvibrissae. (b) Photo of a rat probing a coin with its microvibrissae. (c) Photo of rat macrovibrissae (whiskers). (a) Modified from Brecht et al. (1997) and Ebner and Kaas (2015); (b and c) adapted from Deschênes et al. (2012).

initially using their long macrovibrissae, but then approach it and palpate it extensively with their microvibrissae, nose, and lips in order to identify the precise object shape[13] (Grant et al. 2012; see also figure 5.8b). As with the visual fovea, virtually nothing is known about the neurophysiology of the tactile fovea, the microvibrissae, and hence we will focus here on the macrovibrissae, the whiskers (figure 5.8c).

Another notable aspect of natural whisking behavior is that rodents tend to touch object very gently, with many whiskers. This ultragentle touching was termed "minimal impingement" by Tony Prescott (Mitchinson et al. 2007). It is thus likely that the strong, nongentle whisker deflections that are typically applied to single whiskers in studies of whisker neurophysiology are quite unnatural and nonphysiological. Instead, in real-life whisking behavior, the active positioning of the entire whisker array may be decisive in shaping whisker contacts and neuronal responses. This can be investigated only in naturally behaving rodents. Sadly, despite the thousands of studies on the mouse and rat barrel cortex, very little is known about the natural whisking behavior of rodents, as most studies are performed under head fixation, which reduces the frequency of whisking and creates a poor and abnormal behavior compared to the rich natural whisking behavior seen in freely moving animals. Even less is known about neural responses in naturally whisking rodents. But this is starting to change.

The orderly organization of whiskers on the face (figures 5.8a and 5.9a) is reflected in an orderly organization of columns, or "barrels," in layer 4 of the whisker somatosensory cortex (figure 5.9b); note that the barrels are located adjacent to the "micro-barrels," which correspond to the microvibrissae. Each barrel corresponds to one primary whisker, and they are arranged topographically one-to-one, like the whisker array (compare figures 5.9b and 5.9a). This crisp columnar arrangement, which offers many practical experimental advantages, was a key reason for the rapid growth of research on the rodent somatosensory system over the last half century, with particular emphasis on the barrel cortex. It is important to

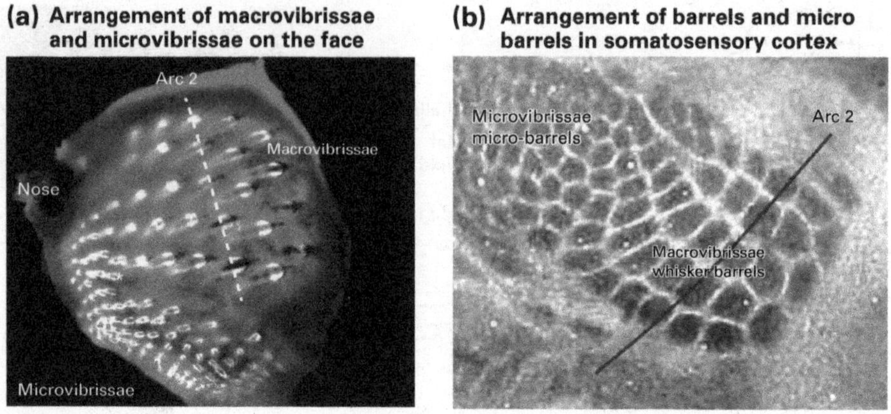

Figure 5.9
The arrangement of the barrels and microbarrels in layer 4 of the whisker somatosensory cortex of the rat (b) matches the arrangement of the macrovibrissae (whiskers) and microvibrissae on the rat face (a). The barrels that correspond to one arc of vibrissae (arc 2) are marked. Adapted from Deschênes et al. (2012).

emphasize, however, that substantial processing occurs subcortically: the cortex and subcortical somatosensory nuclei are tightly linked via multiple cortical-subcortical loops (Kleinfeld et al. 2006), but since the large majority of the research has been done on the rodent barrel cortex, this section will focus primarily on this region.

Most of the vast physiological research on the rodent barrel cortex was classically done in nonbehaving anesthetized animals (Curtis & Kleinfeld 2006). This focus on anesthetized animals came despite the fact that rodents display rich active behaviors with their whiskers (Brecht et al. 1997; Grant et al. 2012). Thus, when rodents explore the environment, they rhythmically move their whiskers in large-amplitude movements, called "whisking," typically at frequencies of 5–12 Hz in rats and 8–15 Hz in mice; and this exploratory whisking is often accompanied by synchronous active olfactory inhalation, or "sniffing" (figure 5.10). When the rat touches an object, it often reduces the amplitude and increases the frequency of whisking to 15–25 Hz, a pattern often called "foveal whisking" (Berg & Kleinfeld 2003). Finally, as already mentioned, the rat (or mouse) moves its chin and snout to palpate the object in detail with its high-acuity, fovea-like microvibrissae (figure 5.8b). Surprisingly, very little is known about the implications of these rich whisking behaviors on the neural codes for touch in the whisking system—however, as will be described next, the few studies that have been done on this topic show that the consequences are quite dramatic.

But before we go on to describe these studies, one may ask a devil's advocate question: Since all the whisking behaviors described here were characterized in laboratory rats that were born in the lab and raised in small, impoverished, rectangular cages, without ever experiencing the rich outdoors, are their whisking behaviors even relevant for understanding the normal, natural behaviors of wild rodents? An interesting study suggests that, surprisingly, the answer might be "yes." In that study (Polley et al. 2004), Frostig and colleagues compared the functional representation in the barrel cortex of rats that were raised in standard, small laboratory cages against animals that were introduced as adults (at age 90 days) into a large naturalistic habitat and lived inside it for four to six weeks. This naturalistic habitat included many objects, as well as other rats, and importantly, it also contained a

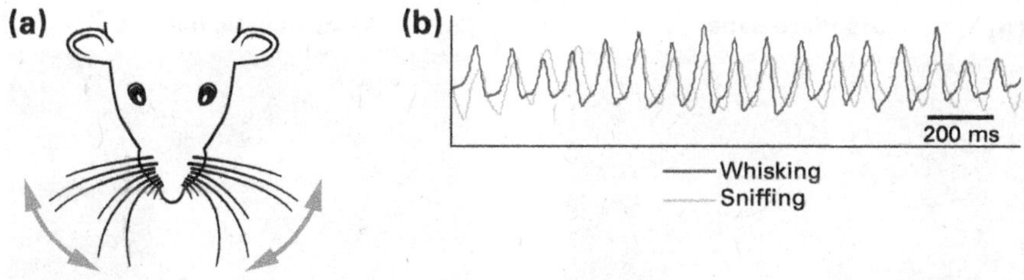

Figure 5.10
Synchronization of whisking and sniffing. (a) Rats move their whiskers periodically to actively explore the environment. (b) When a rat is inspecting an object of interest, its somatosensory-whisking and olfactory-sniffing cycles become phase-locked. (b) adapted from Deschênes et al. (2012).

thick layer of topsoil. The results were quite striking. First, the authors reported that "rats were observed to dig complex, interlocked subterranean tunnel systems that spanned the entire diameter of the [environment]" (Polley et al. 2004); see figure 5.11b (top) for an example of such a tunnel. This in itself is quite an amazing finding because these laboratory rats were bred for > 100 generations indoors and have never seen sunlight or moonlight or natural soil for > 100 generations—and yet, when introduced to topsoil for the first time, they knew what to do: they dug beautifully complex tunnel systems. Paraphrasing the old proverb: You can take the rat out of the wild, but you cannot take the wild out of the rat . . .

The second finding in the study by Frostig and colleagues was equally illuminating. They used intrinsic optical imaging to measure the functional representation in the barrel cortex, and found that the response fields in the barrel cortex were dramatically shrunk and weakened after exposure to the naturalistic habitat for a few weeks, exhibiting a sharpening or refinement of the cortical map (figure 5.11b versus figure 5.11a: bottom row). Such extremely strong plasticity in adult animals is never seen in standard laboratory experiments. This plasticity was likely caused by the extensive whisker use that these rats did while foraging and navigating underground by actively palpating the walls of the tunnel systems that they have dug, as well as palpating the various objects and many other rats that populated the naturalistic-habitat environment. The study by Frostig and colleagues indicates that, on one hand, building tunnel systems (burrows) and using whiskers for it are innate and genetically driven behaviors in rodents (see also Weber et al. 2013), and hence it may be legitimate to study whisking in laboratory rats. On the other hand, the dramatic plasticity observed in the rat somatosensory system upon exposure to a naturalistic environment (figure 5.11) suggests that neurobiological findings may *differ* among animals housed in impoverished cages and tested with highly artificial stimuli, as compared to animals that were housed and tested in enriched, complex, naturalistic environments. Notably, whisking behaviors were never studied at any level of detail in wild rats and were not compared to laboratory rats. Therefore, in the study of Frostig and colleagues, we do not really know how similar were the behaviors (and brain responses) of the laboratory rats that were used in that study, as compared to wild rats that would be tested in the same naturalistic environment. In fact, only recently did researchers start quantifying whisking behaviors systematically in laboratory rats. Such studies have not been conducted yet on wild rats. These fundamental gaps in knowledge call for a detailed behavioral characterization of whisking behaviors in laboratory-born versus wild-born rats, both indoors and outdoors.

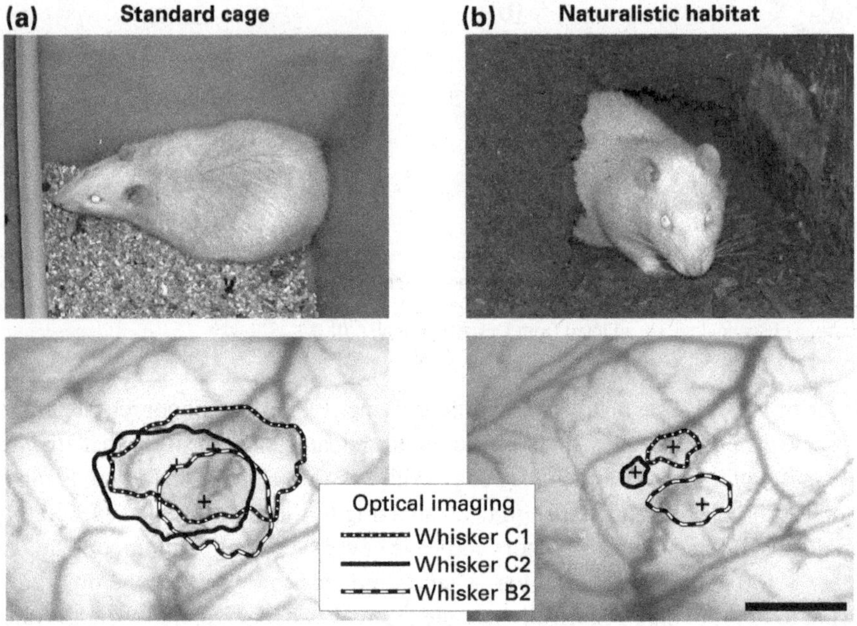

Figure 5.11
Effects of exposure to a naturalistic habitat. Polley et al. (2004) compared rats living in a standard laboratory cage (a, top) versus rats that were introduced as adults (at age 90 days) to live in a complex naturalistic habitat with topsoil (b, top). They found that first, rats spontaneously dug complex tunnel systems when provided with topsoil (b, top). Second, exposure of a few weeks to this rich, naturalistic habitat led to dramatic shrinking of the whisker representation in the somatosensory barrel cortex (b, bottom) compared to the control condition (a, bottom). The curved outlines denote the cortical area activated when experimenters touched a single whisker (C1, C2, or B2), as measured via intrinsic-signal optical imaging. Note that in the control animals (a, bottom) cortical activation fields overlapped and covered several barrels; the authors also reported on electrophysiological recordings showing that individual neurons in the upper cortical layers of control animals responded to multiple whiskers (median: six whiskers). By contrast, in the animals with naturalistic experience (b, bottom), there was no or little overlap between cortical activation fields as measured with imaging, and individual neurons typically responded to only one or two whiskers (median: two whiskers). The scale bar, 1 mm, is the same for both bottom panels. High-resolution images of top panels—courtesy of Ron Frostig; the top-right panel was adapted from Frostig (2006). The bottom panels were adapted from Polley et al. (2004), and were measured in different animals.

Further, the results of Frostig and colleagues have two important implications: First, they highlight the importance of using enriched housing for rodents rather than the sensory-deprived standard cages. Neuroscientific results are fundamentally different in such sensory-deprived cages, as opposed to enriched, naturalistic environments. Second, these results also raise the urgent need for neurobiological investigations of sensory systems in animals running and freely behaving in real-world naturalistic environments.

So what kinds of information does the whisking system convey about objects in the environment? Whiskers can measure the *location* of objects in 3D, their *shape* and *softness*—and rats were even shown to discriminate surface *texture* via whisking (Guić-Robles et al. 1989). The 3D location of objects relative to the head was shown to be encoded in the rat's whisking system by a triple coding scheme: The *identities* of activated sensory neurons represent the vertical coordinate of the object (a **labeled-line code** according to the whisker row); the *timing* or phase of neuronal firing, relative to the whisking cycle, encodes the horizontal object coordinate (**temporal code**); and the *firing rate* encodes the third coordinate—the radial distance from the whisker base (**rate code**) (Ahissar & Knutsen 2008).

But what are the elements of information that allow the rat to discriminate *texture*? As previously mentioned, during natural behaviors, rats rely heavily on their small microvibrissae for texture discrimination, but in laboratory experiments on texture discrimination, the use of microvibrissae is usually prevented (e.g., by removing them or by head-fixing the rats, which prevents normal palpation) to force the rats to rely exclusively on their long whiskers. This clearly limits the display of natural behavior, and therefore, future work will need to be done on the neurophysiology of the microvibrissae during natural texture discrimination (and other tasks). That said, long whiskers can also convey information about texture: When a rat sweeps its whiskers across a rough surface, its whiskers get caught and then released in micromovements, which were termed "stick-slip events"—or "slips" for short. Feldman and colleagues have studied the neural coding of such slip events in the barrel cortex of freely moving rats that performed a texture-discrimination task (Jadhav et al. 2009). They found that neurons responded in a highly temporally precise manner to the ultrashort slip events. Because multiple neurons responded with such precise temporal specificity, this constituted a *synchrony code* for slips. The authors demonstrated that as few as about 100 neurons were enough to decode the occurrence of a slip with >95 percent accuracy. This finding of a precise temporal code or synchrony code for slips in the barrel cortex is important because it was previously shown that the temporal statistics of slips varies with texture identity (Ritt et al. 2008; Wolfe et al. 2008). And indeed, Feldman and colleagues found that individual neurons fired differently for different textures; that is, for different grades of sandpaper (Jadhav et al. 2009). Thus, natural whisking transforms spatially structured stimuli (textured surfaces) into a temporally discrete stream of ultrashort events (stick slips), which are encoded by a precise temporal code in the barrel cortex and can reliably represent texture. Importantly, the natural statistics of slip events occur only when the rat is naturally whisking: This complex natural stimulus is very difficult to reproduce by passive stimulation, which means that if one wants to study the encoding of textures in the rodent whisker system, this must be done in behaving animals that are *actively* whisking.

There are at least *two* different fundamental reasons why the rodent whisking system cannot be adequately studied in a nonbehaving animal, using passive stimulation—and, therefore, why research must be done on actively behaving rats and mice. First, the biomechanical properties of the stimulus cannot be physically mimicked by externally applied forces: The physical forces on the follicle at the base of each whisker are very different in the passive case, when the follicle muscles are relaxed, versus the active case, when the muscles are contracted; therefore, there is a fundamental difference already at the periphery between passive and active sensing. Further, there is another key aspect of this problem: *whisker curvature* is another crucial sensory variable that cannot be mimicked with only external forces. Quist and Hartmann (2012) showed that the curvature of the whisker can dramatically affect the axial forces and bending moments in the follicle at the base of the whisker, where the mechanoreceptors are situated (i.e., the whisker curvature strongly affects the basic sensory information that is available to the brain). Moreover, it was shown that neurons in the first station of the whisking system, the trigeminal nucleus, are highly sensitive to axial forces (Stüttgen et al. 2008). Importantly, the full shape of an untrimmed whisker conveys reliable information about the *3D location* of the encountered object because the whisker is being bent in a unique manner in every 3D location where it touches an object (Bagdasarian et al. 2013). Thus, the morphological shape of the whisker conveys very rich sensory information based on biomechanics alone; this can be thought of as "morphological coding" (Bagdasarian et al. 2013). Notably, it is impossible to fully mimic natural whisker shapes and curvatures by applying external forces using standard laboratory techniques (Bagdasarian et al. 2013). Unfortunately, however, many neurophysiological studies on the whisker system are conducted with some of the whiskers being trimmed (to facilitate video tracking of the remaining, untrimmed whiskers), thus missing entirely the important contributions of the whisker shapes across the entire array of whiskers (Quist & Hartmann 2012). Taken together, this means that to fully study the neurobiology of sensory transformations in the rodent whisker system, one should best study it on untrimmed whiskers, in actively behaving animals.

Second, neuronal activity in the whisking system changes dramatically in the active versus the passive condition. A study by Ahissar and colleagues (Szwed et al. 2003) showed that already in the sensory periphery, at the trigeminal ganglion, many neurons encode movement-related variables: they found that 37 percent of the neurons were "whisking cells" or "whisking/touch cells," whose whisking-related activity could not be studied in the passive condition, of course. They also found many "touch cells," which did not directly encode whisking movements, but rather encoded contact, steady pressure, or detachment—but, importantly, the activity of these touch cells in the active condition was only poorly predicted by their activity in the passive condition. Another study (Campagner et al. 2016) found that the responses of trigeminal-ganglion neurons in freely whisking rats were well predicted by the rotational forces (moments) acting on the whisker, and this was true both during whisking in the air and upon touch. Importantly, such rotational forces are easy to *measure* using high-speed video cameras but are exceedingly difficult to *re-create* using passive stimulation (Campagner et al. 2016). Both these studies concluded that for the very periphery of the whisking system, it is crucial to study it under naturalistic active-whisking conditions (Szwed et al. 2003; Campagner et al. 2016).

This point becomes even more evident higher in the sensory processing pathway—in the barrel cortex. Carl Petersen and colleagues (Ferezou et al. 2006) used a fiber-optic bundle to conduct the first optical imaging recordings from the barrel cortex of freely moving, freely behaving mice. They examined the neural responses to passive stimulation, where the stimulation was done via a magnet that pulled a small piece of metal glued to one whisker, and compared it to active touch of an object by the actively whisking mouse. When they passively stimulated the whisker while the mouse was awake but sitting still, they observed a strong response (figure 5.12a), which could be interpreted as a startle response (Curtis & Kleinfeld 2006). However, when they passively stimulated the whisker while the mouse was actively whisking and exploring the environment, they observed almost no response at all (figure 5.12b). By contrast, when the mouse actively touched an object during whisking, the response was extremely strong again (figure 5.12c). These differences could be caused by different whisking patterns in the exploration mode versus object-detection mode. Alternatively, these large differences may indicate that sensory responses in the barrel cortex are very strongly modulated by nonsensory information, such as the *behavioral state* of the

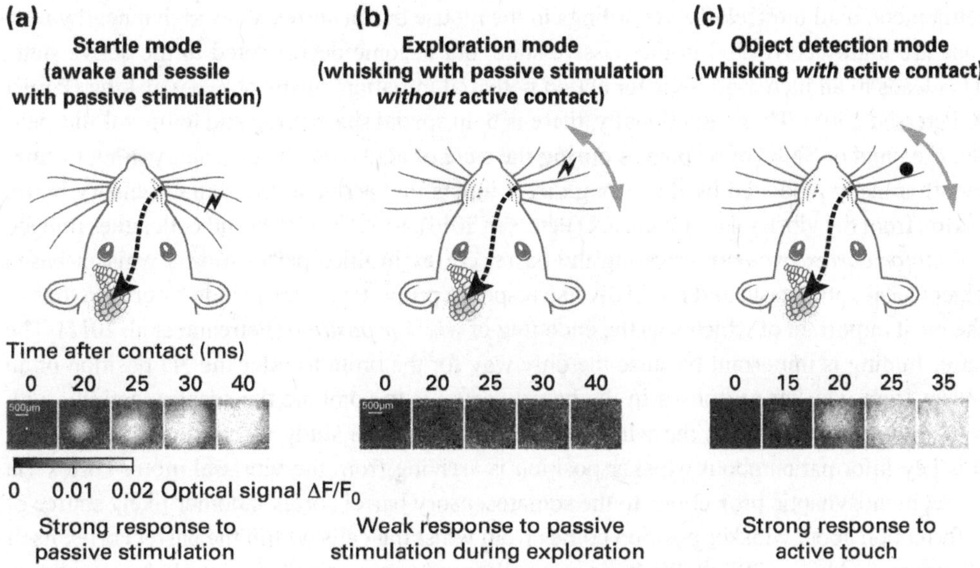

Figure 5.12
Imaging activity in the somatosensory barrel cortex of mice running in an arena under three behavioral states (a–c). Adapted from Curtis and Kleinfeld (2006), based on the original study (Ferezou et al. 2006).

animal (compare figures 5.12a and 5.12b) and *expectation*—that is, whether an actively exploring animal expects or does not expect to encounter a sensory stimulation at a certain location (compare figures 5.12c and 5.12b).

Nonsensory modulation of neural activity in the barrel cortex also was demonstrated by Michael Brecht and colleagues, who studied the representation of *social touch* in the barrel cortex (Bobrov et al. 2014; Lenschow & Brecht 2015). They found that a rat's whisking dynamics were very different when the animal was socially palpating another live animal, as opposed to palpating a stuffed rat. The spiking responses were also very different in the social context—in particular, the neurons started responding *before* the touch itself, indicating the existence of nonsensory signals in the barrel cortex, such as expectations. Moreover, in females, the neural responses were strongly affected by whether they were in estrus or not. Further, during social whisking, the membrane potential modulations both after and before whisker-touch were much stronger to a live rat than to a stuffed rat. Together, these results indicate that the differences in neuronal responses between social touch and regular touch are likely driven by enhanced top-down inputs during social touch: either enhanced *arousal*, *attention*, or *expectation* of contact, or perhaps a *multisensory* pheromonal effect—or all of the above (Lenschow & Brecht 2015).

Even if we ignore the complex modulatory effects of expectation or multisensory input, and consider just the contribution of whisker movement per se—it is clear that the *motor* act of moving the whiskers can dramatically affect responses in the barrel cortex. Several studies have shown that during active touch, as compared to passive touch, cortical responses become weaker in amplitude but are spatially more specific—that is, fewer cortical barrels respond to a touch by a given whisker (Fanselow & Nicolelis 1999; Hentschke et al. 2006). This spatial refinement of responses during the active state is also accompanied by temporal refinement: dual intracellular recordings in the mouse barrel cortex showed that nearby neurons are highly correlated in the passive state, but become decorrelated in the active state. This leads to an increased SNR for action potential initiation during active whisking (Poulet & Petersen 2008). Thus, functionally, there is both spatial sharpening and temporal sharpening (i.e., higher SNR) of responses during the state of active touch. This sharpening (refinement) is likely mediated by the strong direct inputs that arrive at the somatosensory barrel cortex from the vibrissal motor cortex (Petersen 2014). Karel Svoboda and colleagues imaged the *motor cortex axons* innervating the barrel cortex in mice performing a whisker-based object localization task, and found diverse response properties in these motor-cortex axons—the most important of which was the encoding of *whisker position* (Petreanu et al. 2012). The latter finding is important because the only way for the brain to infer the 3D position of an object from whisker responses in the barrel cortex is to combine the sensory activity with reliable information about the whisker position. Svoboda's study nicely demonstrated that this key information about whisker position is arriving from the vibrissal motor cortex via direct monosynaptic projections to the somatosensory barrel cortex (another likely source of information about whisker position comes from whisking cells within the barrel cortex itself (Knutsen & Ahissar 2009)). Svoboda and colleagues summarized their study by suggesting that the motor-sensory whisking loop could integrate haptic (touch) information across space and time, which is crucial for object recognition and somatosensory navigation (Petreanu et al. 2012). These motor-sensory interactions could be studied only in a behaving animal.

In summary, extensive evidence has accumulated over recent years that the entire whisking system operates completely differently in behaving, actively whisking rodents compared to passive stimulation. These dramatic differences are observed starting from the biomechanics of the whisker hairs, through the neural activity of the mechanoreceptors and first-order neurons in the trigeminal ganglion, and ending with major differences in neural activity observed in the barrel cortex. But why should we even be surprised that this is the case? After all, although a rat's whiskers may sometimes be passively touched by a moving object, it is clear that the major use of the whisking system in real life comes during active exploration and palpation. The world does not usually move toward a stationary rat; instead, a rat has to locomote its body, move its head, and actively whisk with its whiskers to map the spatial structure of its environment and the objects within it. In other words, the rat has to *actively pick up information* about the shapes, textures, and relative locations of surfaces, objects, landmarks, and other rats. If the animal will not actively explore space and palpate objects, there will be hardly any information at all impinging on its whiskers. The whisking system is thus, fundamentally, an *information-seeking system* that is busy with *active sensory acquisition* (Gibson 1962; Prescott et al. 2011). It is, therefore, no wonder that those "sensory" areas in the rodent brain that are responsive to whiskers are very far from being purely sensory—instead, they are part of a *purposive motor-sensory system* that is strongly affected by behavioral state, social context, expectations, intentions, and possibly navigational goals. It is also no wonder that in recent years, the research community that studies the brain's whisking system is gradually moving away from using reduced preparations, and instead has launched a major effort toward fine-grained behavioral characterization and neural recordings in freely moving, naturally behaving rodents.

5.4 Olfaction and Sniffing

Olfaction in mammals is another fascinating example of an active-sensing system where the motor component is absolutely essential for sensory perception. Diffusion is too slow to allow odor molecules to traverse the nasal cavity: it would take several days for most odor molecules to reach the olfactory receptors in the nose based on diffusion alone (Dusenbery 1992). Therefore, the physics of diffusion dictates that, in almost all circumstances, animals have to inhale the odor molecules actively in order to smell them. This process of active odor inhalation, which is typically performed rather periodically, is called "sniffing" (Welker 1964; Wachowiak 2011). The alternative strategy for actively sampling odor molecules is to run very fast to create constant airflow into the nose. Thus, bird-dogs during outdoor hunting were shown to exhibit one of two olfactory-search strategies: when tracking ground scents, these dogs exhibited rapid periodic sniffing near the ground; but when the dog was searching for game with its head poised high up, it actually avoided active sniffing and instead relied on the strong airflow created through its fast running (Steen et al. 1996). However, most mammals—including rodents and humans—do not use the running strategy; rather, they rely on sniffing for active olfaction (Mainland & Sobel 2006; Wachowiak 2011).

Several lines of evidence suggest that sniffing is crucial for olfactory perception. First, humans injected with an odorant into their bloodstream could not detect the odorant when they did not actively sniff, but they did perceive the odor upon sniffing, despite the fact that no odor was actually present in the nostrils, but rather reached the olfactory epithelium through the bloodstream (Bocca et al. 1965; Mainland & Sobel 2006). This directly demonstrates that the act of sniffing is fundamental for olfactory perception. Second, an ingenious experiment by Sobel and colleagues showed that humans can scent-track an olfactory trail, and that internostril comparisons during sniffing are essential for scent-tracking (Porter et al. 2007). When subjects wore a device that mixed the air coming into the two nostrils, their scent-tracking abilities deteriorated compared to subjects who wore the same device that did not do this air-mixing. This study suggested that sniffing is crucial for scent-tracking in humans, and moreover, that *independent* olfactory inputs into both nostrils are essential for spatial localization using olfaction. Third, rodents actively modulate the rate of sniffing, increasing it from the basal respiratory rate of about 1.5–5 Hz to a higher active-sniffing rate of about 6–9 Hz when encountering a new odor or examining an object of interest (see figures 5.13a and 5.13b) (Kepecs et al. 2007). Similar increases in sniffing rate occur also in humans. This increase in the rate of olfactory active sensing is very similar to the increased rates of active sensing observed in other sensory systems when encountering an object of interest, such as in whisking, vision, and echolocation (Geva-Sagiv et al. 2015). Fourth, rats are able to perceive odors within one sniff cycle, suggesting that the sniff is the minimal discrete unit of olfactory processing (Uchida & Mainen 2003; Uchida et al. 2006). This does not mean, of course, that animals cannot integrate information across multiple sniffs, or that such temporal integration does not play a role in some circumstances; but it does suggest a unique role for individual sniffs. Fifth and finally, as we will see, neurons in the olfactory bulb encode information via a phase code relative to the sniff cycle—again highlighting the importance of sniffing. The sniff thus seems to play a key processing role, being the basic

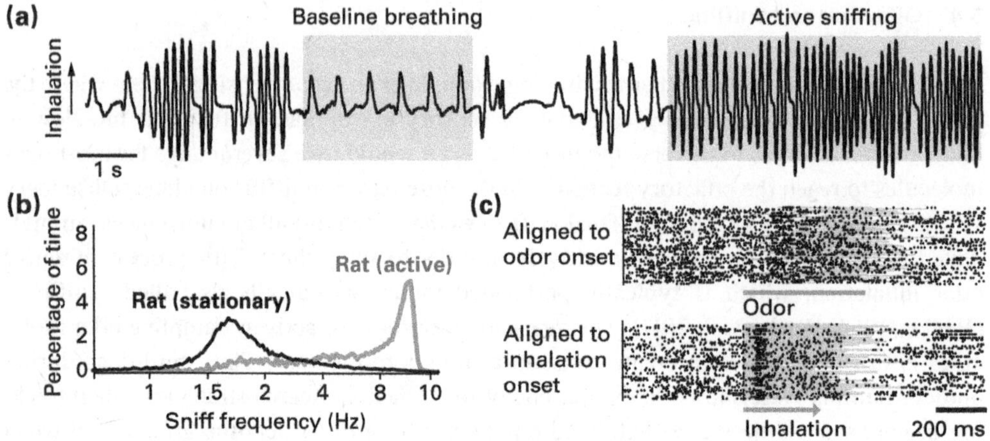

Figure 5.13
Active sniffing in rodents and the locking of olfactory bulb neurons onto the sniffing onset. (a–b) When rodents actively inspect an object of interest, they increase the rate of sniffing. (c) Example of an olfactory bulb neuron that showed only weak response to an odor when its activity was aligned to odor onset (top), but showed much stronger neuronal response when it was aligned to the first inhalation after the odor onset (bottom). This illustrates the importance of considering active sniffing and inhalations in olfactory research. (a–b) adapted from Wachowiak (2011), (c) adapted from Shusterman et al. (2011).

packet of information in olfaction, perhaps similar to the role of a single whisk in the whisking system or a single saccade in vision.

Olfactory sniffing in rodents is tightly coupled to whisking, and the two systems share many functional similarities. Deschênes, Moore, and Kleinfeld (2012) reviewed in detail the similarities between these two orofacial rhythms—sniffing and whisking. Two aspects are particularly interesting. First, during active exploration of the environment or when examining an object, rats typically both sniff and whisk at the object of interest, and the sniffing and whisking are often synchronized and *phase-locked*, as shown in figure 5.10b (see also Fonio et al. 2016). Second, Kleinfeld and colleagues found that the pre-Bötzinger complex in the brainstem serves as a single master pacemaker that controls both sniffing and whisking (Moore et al. 2013). However, very little is known to date about how the tight coupling between these two active-sensing systems affects the animal's behavior and the neural processing in either of these systems—and about their multisensory merging downstream.

Furthermore, at the neural-coding level, both the olfactory and whisking systems exhibit a similar temporal code (phase code), whereby the phase of the neuronal response relative to the behavioral oscillation is encoding the properties of the sensory stimulus. For whisking, the phase of the neuronal response relative to the whisking cycle encodes object horizontal location, constituting a phase code for object location (Knutsen & Ahissar 2009). For sniffing, Dmitry Rinberg and colleagues showed that information about the odor identity is encoded by the phase of the neuronal response relative to the sniffing cycle—again, an example of a phase code. Moreover, they showed that this phase information is also decoded by downstream circuits to determine the animal's sensory perception and behavior. Specifically, they conducted the following series of experiments: First, they compared the activity of mitral cells in the olfactory bulb of behaving versus anesthetized mice, and found that the neuronal responses are much weaker and sparser in the awake behaving animal, as compared to anesthesia (Rinberg et al. 2006). This suggests, again, that the anesthetized brain is quite different from the awake brain. Second, they found that the seemingly weak responses to odors are in fact much stronger when time-aligning them not to odor onset, but to inhalation onset (figure 5.13c; compare bottom to top). Thus, the mitral cells were strongly locked to the sniff-cycle phase, with about 10-ms precision. Moreover, the precise phase and temporal profile of this phase-locked response were found to encode the stimulus identity (Shusterman et al. 2011). This high 10-ms temporal resolution may seem surprising for the seemingly slow olfactory system, but it may allow responding rapidly to natural olfactory stimuli, which are very transient outdoors, arriving in intermittent plumes (Dusenbery 1992); and this high temporal resolution also may be required to compute odor direction via internostril comparisons (Porter et al. 2007).

Third, Rinberg and his colleagues examined whether the mice can behaviorally decode (read out) this phase information to guide behavior. They developed a mouse line that expressed ChR2 in olfactory-bulb sensory neurons, which allowed for optogenetically stimulating the neurons at precise phases relative to the sniff cycle, while controlling for other aspects of the stimulus (Smear et al. 2011). They trained mice to discriminate between light stimuli delivered to the olfactory bulb at different phases—mimicking endogenous neuronal activation at various phases of the sniff cycle—and found that the mice were able to discriminate temporal (phase) difference down to about 10 ms. This behavioral-precision of 10 ms was very similar to the locking precision revealed by the electrophysiological recordings. Crucially, this study causally demonstrated that mice can indeed read out the phase code of olfactory neurons relative to the sniff cycle. Needless to say, these fundamental aspects of olfactory neural coding—precise phase coding, and sparsity—could be elucidated only in actively sniffing animals. It would be important to study this further in mice that are not head-fixed but are totally free to move and behave because natural sniffing and head-movement behaviors will likely reveal additional important aspects of active sensing in olfaction, which are masked by head fixation.

Future work will also need to determine whether—similar to the visual and whisking systems—olfactory neural processing is influenced by top-down effects, such as attention or expectation. Another key question is how do mammals modulate their sniffing patterns during natural active olfactory exploration; some data on this were already collected, but much more needs to be done. It would also be important to answer the following questions: Why do mammals change their sniffing patterns the way they do during ongoing natural behavior? What kinds of different functional information is the animal trying to acquire by these active modulations of sniffing? To date, this is unknown. In any event, the picture that emerges from the studies discussed here is that one *must* consider natural olfactory behaviors to understand fully the neurobiology of olfaction.

5.5 Echolocation

Echolocation in bats and toothed whales (including dolphins) is another prime example of an active-sensing system: The animal has to actively emit a sound to receive an echo, which carries rich sensory information to the animal. Therefore, the active motor production of echolocation signals is absolutely necessary for this system. Here, I will focus on bat echolocation (or biosonar). Compared to other active-sensing systems, bat echolocation offers three major advantages as a model system for studying active sensing. First, there is a clear mathematical theory—the theory of sonar and radar—which can explain quantitatively why bats change their echolocation calls the way they do; this theory provides a clear mathematical framework for understanding the kinds of sensory information that bats can acquire using one type of biosonar call or another (Skolnik 2001; Peebles 2008). We do not have such a mathematical theory for other active-sensing systems, such as vision, whisking, or olfaction. The second advantage of studying echolocation is an experimental one. We can use microphone arrays to characterize the full sensory behavior of a freely behaving, freely flying bat. For example, we can measure the full spectrotemporal properties of the emitted echolocation calls—and often also the returning echoes—and in addition, we can characterize precisely when, where, and in which direction the flying bat emitted each individual sonar call (Ghose & Moss 2003; Ulanovsky & Moss 2008; Moss & Surlykke 2010; Yovel et al. 2010). By contrast, it is much more difficult to measure eye movements in freely moving monkeys or to track all the individual whiskers of a rat as it freely explores a complex environment (because the thin whiskers go in and out of the plane of focus of the camera), although recent technological advances are starting to allow it. The third advantage is the high diversity of echolocation call structures across different bat species, which yields a rich ground for comparative studies of this active-sensing system (Schnitzler et al. 2003; Siemers & Schnitzler 2004; Jones 2005; Jones & Holderied 2007; Ulanovsky & Moss 2008).

However, while much is known about the processing of echolocation sounds in the bat auditory system (Suga 1989; Ulanovsky & Moss 2008), much less is known about the motor-sensory loops in the brain that control the motor changes in echolocation call production based on the incoming sensory input. Indeed, bats exhibit large changes in their echolocation calls based on the sensory inputs (echoes): they change the spectrotemporal acoustic properties of their calls very dynamically and rapidly according to the incoming echoes in order to optimize sensory acquisition. For example, when an insect-eating bat hunts an insect, it undergoes transitions among three phases of echolocation (see figure 5.14a): the "search phase," which optimizes detection; the "approach phase," which localizes a target; and then the "attack phase" which occurs just before reaching the target. The transitions among these phases are characterized by a dramatic increase in the biosonar pulse emission rate (providing higher sensory update rates), and an increase in the pulse bandwidth (which improves the resolution of distance measurement (Skolnik 2001)).

Several reviews have discussed in detail these echolocation behaviors in bats, the mathematics of bat biosonar, and the neural basis of auditory processing of echolocation (Suga 1989, 1990; Schnitzler et al. 2003; Jones & Holderied 2007; Ulanovsky & Moss 2008; Yovel et al. 2011; Geva-Sagiv et al. 2015). In brief, as described in chapter 2, in the 1970s Nobuo Suga and colleagues discovered "delay-tuned neurons," which were tuned to a particular pulse-echo delay—these neurons may thus represent the physical distance to the object (Suga et al. 1978; O'Neill & Suga 1979; Suga & O'Neill 1979). Moreover, Suga and colleagues found neurons that preferred sequences of pulse-echo pairs where the time delay between pulse and echo became progressively shorter, matching the behavioral increase in call rate and decrease in time delay as a bat is tracking an insect and closing in for the capture. They called these cells "tracking neurons" (O'Neill & Suga 1979). These neurons thus provide an amazing matching of sensory processing in the brain to the behavior of the animal, in line with the expectation from an active-sensing system.

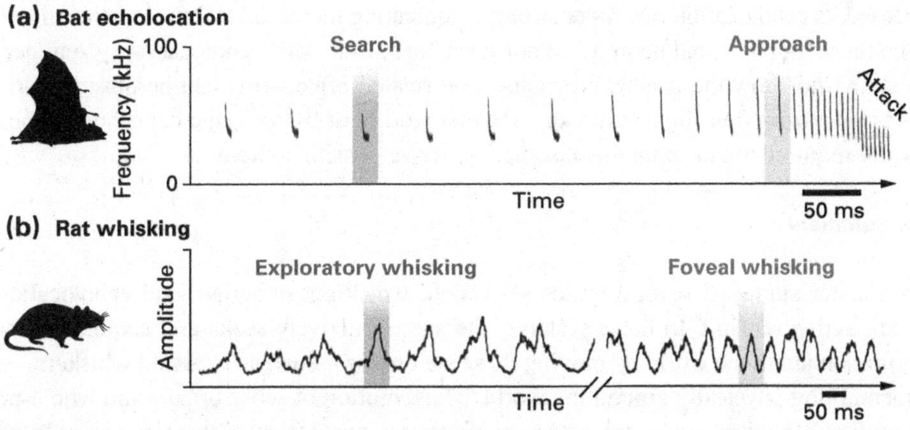

Figure 5.14
Comparing bat echolocation and rodent whisking. (a) Behavioral phases of bat echolocation: "search phase," in the absence of targets, optimizes detection; "approach phase" localizes a target, which enables approach; and "attack phase" is emitted just before reaching the target. (b) Rats increase the rate of whisking (and sniffing) when approaching an object ("foveal whisking"), akin to bat echolocation. (a) Adapted from Schnitzler et al. (2003). (b) Adapted from Berg & Kleinfeld (2003). Figure arranged similarly to Geva-Sagiv et al. (2015).

Notably, however, most studies on the neurobiology of bat echolocation have been conducted on head-fixed bats that were actually *not* echolocating actively during the experiment, but were passively listening to presented pairs of pulse and echo. Thus, Suga's paired-pulse paradigm was both an immense success, as it delivered incredible insights on the bat auditory cortex, but also a tragedy, as it did not study real echolocation behavior. To date, only a handful of studies recorded neural activity in freely echolocating bats from either the auditory system or vocal premotor centers (Kawasaki et al. 1988; Sinha & Moss 2007; Kothari et al. 2018; Wohlgemuth et al. 2018)—and only one study recorded echolocation-related neurons in freely flying bats (Kothari et al. 2018). The latter study found that when bats used their echolocation to inspect objects, the spatial tuning of auditory neurons in the superior colliculus became more sharply tuned. This finding highlights the importance of studying the neurobiology of echolocation in freely flying, actively echolocating bats. However, no such studies were conducted to date in the auditory cortex of bats. Thus, to this day, practically nothing is known about the neural basis of echolocation in bats during actual echolocation behaviors. Therefore, even though a strong ethological motivation has underlain the study of bat echolocation over many decades, the field is still awaiting experiments to be conducted during actual ethologically relevant echolocation behaviors, such as insect-hunting, collision-avoidance, or landing maneuvers.

As detailed in chapters 3 and 4, we have recently developed methodologies to do exactly such experiments, including wireless electrophysiology methods and wireless on-bat audio recordings. These methods were used in a recent study (Sarel et al. 2022), where we conducted experiments in pairs of bats flying in a 135-m-long tunnel. We found that when bats switched from navigation to collision-avoidance behavior (i.e., not colliding with the other bat), hippocampal neurons exhibited a dramatic and very rapid switch, as fast as 100 ms, from representing the bat's own position to representing its position together with the distance to the other bat. Interestingly, this neuronal switch was more pronounced when the bat increased its echolocation rate more strongly, indicating increased attention to the other bat. While these hippocampal neurons do not form part of the bat's echolocation system per se, this study did show the dramatic echolocation-related effects that can be observed in the bat's brain during free-flight behaviors. Further studies of freely flying bat echolocation are urgently required to elucidate this fascinating active-sensing system.

5.6 Summary

This chapter surveyed several systems—vision, whisking, olfaction, and echolocation—that are active sensing. In these systems, the subject actively seeks and acquires sensory information about the world by moving its sense organs—such as eyes and whiskers—and by locomoting physically around the world. Thus, motion of sense organs and whole-body locomotion behaviors are an integral part of sensory perception in these systems. In other words, if we consider the brain-behavior-world interactions depicted in figure 1.3 of chapter 1, then there are almost no direct sensory inputs from the world to the brain: the direct arrows connecting world → brain are almost nonexistent. Instead, almost all sensory information from the world passes through the behavior step (i.e., active-sensing behaviors) to reach the brain. Natural behaviors are a fundamental aspect of these sensory-acquisition systems. Therefore, to fully understand natural sensory perception, we cannot study it only passively, in isolation. Instead, we must allow the subject to do what these exquisite systems have evolved for—to acquire sensory information actively. This should be done by freeing the animal to behave naturally: move about, use its sense organs, and explore the world.

Specific examples of the importance of this approach were discussed, involving vision, whisking, and olfaction. There are many additional active-sensing systems in the natural world, which I have not discussed here for the sake of brevity. Some of these systems have a wonderful history of behavioral and neurobiological studies, including electrolocation in weakly electric fish (Heiligenberg 1991b; Krahe & Maler 2014) and the somatosensory tentacles of star-nosed moles (Catania 2006). Even the sense of hearing, which is classically considered a passive sense (except in echolocation), incorporates some aspects of active sensing, in the form of active movements of the head and ears, which actively scan the acoustic environment, as can be appreciated by watching a cat as it intently listens and moves its ears to localize sounds actively in the environment. Further, certain classes of neurons in the cat's dorsal cochlear nucleus—a key station in the ascending auditory pathway—exhibit strong somatosensory responses to ear movements (Young et al. 1995; Nelken & Young 1996), which perhaps reflects the active-sensing nature of hearing.

Interestingly, there are notable parallels between many active-sensing systems. First, there is strong behavioral *rhythmicity* in many of these sensory systems, including bat echolocation, rodent whisking and sniffing, primate saccades, and fish electrolocation (Heiligenberg 1991b; Berg & Kleinfeld 2003; Kepecs et al. 2006; Ulanovsky & Moss 2008; Jutras et al. 2013). In addition, most of these active-sensing systems exhibit an increase in sensory update rate near objects of interest, such as an increase in the bat echolocation rate (figure 5.14a) or in the rodent whisking and sniffing rate (figure 5.14b) (Geva-Sagiv et al. 2015). Several other functional parallels exist (Gibson 1962; Nelson & MacIver 2006; Geva-Sagiv et al. 2015). Future work is needed to understand whether these functional parallels between the various active-sensing systems are accompanied by parallels at the computational/algorithmic levels or by implementational parallels among the relevant brain circuits.

Another future challenge is to study the multisensory integration among different active-sensing systems. Many active-sensing systems are often synchronized, as discussed in this chapter for rodent whisking and sniffing (figure 5.10b)—but this is also true for vision and hearing in humans (Gruters et al. 2018), as well as for echolocation and somatosensation in bats. The echolocation calls and wingbeats are strongly synchronized in flying bats (Speakman & Racey 1991; Holderied & von Helversen 2003; Holland & Waters 2005), which in turn synchronizes the auditory information arriving via echolocation with the somatosensory air-flow information arriving via the wings (Sterbing-D'Angelo et al. 2011). The functional implications of these synchronizations remain to be elucidated, for both behavior and neural processing. But one thing is clear: to elucidate the interplay and synchronization of information flow across sensory systems, it is crucial to perform studies in freely behaving animals.

Taken together, active-sensing systems provide a strong case study for the importance of investigating the brain during the natural behaviors of the animal. That is, the natural neuroscience approach.

6

Sensory Coding of Natural Stimuli

Sensory neuroscience has traditionally used very simple stimuli. Thus, most studies of the auditory system have used pure tones, white noise, or clicks, while most studies of the visual system have used spots, bars, or gratings. These controlled, simple stimuli were used in order to understand basic sensory processing mechanisms in the brain, with the hope that these same mechanisms could be extended in a rather straightforward manner to understand the neural processing of natural sounds and natural images. This hope is what guided Hubel and Wiesel's bottom-up research program on the visual cortex, as well as the research programs of their contemporaries studying the auditory cortex, such as Gerstein, Kiang, Goldstein, and Abeles. However, as I will discuss here, this hope has largely failed to materialize because, in most cases, the responses of sensory neurons to pure tones or drifting gratings are poor predictors of their responses to natural sounds or images. Experimental findings from recent years have made it clear that the extension from simple stimuli to complex, natural stimuli is anything but straightforward—and might not be possible at all. The reason for this failure is that the reductionist approach to sensory neuroscience has traditionally assumed that (1) sensory neurons are linear (or weakly nonlinear), (2) sensory processing is time invariant and does not depend on stimulus history, and (3) sensory processing is largely feedforward. However, none of these assumptions is correct. In fact, we know today that sensory processing is highly nonlinear, highly history dependent, and strongly affected by recurrent feedback. All these factors argue against the sole use of simple stimuli, such as visual gratings or auditory pure tones, for elucidating sensory systems. I will therefore posit that the best approach would be a *combined* approach: to use simple stimuli for studying basic circuit mechanisms, but also to employ more complex stimuli—both natural stimuli and artificial complex stimuli—in order to elucidate the more complex brain mechanisms that are highly adapted for processing natural sounds and images.

The focus in this chapter will be mostly on two senses—audition and vision—because most of the data on natural stimuli come from these two senses. However, beautiful examples also exist in other sensory systems, and I will discuss some of them in this chapter. In fact, this entire chapter should be read not so much as a review of specific sensory systems, but rather as a general call to study *all* the sensory systems using the combined approach, where the same neurons should be probed using both simple and complex naturalistic stimuli.

Finally, I will briefly review three research traditions: *scene analysis*, *sensory ecology*, and *ecological psychology*. These distinct traditions have very different historical roots, and different research-communities follow them, but they all support the same notion: the importance of studying sensory perception using natural stimuli and during natural behaviors. Here again, I will focus mostly on audition and vision in order to illustrate how different traditions study these sensory systems with an eye (or ear) on the real world.

I will start with audition, the sensory system where I took my first steps in neuroscience (Ulanovsky et al. 2003, 2004).

6.1 Audition: Natural Sounds and Natural Soundscapes

It is difficult to define precisely what "natural sounds" are. First, this can differ among species: For a polar bear, natural sounds might be the sounds of wind and waves and breaking ice. For a human being, these might be the sounds of the savanna where *Homo sapiens* evolved, the sounds of human speech, or the more recent sounds of car honks and cell-phone ringtones. Second, these may differ among environments for the same species: Rats, which lived for millions of years in grasslands and forests, have adapted over the last millennia to live in our urban environments—so perhaps car honks and ringtones are "natural" for them too? And what about the laboratory rat: are his natural sounds the sounds of air conditioning and floor cleaning by animal-care staff?

This raises the question: Which sounds should we choose as stimuli if we wish to understand the neural processing of natural sounds? There are three answers typically given to this question, which are not necessarily mutually exclusive (Nelken 2017). First, there is an answer based on *evolution*: we should focus on sounds that are important for natural selection. For example, social communication calls produced by conspecifics, as well as sounds produced by predators and prey, but also sounds generated by environmental dangers, such as those of a distant thunderstorm. The assumption is that through evolution, such sounds must have become prominently represented in the auditory system. Second, there is an answer based on *behavioral relevance*: the auditory cortex is highly plastic and shows particular plasticity for salient or behaviorally important stimuli (Irvine 2017). Therefore, over the lifetime of an animal, those sounds that are important and relevant for that animal may become strongly represented in its auditory circuits, and therefore should be studied. This again points to social communication calls, which exist even within the impoverished world of the laboratory mouse or rat—but also perhaps include the sounds of the animal-care technicians as they bring food to the cage. The latter sounds might be irrelevant for natural selection and evolution, as the survival and reproduction of the laboratory mouse do not depend on whether he hears these sounds—but they are certainly subjectively important and relevant for him. The third answer argues that *natural stimulus statistics* is crucial for the neurobiology of the auditory system, regardless of their subjective relevance or evolutionary importance. This approach suggests that auditory circuits (at least in the auditory cortex) develop during ontogeny to extract the statistical regularities present in the natural soundscapes surrounding the animal, or its "acoustic biotope." This answer thus proposes to study the auditory system using some statistically representative selection of natural sounds (Smith & Lewicki 2006; Nelken 2017). Similarly, these three answers apply to other sensory systems, where we may ask what a "natural visual scene" or a "natural odor" is, as will be discussed later in this chapter.

What do classical models of the auditory system predict regarding neuronal responses to natural sounds? Overall, these models posit that auditory processing relies on linear summation of the sound energy impinging within the neuron's frequency tuning curve (Schnupp et al. 2012). The assumption of linearity means that if we present two sounds to the animal, A and B, and measure the responses of a neuron to these sounds, $r(\mathrm{A})$ and $r(\mathrm{B})$, then the response of the same neuron to a combination of these two sounds should be $r(\mathrm{A} + \mathrm{B}) = r(\mathrm{A}) + r(\mathrm{B})$. Thus, if true, these models should be able to explain neuronal responses to natural sounds as linear combinations of responses to the simple sounds that make up the natural sound. However, as I will discuss next, several lines of evidence in fact suggest that this does not work: such simple linear models *cannot* explain the responses of auditory neurons to natural sounds in a satisfactory manner.

The first evidence against these linear models comes from studies showing that the responses of auditory neurons to pure tones are poor predictors of the responses of the same neurons to natural sounds. Machens, Wehr, and Zador (2004) demonstrated that two neurons in the primary auditory cortex of rats can respond very differently to the same natural sound—even though these same two neurons responded very similarly to tones (i.e., exhibited similar frequency tuning to pure tones). But it is not only cortical neurons that are highly nonlinear—even peripheral neurons in the auditory brainstem are highly nonlinear. This was shown by Roberts and Portfors (2015), who demonstrated that auditory brainstem neurons can respond strongly to natural sounds at a frequency at which they do not respond to pure tones *at all*. They presented mouse ultrasonic social vocalizations with frequencies of 50–100 kHz and recorded responses to these vocalizations in the brainstem dorsal cochlear nucleus of the mouse from neurons with pure-tone selectivity at frequencies of 10–20 kHz. Because the frequencies of the social vocalizations did not overlap at all with the frequency tuning curve of the neurons, the classical linear models would suggest that there should be no neuronal response whatsoever to these vocalizations. Surprisingly, however, many of these vocalizations evoked a strong response. Roberts and Portfors then showed that this unexpected neuronal response happens because of *cochlear nonlinearities*, whereby lower frequency distortion products are created at the output of the cochlea—and these lower frequencies, which do not exist in the original stimulus, drive strong responses in neurons tuned to low frequencies. This study, as well as a similar one on the inferior colliculus (IC; Portfors & Roberts 2014), have powerfully illustrated that peripheral sensory-transduction mechanisms are highly nonlinear, which makes it difficult to use simple stimuli and classical linear models to understand the neural processing of complex natural stimuli.

Second, the responses of auditory neurons to natural sounds are poor predictors for the responses of the same neurons to *other* natural sounds. The study by Machens et al. (2004) demonstrated this point very nicely. They used a diverse set of natural sounds—ranging from animal vocalizations and the sound of a blowing wind to Jimi Hendrix playing the guitar—and measured the subthreshold responses of rat auditory-cortex neurons to these sounds by conducting whole-cell patch-clamp recordings. Then they computed the spectrotemporal receptive field (STRF) of these neurons: a linear spectrotemporal estimate of the temporal responses of the neuron to pure tones at all possible frequencies.[1] They used the

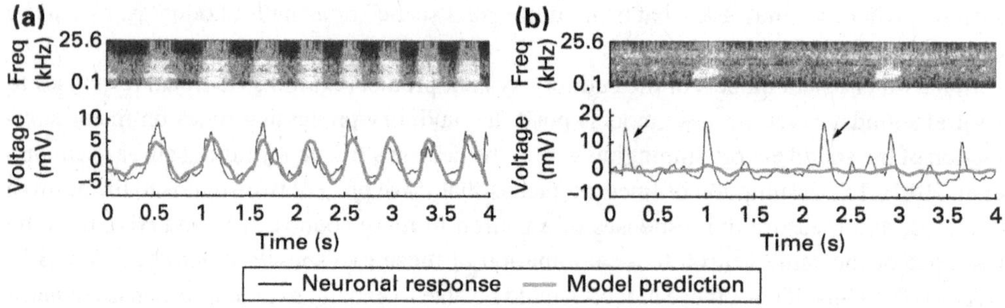

Figure 6.1
Linear STRF models usually do not satisfactorily predict the responses of primary auditory cortex (A1) neurons to natural sounds. Top: spectrograms of natural sounds presented to rat A1 neurons (white denotes high intensity). Bottom: experimentally measured subthreshold responses (black) and responses predicted based on the linear STRF model (gray). Adapted from Machens et al. (2004). (a) A nontypical A1 neuron, in which the linear STRF-based prediction matched well the actual neural response (compare the gray and black lines). (b) A more typical neuron, in which the linear STRF model failed to predict the actual neural response. In this neuron, there were two failure modes of the linear model. First, it failed to predict the magnitude of the neural response. Second, it completely missed the very occurrence of many of the strongest responses (some of these misses are marked by arrows).

STRF of each neuron to predict the expected responses of that neuron to new sounds, and then compared these linear predictions to the actual neuronal responses. In some cases, the predictions were remarkably good: figure 6.1a shows an example of a neuron whose actual response (bottom: thin black line) was very well predicted by the linear STRF estimate (thick gray line). However, in most cases, the linear predictions were very poor: figure 6.1b shows another neuron, in which the linear model failed to predict not only the magnitude of the neural response, but even the occurrence of many of the strongest responses (arrows). The prediction success for this neuron was just 8 percent—very close to the population average prediction success, which was 11 percent, a remarkably low success rate (Machens et al. 2004).

Machens et al. were able to rule out simple static nonlinearities—such as saturation and rectification (a nonmonotonic input-output relation)—as an explanation for the dramatic failure of the linear model. Hence, they proposed that more complex nonlinearities must be at play. Other studies, which estimated the STRFs of auditory-cortex neurons using non-natural, but complex artificial sounds—such as random chords and auditory ripple stimuli—have reported somewhat better predictions of the linear STRF, ranging from a prediction success of about 29 percent in rats and mice (Sahani & Linden 2002) to about 50 percent for linear estimations of spatiotemporal auditory receptive fields in ferrets (Schnupp et al. 2001). These differences may stem from interspecies differences (nonlinear cortical processing in rodents versus linear processing in ferrets) or, more likely, from differences in the extent and types of nonlinearities engaged by each stimulus set. Machens et al. suggested that the dramatic nonlinearities revealed by natural sounds may stem from the unique properties of natural sounds: their long-term correlations (both temporal and spectral), and their nonstationarity (changes over time). The long-term correlations and nonstationarity of natural sounds may engage cortical mechanisms of forward suppression (Brosch & Schreiner 1997), stimulus-specific adaptation (Ulanovsky et al. 2003, 2004) and spectral interactions (Nelken 2017), all of which are highly nonlinear and dynamic phenomena that are not captured by classical linear STRF models.

An attempt by Ahrens et al. (2008) to explicitly incorporate some of these nonlinearities into STRF models resulted in a substantial improvement in the average prediction success, from 30 percent to 42 percent (a 1.4-fold increase). This improvement offers unique insight into the physiology of cortical processing because it may point to *which* nonlinear processes might be at play. Furthermore, extending the approach of Ahrens et al. (2008) offers an exciting avenue for future research. In order to systematically understand the nonlinear processing of natural sounds in the cortex, auditory neurons should be probed with combinations of natural sounds and complex-yet-unnatural sounds, and their responses should be fitted with explicit models of various nonlinearities. First steps in this direction were taken by a number of studies, including Williamson et al. (2016), which revealed a host of strong nonlinearities.

A very nonintuitive and dramatic example of a cortical nonlinearity comes from several studies by Israel Nelken and colleagues, who showed that the responses of neurons in the cat's primary auditory cortex to complex sounds are extremely sensitive to small perturbations in the acoustic input—a sensitivity that cannot be explained by any of the current models of cortical processing. In a first set of studies, they used natural bird chirps as stimuli and found that even the smallest perturbation that they made to the acoustic stimulus, such as removal of the (very weak) background noise, dramatically modified the neuronal responses in the cat auditory cortex (Bar-Yosef et al. 2002; Bar-Yosef & Nelken 2007). Specifically, they separated the natural sound mixture (chirp+noise) into its two components and found that the majority of the neurons were responsive mostly to the *weak* background noise, not to the cleaned tonal chirp, which was by far the most acoustically dominant component, being louder by 29 dB than the background noise. Figure 6.2a shows an example of such a neuron, which responded similarly strongly to the weak background noise and to the original, natural chirp (which also contained the background noise)—but did not respond at all to the cleaned-up chirp. Figure 6.2b shows the population analysis for one animal, demonstrating that across the neuronal population, the weak background noise elicited large neuronal responses, while the main cleaned-up chirp elicited weaker responses, even though it contained much more acoustic energy. In a complementary set of studies, in which they used artificial mixtures of pure tones and amplitude-modulated noise—where now the broadband noise was the acoustically dominant component and the tonal component was the weak one—they found a similar result: the responses to the strong noise were dramatically modified by the addition of the very weak tone (Nelken et al. 1999; Las et al. 2005). None of the current linear models of spectrotemporal integration by auditory neurons can explain such extreme sensitivity to small acoustic perturbations because all these models predict that practically no change in neuronal responses should occur when adding a very weak sound—in stark contrast to the experimental results.

Following these findings, Nelken and Bar-Yosef (2008) proposed that perhaps the job of the primary auditory cortex is *not* low-level feature extraction, as commonly assumed, but rather a high-level representation of *acoustic objects,* and that in complex mixtures of sounds, this may actually entail encoding the weak acoustic background: the nondominant acoustic object. They argued that the reason for the dramatic cortical overrepresentation of the weak background is that for a cat, the background sounds might be particularly important, as they may contain rich information about prey (e.g., rodents), distant conspecifics (other cats), and approaching predators. It remains a challenge for future research to further

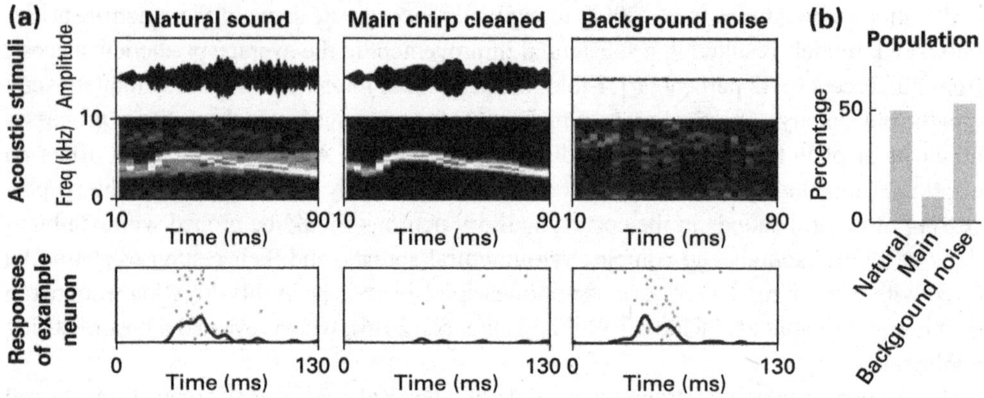

Figure 6.2
A1 neurons respond more strongly to weak background noises than to loud foreground sounds, exhibiting a profound nonlinearity that cannot be explained by any of the classical models of sensory responses in A1. Adapted from Bar-Yosef and Nelken (2007). (a) Responses of an example neuron to natural sounds (bottom row: the neuron's spike-raster [dots] and the average response [black line, PSTH]; top two rows: sound waveform and spectrogram). Left: Original sound, consisting of a bird chirp on top of a weak acoustic background noise (wind, rustling leaves). Middle, the cleaned-up bird chirp only, with the background noise removed. Right, acoustic noise only (also excluding the echoes). Surprisingly, the responses of this neuron to the original natural sound (left) were much more similar to its responses to the weak background noise (right) than to the strong main chirp (middle), suggesting that this neuron was driven very nonlinearly by the weak background noise. (b) This dramatic nonlinearity holds at the level of the neuronal population in A1. Shown here are population responses for all the neurons recorded from one animal, for the same stimuli as in (a). The weak background noise elicited many strong neuronal responses (right bar), while the loud main chirp alone elicited very few large responses (middle) even though it contained much more acoustic energy.

explore this idea—and in particular, to understand how the auditory cortex may isolate and bind together the relevant portions of the incoming acoustic waveform into one "object." Elucidating these mechanisms will likely require the use of naturalistic sound-mixtures—something rarely done in auditory research.

Third, another piece of evidence against the classical linear models comes from findings that spectrotemporal context can strongly affect the cortical processing of natural sounds. This violates a basic assumption of classical linear-processing models—namely, that neuronal responses are memoryless and do not depend on stimulus history. It is a dramatic violation because one of the hallmarks of natural sounds—such as animal vocalizations, human speech, and music—is that each sound element appears within a complex spectrotemporal context. Nelken's studies with the bird chirps already demonstrated that auditory neurons are extremely sensitive to such context, such as to background noise (figure 6.2). A subsequent study by Linden, Sahani and colleagues has set out to systematically quantify these effects (Williamson et al. 2016). Using complex, random-chord stimuli, they examined how the local spectrotemporal context affects neuronal responses in the auditory cortex to a sound element. They found that a coincident sound at a different frequency enhanced the response, while a preceding sound at a similar frequency suppressed the response. The suppressive effects occurred for time intervals up to about 250 ms. Similar timescales of suppression have been previously reported (Brosch & Schreiner 1997; Wehr & Zador 2005; David & Shamma 2013). Multiple other studies, which used pure tones, have demonstrated that the responses of auditory-cortex neurons are extremely sensitive to the preceding sound history even over *longer* timescales. In the cat primary auditory cortex, for example, the firing rate in response to the *same* physical sound differs threefold, on average, depending on which sounds preceded it, and it can differ more than tenfold in some neurons (Ulanovsky et al. 2003). The response to a tone is usually suppressed if it is preceded by many sounds with a similar frequency, and it is enhanced if the preceding sounds have a different frequency—a phenomenon often referred to as "stimulus-specific adaptation" (Ulanovsky et al. 2003, 2004; Rubin et al. 2016). These temporal-context history effects occur over multiple timescales, ranging from a few milliseconds up to tens of seconds, and even minutes (Ulanovsky 2004; Ulanovsky et al. 2004; Yaron et al. 2012). The very long timescales from seconds to minutes suggest that these stimulus-history effects are relevant to long natural sound sequences, such as social vocalizations, song, speech, and music. They also imply that it's practically impossible to conduct an auditory-neurophysiology experiment in the absence of strong history effects. To date, almost no studies have examined how temporal context may affect the responses to *complex* sounds at timescales longer than a few hundreds of milliseconds. One notable exception was Asari and Zador (2009), who found temporal-context history effects lasting up to 4 s when they recorded from auditory-cortex neurons of rats presented with long sequences of natural sounds. Asari and Zador achieved a twofold improvement in STRF model performance when incorporating such a 4-s history. Importantly, this marked improvement was still quite far from a perfect performance, suggesting that the neurons exhibited a nonlinear long-lasting history effect. Thus, the dramatic failures of linear STRF models in explaining neuronal responses to natural sounds may arise largely from such unaccounted nonlinear history effects. Future research should investigate the mechanisms underlying these history effects, and how one should build predictive models that can account for them.

Fourth and finally, a number of studies have reported selective neural responses to species-specific vocalizations in various auditory brain regions. This was reported in mammalian species, such as monkeys, guinea pigs and bats, and in nonmammalian species, such as songbirds and frogs (Wollberg & Newman 1972; Suga et al. 1978; Suga 1989; Feng et al. 1990; Doupe & Konishi 1991; Esser et al. 1997; Margoliash 1997; Tian et al. 2001; Grimsley et al. 2012). Notably, however, there was a debate as to whether such neuronal responses represent a truly unique selectivity for vocalizations or whether they perhaps reflect some nonlinear selectivity to complex sounds in general, but not necessarily vocalizations (Theunissen & Elie 2014; Nelken 2017). Of particular interest in this respect is Wang and Kadia (2001), which reported a surprising interspecies difference in how marmoset monkey vocalizations are processed in the primary auditory cortex of marmosets versus cats. They presented a set of natural marmoset twitter calls and their time-reversed versions (which maintain the same spectrum but lose the natural temporal ordering), and they found that the average response of marmoset auditory-cortex neurons was almost threefold stronger to the natural call. By contrast, no difference between the responses to natural and time-reversed calls was found for the same set of marmoset twitter calls when recording in the auditory cortex of cats.

This enhanced selectivity for marmoset vocalizations in the marmoset auditory cortex, but not in the cat auditory cortex, is *not* predicted by the response properties of the auditory cortex neurons to simple stimuli, which are very similar between marmosets and cats. The experiment by Wang and Kadia was done with the animals under anesthesia, so it is possible that the results in awake animals would be different. Despite this caveat, this interspecies difference is interesting and may suggest complex processing mechanisms that enhance the responses of marmoset cortical neurons to marmoset social vocalizations, which are important natural stimuli for marmosets, both behaviorally and evolutionarily. This result emphasizes that the study of cortical mechanisms responsible for encoding communication sounds must be based on behaviorally relevant vocalizations. This is especially true for awake animals, where auditory cortex neurons were shown to dramatically alter their response properties according to the behavioral relevance of the sound (Fritz et al. 2003, 2007).

For a detailed discussion of the various complex mechanisms that may underlie the dramatic differences in responses to natural sounds versus artificial sounds, there are several excellent reviews (Wang 2000; Mizrahi et al. 2014; Nelken et al. 2014; Theunissen & Elie 2014; Nelken 2017). I will turn now to another line of evidence for why the auditory system should be studied using natural sounds. This evidence comes from studies demonstrating that information coding by auditory neurons is optimized for sounds with natural stimulus statistics.

The first of these studies was conducted by William Bialek and colleagues, who compared the responses of bullfrog auditory nerve fibers to two kinds of stimuli: artificial broadband white noise, and naturalistic noise with an amplitude spectrum that was identical to the frog's communication call (Rieke et al. 1995). They found that neuronal information rates about the stimulus (the information that the spike-train transmits about the stimulus, in bits per second) increased dramatically for naturalistic noise, exhibiting an average 5.4-fold increase in the information rate compared to artificial noise. Remarkably, in some neurons, the coding efficiency for naturalistic sounds reached 90 percent, meaning that the neuron conveyed 90 percent of the fundamental theoretical limit on the information rate, set by the physical properties of the spike train. Figure 6.3a shows an example of such a neuron, which exhibited a much higher coding efficiency for naturalistic sounds than for artificial white noise. The authors concluded that the coding properties of the auditory nerve are matched to the structure of natural sounds. They argued that these results cannot be explained by linear processing; instead, they suggested that nonlinearities in auditory processing increase the information rate and coding efficiency for naturalistic stimuli, bringing them very close to the fundamental theoretical limits on information transfer. This result again highlights the likely functional importance of the strong nonlinearities in the auditory system and argues against the use of simple stimuli and linear models for elucidating auditory processing.

A theoretical study by Smith and Lewicki (2006) has shown that **efficient coding theory** can predict the shape of cochlear auditory filters and the response properties of auditory-nerve fibers. Their model learned a set of independent auditory filters from an ensemble of input sounds. Interestingly, when the model was presented with only animal vocalizations or only environmental sounds, the filters that it learned did *not* resemble physiological filters—whereas when it was presented with a realistic ensemble of *both* animal vocalizations (recorded from multiple species) *and* environmental sounds (such as wind and breaking twigs), the model has learned filters that strongly resembled physiological cochlear filters. Further, these filters represented the natural sound ensemble much more efficiently than did classical models. This suggests a remarkable adaptation of the auditory system for efficient representation of ensembles of natural sounds.

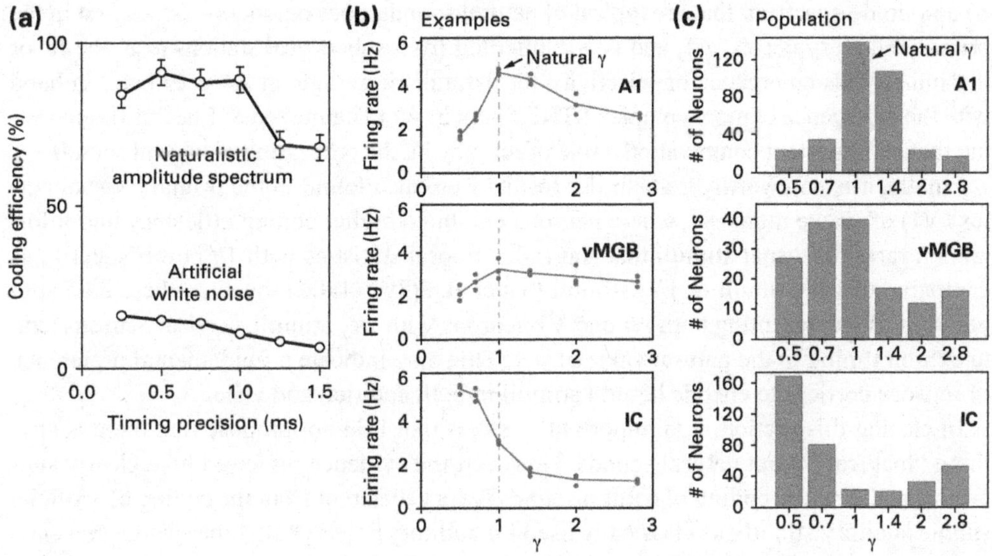

Figure 6.3
Auditory neurons respond optimally to naturalistic sounds. (a) Example neuron from a bullfrog auditory nerve. This neuron carried much more information (in bits per second) about the naturalistic stimulus presented to it (top curve) than for artificial white noise (bottom curve). Surprisingly, the coding efficiency for this neuron reached 90 percent for the naturalistic sound—that is, it carried 90 percent of the theoretically possible information rate. This neuron thus optimally encoded the naturalistic sound. Based on Rieke et al. (1995) and adapted from Bialek (2012). (b) and (c) show neuronal responses of ferret auditory neurons to sounds with a $1/f^{\gamma}$ amplitude spectrum, where γ is a parameter that was manipulated by the experimenters; γ=1 corresponds to natural sounds. Adapted from Garcia-Lazaro et al. (2011). (b) Examples of three neurons recorded in three successive stations along the ascending auditory pathway: inferior colliculus, IC (bottom) → auditory thalamic nucleus vMGB (middle) → primary auditory cortex, A1 (top). The neuronal responses are shown versus γ: note that in A1, but not in the lower auditory stations, the responses peaked strongly for the natural value of γ=1. (c) Population distribution of preferred values of γ across all the neurons recorded in IC (bottom), vMGB (middle) and A1 (top). A1 neurons exhibited a strong preference for the natural value of γ=1, suggesting that A1 neurons might play a particular computational role in processing natural sounds.

Other studies have focused on the observation that natural sounds,[2] including speech and music, exhibit a $1/f$ power law in their temporal envelope spectrum (Voss & Clarke 1975; Theunissen & Elie 2014). That is, the amplitude of the sound envelope, A, varies with frequency, f, as follows: $A(f) \sim 1/f^{\gamma}$, with $\gamma \approx 1$. Interestingly, human listeners judged music produced with a $1/f$ amplitude-spectrum (γ=1) to be more pleasant than either white noise (γ=0), which was judged to vary "too randomly," or $1/f^2$ noise (γ=2), which was judged to vary too slowly and to be "too correlated" (Voss & Clarke 1975, 1978). Schnupp and colleagues examined how neurons along the ascending auditory pathway respond to $1/f^{\gamma}$ sounds with varying values of γ (Garcia-Lazaro et al. 2006, 2011). They found many neurons in the ferret primary auditory cortex (A1) that were tuned to γ=1, exhibiting a higher firing rate for γ=1 than for lower or higher values of γ (figure 6.3b, "A1"). The results were different in the auditory thalamus, where neurons did not show strong preference for a particular value of γ (figure 6.3b, "vMGB")—and they were even more different in the inferior colliculus, where neurons preferred fast-changing stimuli with γ=0.5 (figure 6.3b, "IC"). The same tuning to γ=1 in A1 was also seen in the population results summarized in figure 6.3c. These results thus demonstrate that the preferential coding for naturalistic sounds—that is, sounds with a

$1/f$ amplitude-spectrum that are typical of natural soundscapes outdoors—arises first in the primary auditory cortex, A1, and is *not* inherited from subcortical stations (e.g., the IC or thalamus). This emergence of selectivity for naturalistic sounds in A1 goes hand in hand with the emergence of more complex STRF filters in A1 (Theunissen & Elie 2014), suggesting that an important computational role of A1 may be the processing of natural sounds.

Finally, it is noteworthy that similar results were also found in the primary visual cortex (V1) of awake monkeys, where neurons exhibited higher coding efficiency and information rates for visual stimuli that showed temporal statistics with $1/f$ power spectra, as compared to $1/f^0$ stimuli or $1/f^2$ stimuli (Yu et al. 2005). Taken together, these data suggest that when presenting both A1 and V1 neurons with $1/f^\gamma$ stimuli, cortical neurons tend to exhibit tuning to the natural value of $\gamma \approx 1$. This may indicate a fundamental preference of sensory cortices to encode natural stimuli in both audition and vision.

In closing this section, it is important to stress that I do not propose that there is anything "magical" about natural sounds. However, the evidence reviewed here clearly suggests that the neural coding of natural sounds is very different from the coding of artificial simple auditory stimuli, as classically used for auditory research. In other words, the classic bottom-up program of sensory neuroscience—which seeks to use simple stimuli in order to elucidate (eventually) the neuronal processing of complex sounds—has severe fundamental limitations. These limitations stem from the strong nonlinearities, massive top-down feedback, context dependence, and history dependence of the auditory system.

To overcome them, I propose to take a top-down approach, using top-down knowledge to choose relevant natural sounds. It remains an open question whether the differences between responses to simple sounds versus natural sounds stem from the *naturalness* or from the *complexity* of natural sounds—in other words, whether there is a real difference between using a truly natural sound versus using an artificial sound of similar complexity to that of the natural sound (Nelken 2017). Advantages of using natural sounds include their higher behavioral relevance and their adaptive value for evolution. On the other hand, complex artificial sounds offer clear experimental advantages, as they can be parameterized and controlled. Thus, one approach would be to use batteries of complex artificial sounds as well as real natural sounds, and systematically compare neuronal responses to these two classes. To ensure the acoustic similarity of the artificial sounds to the natural sounds, one could use natural sounds to define smaller subspaces of the full possible stimulus space, and then generate ensembles of complex artificial sounds that uniformly cover these restricted subspaces. Another approach would be to take fully natural sounds and start to systematically simplify them. In either case, it is clear that an intensive future effort is needed in order to understand the neural coding of natural sounds.

6.2 Vision: Natural Images

Studies of the visual system using natural images have suggested a very similar picture to that reviewed in this chapter for the auditory system, including the following points:

1. The responses of V1 neurons to simple stimuli (such as bars, gratings, or spots) are usually rather poor predictors for their responses to natural movies.

2. V1 neurons are extremely sensitive to the spatial and temporal context present in natural scenes, and they exhibit various forms of strong nonlinearities and temporal-history effects.
3. Efficient coding theory can predict the shape of V1 receptive fields based on natural image statistics.
4. Natural images maximize information transmission in V1 neurons.

This large body of work on the neurobiology of natural vision has been summarized in a number of excellent reviews (Simoncelli & Olshausen 2001; Kayser et al. 2004; Carandini et al. 2005; Olshausen & Field 2005). For the sake of brevity, I will not attempt to review it all here; rather, I will refer readers to these reviews. Instead, I will briefly highlight here four points that are of particular interest, and then move on to discussing a related issue: the interaction between natural images and natural eye movements.

First, the understanding of 3D visual scenes requires context information and higher-level knowledge, which is absolutely essential for disambiguating the 3D layout of surfaces and 3D objects, because the 2D image itself simply does not contain enough information to do so (Lewicki et al. 2014). The 3D binocular visual information is already encoded by neurons in early visual cortices (including V1) and is widespread throughout the cortical visual hierarchy (Parker 2007; La Chioma et al. 2019). Some of this binocular information may be transmitted into early visual areas through top-down feedback from higher visual cortical areas. It is therefore crucial to study both bottom-up and top-down visual processing together. Olshausen and Field (2005) summarized it as follows:

> It is now time to dig deeper, using richer, ecologically relevant experimental paradigms, and developing theories that can help to elucidate how the cortex performs the computationally challenging problems of vision. Reductionism does have its place, but it needs to be motivated by functionally and ecologically relevant questions. . . . Given the extensive feedback projections to V1, in addition to projections from pulvinar and other sources, it seems unlikely that we will ever understand V1 in isolation. Thus, our investigation must also be guided by how V1 fits into the bigger picture.

This challenge has largely remained unaddressed.

Second, along the same lines, it is likely that higher visual areas, from V2 and beyond, are tuned to analyzing natural images even more strongly than V1. This was suggested by a number of recent studies from the lab of Tony Movshon, who showed that V2 neurons respond particularly vigorously to images with naturalistic textures, unlike neurons in V1 (Movshon & Simoncelli 2014; Ziemba et al. 2019). Thus, while the aforementioned reviews called for using natural or naturalistic images when studying V1, this is clearly even *more* imperative when going beyond V1. An extreme example of this point comes from face-selective neurons in primate inferotemporal (IT) cortex—a high visual area well beyond V1. These face cells respond extremely selectively to faces compared to nonface objects or body parts (Bruce et al. 1981; Hesse & Tsao 2020). Remarkably, these neurons are organized anatomically in a beautiful system of face patches, with each patch exhibiting distinct specializations for extracting view angle, gaze direction, identity, and other elements (Tsao et al. 2006; Hesse & Tsao 2020). The intricate selectivities of these neurons cannot be studied using simple stimuli, such as oriented bars—rather, they must be studied using the natural stimuli themselves.

Third, another evidence for the importance of studying natural visual stimuli comes from a series of studies by Dale Purves and colleagues, which demonstrated that the statistics of natural images strongly affect visual perception in humans. For example, the perceived length of a line has long been known to depend on the line orientation—a puzzling illusion that remained unresolved until Purves showed that this illusion can be explained by the statistical relationships between the length of lines on the retina and the 3D lengths of their real-world sources. By measuring the 3D structure of natural scenes using a 3D laser scanner, they showed that the average length of lines in the real 3D world—for example, the length of tree branches—is related systematically to the orientation of their 2D (retinal) projections in a manner that matched the psychophysical illusion extremely well (Howe & Purves 2002). Specifically, they showed that the 2D length of natural *vertical* lines (i.e., the length of their 2D projection, as would be seen on the retina) tends to match their true 3D length because tree trunks stand vertically, parallel to the plane of their 2D image. By contrast, the same 2D length for *diagonal* natural lines strongly underestimates their true 3D length because diagonal lines in a 2D image often correspond to diagonal or horizontal branches that are *not* parallel to the 2D plane, or to horizontal receding roads in 3D, whose true 3D length is much longer than their perceived 2D projection due to perspective. The study, however, went well beyond this basic intuition, showing a remarkable quantitative fit to human psychophysics. This study powerfully illustrated that our visual system is strongly tuned to process real-life, 3D natural scenes, which contain real 3D surfaces.

Fourth and finally, a theoretical study was able to explain the tuning curves of monkey medial temporal (MT) neurons to 3D motion as arising from the projective geometry of real-world 3D motion onto the 2D surface of the retina (Bonnen et al. 2020), thus connecting natural vision with natural motion. Overall, it is clear that much more effort must be devoted to studying the neural representations of natural stimuli in the visual system, and in particular natural 3D surfaces and natural 3D motion. Remarkably little neurophysiological research has been done on this to date.[3]

While the studies discussed here focused on the *spatial* structure of natural images and natural scenes, and how this structure affects neural coding and perception, there is another important aspect of natural vision—namely, the *temporal* structure of eye movements, and in particular small fixational eye movements, which already have been discussed in chapter 5. How do the spatial and temporal aspects of natural vision interact? Michele Rucci and colleagues have investigated this question and found that natural images contain more power at *low* spatial frequencies (figure 6.4, dashed line), while fixational eye movements serve to enhance the power of *high* spatial frequencies (via the space-to-time transformation discussed in chapter 5: see the dotted line in figure 6.4). Remarkably, these two opposing tendencies precisely counterbalance each other so that the information arriving at the retina ends up having a white spectrum (figure 6.4, gray solid line) (Kuang et al. 2012; Aytekin et al. 2014; Rucci & Victor 2015). A signal with a white spectrum (temporally decorrelated signal) is known to be optimally efficient in terms of information transfer (Rieke et al. 1996). Therefore, this "behavioral whitening," achieved through eye movements, could enable a more efficient neural code and may have other major implications for coding in the visual system.

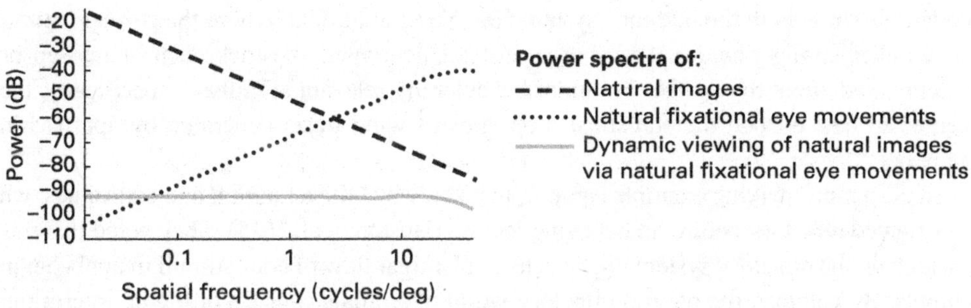

Figure 6.4
The spectrum of fixational eye movements precisely counterbalances the spectrum of natural images, effectively whitening the input to the visual system. Black dashed line: Spatial-frequency spectrum of natural images. Black dotted line: Spectrum of fixational eye movements measured in one subject using a particular range of temporal frequencies (the results were similar in other subjects and other temporal frequencies). Gray solid line: Effective input to the visual system, assessed by multiplying the dashed and dotted curves, showing the spatial spectrum of natural images as viewed via natural fixational eye movements. The flat shape of the gray solid curve shows that natural fixational eye movements serve to whiten the inputs to the visual system, which may yield substantial computational advantages. Adapted from Aytekin et al. (2014).

Consequently, when we think about "natural stimuli" in the visual system, we should always remember that this naturalness has two components, which interact mutually—the spatial structure of natural visual scenes and the temporal structure of natural eye movements—and these two must be studied *together*. Therefore, the "perfect" neurophysiological experiment on natural vision would be to record neurons from various stations of the visual system and to present the monkey with natural images, allowing the animal to freely move its eyes and scan the image, all while precisely measuring its eye movements, including fixational eye movements. In other words, the largely separate approaches of natural stimuli and active sensing should be combined into one approach and studied together within the same experiment. No such experiments have been done to date.[4] It would also be desirable to allow the animal to move freely and look at the stimulus from various angles because, as discussed in chapter 5, self-motion strongly influences the neural activity in visual areas. Thus, visual neuroscience clearly has a long way to go before we understand the neurobiology of natural vision.

6.3 Olfaction and Touch: Odor Plumes and Wind Gusts

Studies of other sensory systems have also suggested extreme adaptation of neural circuits for processing natural stimuli. I will mention just two examples, from the somatosensory and olfactory systems of insects. Rinberg and Davidowitz (2000) presented cockroaches with wind gusts that had a defined spectral shape (power spectrum of the temporal structure of the wind velocity). They found that, surprisingly, the firing rates of somatosensory neurons were invariant to the wind's total power—so long as the wind spectrum was conserved. Conversely, when they manipulated the wind spectrum, neuronal firing changed dramatically, with the firing rates being maximal for narrowband wind stimuli and minimal for wideband stimuli, even though the total power of the wideband stimuli was larger. Importantly, a narrowband wind spectrum is generated naturally by large approaching objects, such as predators. Thus, as in the auditory system (Bar-Yosef et al. 2002), here the stronger stimuli did *not* elicit stronger neuronal responses. Instead, neurons in the cockroach somatosensory system cared about the weaker but more behaviorally relevant stimulus—specifically, they seemed to best encode the statistical properties of wind gusts generated by approaching predators.

An even more striking example came from a study by Gilles Laurent and colleagues, who investigated olfactory coding in behaving locusts (Huston et al. 2015). They noted that most research on the olfactory system uses a constant laminar flow of odor stimuli in nonbehaving animals. By contrast, the natural olfactory world of animals consists of highly intermittent olfactory inputs, where the intermittency has two sources: the turbulent, intermittent non-laminar flow of olfactory filaments (figure 6.5a, left), and the intermittent olfactory-sampling behavior of the animal (e.g., sniffing in mammals or antenna flicking in crustaceans and insects: figure 6.5a, right). To elucidate the independent contributions of these two sources of intermittency, they used two experimental setups. In the first, they removed one of the antennas from a locust and presented the immobilized locust with two kinds of olfactory stimuli: either classical, laminar-flow pulsed stimuli, or turbulent, naturalistic stimuli. The time course of the stimuli was directly measured using an electroantennogram from the disconnected antenna, which was placed next to the locust. They found that the neural-coding principles for the turbulent naturalistic stimuli were generally quite similar to previous findings using highly controlled pulsed stimuli (Laurent 2002): Each olfactory filament elicited a local field potential (LFP) oscillatory bout in the antennal lobe of the insect's brain, and spikes of olfactory projection-neurons phase-locked to these oscillations and encoded the odor identity in a highly precise manner. Moreover, across repeated presentations of olfactory filaments, the neurons' firing rate decreased, but their phase-locking to the LFP oscillations increased, creating sparser but increasingly more structured neuronal responses.

After finding generally similar coding principles for natural stimuli and for controlled-but-unnatural stimuli, Laurent and colleagues turned to investigate how the animal's olfactory sampling affects neural responses (Huston et al. 2015). In their second experimental setup, they allowed the locust to move its antennae freely and measured the antenna movements in 3D. They found that in clean air, the locust slowly moved its antenna in large sweeps (similar to rodent whisking in free air), but when presented with an odor source, the locust shifted its antenna flicks to concentrate around the odor source and also increased the rate of its antenna flicking (figure 6.5b, black line). Strikingly, the increased rate of antenna flicking, 0.8 Hz, precisely matched the 0.8-Hz stimulus rate that was shown previously to optimize olfactory coding in locusts (Geffen et al. 2009). This optimal sampling behavior resulted in a dramatic improvement in the localization of the wind source by a neural decoder (figure 6.5c: small localization errors with odor [black] versus no-odor antenna-flicks [gray]). This study nicely connected the topics of active sensing (chapter 5) with natural stimulus statistics (this chapter), demonstrating that in fact, these topics are inseparable because the active-sensing behavior of the animal affects the stimuli available to the brain, and in turn, the available stimuli affect the animal's behavior: a closed-loop duet. This study also beautifully demonstrated the natural neuroscience approach—highlighting the power of combining natural stimuli and natural behaviors with more traditional, well-controlled, but unnatural stimulation.

Further, these studies in cockroaches and locusts underscore why we have to know the full behavioral repertoire and sensory world of a species—its so-called *Umwelt* (von Uexküll

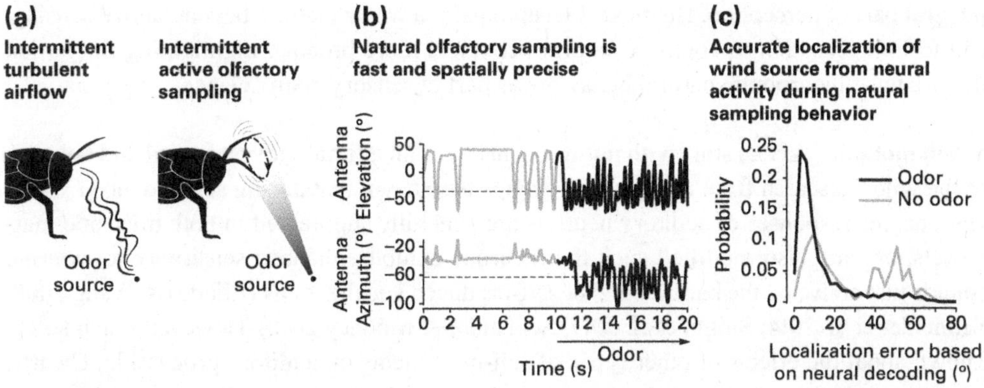

Figure 6.5
Olfactory coding of naturalistic odors in locusts. Adapted from Huston et al. (2015). (a) While typical experiments on olfactory coding use laminar airflow to deliver odors, real-life natural olfactory stimuli are in fact very turbulent and intermittent. Here, the two main sources of this intermittency are illustrated: the natural airflow is turbulent and comes in highly intermittent plumes (left); and animals actively sample the airflow intermittently, either using their antennae (right) or by sniffing, in mammals. (b) The antenna-flicking behavior of a locust changes drastically when the animal is presented with an odor (black lines) compared to without an odor (gray lines): the flicking becomes faster and spatially concentrated around the odor source direction (dashed horizontal lines). (c) Decoding error of wind-source position based on LFP recordings from the antennal lobe in the locust brain, performed during active-sampling behavior. The decoding became very precise when the odor was injected into the wind stream.

1920)—in order to elucidate the sensory neuronal responses of that species. If researchers hadn't wondered about the power spectra of intermittent wind gusts generated by approaching predators, or how locusts move their antennae, it wouldn't be possible to conduct these studies or to understand what these neurons are doing. The topic of active sensing, which was discussed in chapter 5, is an example of the things that we have to know when studying sensory systems. We have to know how animals move their sensory organs, such as their eyes, nose, whiskers, fingers, or antennae. As detailed in chapter 5, understanding these motor actions is absolutely crucial for understanding sensory perception. Perhaps even the movement itself is an integral part of perception (O'Regan & Noë 2001). But the importance of natural behaviors goes far beyond active sensing. As will be discussed in the next section, studying sensory neuroscience requires a comprehensive consideration of the full repertoire of animals' natural behaviors.

6.4 Scene Analysis, Sensory Ecology, and Ecological Psychology

Ultimately, the purpose of sensory systems is to guide behavior. As previously discussed, active sensing links sensorimotor behaviors with perception in a dramatic manner. This link happens for two main reasons. First, movements of the eyes, fingers, antennae, or other sensory organs strongly affect the properties and statistics of incoming sensory streams. In other words, active sensing dramatically affects the key properties of the sensory input itself. And, as we have seen, the brain's sensory systems are highly sensitive to such complex natural stimuli. Second, movements of the sensory organs also generate motor corollary discharges, which interact with the sensory inputs. As discussed in chapter 5, movements of sensory organs strongly affect neuronal responses, and might even form an integral part of perception. But next, I wish to paint a wider picture, beyond active sensing, and to link sensation to complex natural behaviors more broadly, highlighting the notion that we must incorporate natural behaviors as part of sensory neuroscience.

1. Self motion. Let us start with natural hearing as an animal moves around and behaves in the wild. Research from a number of groups has shown that during self-produced vocalizations, the responses of auditory neurons are generally suppressed in both mice and marmosets, and are also modified such that neurons exhibit enhanced sensitivity to external sounds that arrive on the background of self-produced vocalizations (Eliades & Wang 2008; Schneider et al. 2014; Singla et al. 2017; Schneider & Mooney 2018). However, much less is known about the effects of other types of self-movements on auditory processing. Clearly, whole-body locomotion strongly modulates the acoustic input, as well as the relevance of the sounds (e.g., when approaching the sound source versus moving away from it), so we may expect strong movement-related modulations in the auditory system. Indeed, Mooney and colleagues found that the responses of mouse auditory cortex neurons to sounds were strongly suppressed during a wide variety of movements, including locomotion and grooming—an effect that was mediated by direct projections to the auditory cortex from the secondary motor cortex, M2 (Schneider et al. 2014). Interestingly, this suppression had the opposite sign of the motion-induced enhancement of V1 visual responses, which was discussed in chapter 5 (figure 5.4b). A subsequent study (Schneider et al. 2018) showed that this suppression is learned, and serves as a cancelation mechanism ("negative image") for suppressing the predictable acoustic consequences of movement (see also Singla et al. 2017). Another study found different auditory responses of neurons in the dorsal cochlear nucleus of rats when the sound source was rotated relative to the rat, as opposed to when the rat was rotated at the same speed relative to the sound source: the acoustic input was verified to be identical in both conditions, and yet many neurons showed striking differences in their responses (Wigderson et al. 2016). The authors concluded: "The existence of a rich representation of both sound and body orientation so early in the auditory system may be the consequence of a strong evolutionary pressure for computing a representation of sound direction that is invariant to the animal's own movements." Such experiments have hardly been done to date in other regions of the auditory system, outside the dorsal cochlear nucleus and primary auditory cortex; hence, much more work is needed to elucidate the modulation of auditory responses by self-motion.[5]

Further, when an animal moves through the real world, all sounds appear against some acoustic background—a background generated by the animal's own movements, breathing, echoes, or environmental noises. There is no such thing as complete silence in the real world. Combining this observation with the dramatic electrophysiological finding that auditory-cortex neurons are often more sensitive to the *weak* sound, not to the strong sound in a mixture (Nelken et al. 1999; Bar-Yosef & Nelken 2007), it becomes clear that much more research should be conducted to investigate the neural coding of sound mixtures. It is almost inconceivable that there is hardly any systematic research on the key question of how sound mixtures are represented in the auditory system. Isolated, single sounds that suddenly pop up from complete silence—the typical type of stimulus used in auditory research over the past seventy years—are not very natural, certainly not for freely moving, freely behaving animals, which always produce self-generated sounds as they move around. It is, therefore, high time that we direct substantial effort toward studying the neural coding of natural sound mixtures.

2. Sensory history. The explicit effects of sensory history must be studied in much more detail, paying particular attention to its ethological relevance. This is true for any sensory modality. We have already mentioned a number of studies showing that sound history (i.e., temporal context) can strongly affect auditory neuronal responses. These dramatic effects lasted from tens of milliseconds to tens of seconds, or even minutes (Ulanovsky et al. 2003, 2004; Asari & Zador 2009). However, although these studies clearly showed that sound history matters a lot, they still used rather simple stimulation paradigms. But what if we consider specific sound histories that have a particular behavioral relevance to the animal? One such example comes from a study of the responses of bat auditory-cortex neurons to natural versus scrambled echolocation sequences (Beetz et al. 2016). These authors studied delay-tuned neurons in the echolocating bat *Carollia perspicillata*, and presented them either with a natural target-approach sequence recorded from a bat—where the delay between the pulse and the echo progressively decreased over consecutive pulses—or the same sequence, but with a scrambled order of the pulse-echo pairs. They found that for most neurons, the delay-tuning was substantially sharper for the natural sequence than for its scrambled version, indicating a strong effect of natural stimulus history on bat auditory neurons. There are also many examples of ethologically relevant sound sequences in other mammals. For instance, the sound of a car whose loudness increases over time implies something completely different for our personal safety than the same sound of a car that is slowly fading away. Or the gradual change in the forest sounds as a predator creeps closer. Further, long-term memory also plays an important role, and studies in monkeys suggested that long-term auditory memory is stored in the auditory cortex (Fritz et al. 2005). These examples illustrate that auditory processing, sound history, and auditory memory are likely strongly interwoven. However, there have been hardly any neurophysiological studies addressing the triple interactions between long-term auditory memory, auditory working memory, and the temporal-context effects of sound history.

Finally, it is noteworthy that similar history effects and memory effects manifest in the visual system as well. And in vision, too, these effects were hardly studied neurophysiologically.

3. Scene analysis. Arguably, scene analysis in its fullest sense is rarely incorporated into thinking about sensory neurophysiology, in either audition or vision. Lewicki et al. (2014) beautifully reviewed the large gaps between scene analysis in the natural world and typical laboratory studies; I will mention here just a few of the many points brought up in their extensive review. Consider a monkey or a jaguar in a South American rainforest. As the animal moves around the forest, it is constantly bombarded by a barrage of many sensory inputs from multiple sources, and it needs to rapidly understand the 3D layout of the environment: the locations of potential food items, conspecifics, predators, obstacles, safe paths, etc. Monkeys and jaguars, of course, possess highly developed brains, so, for simplicity, it is instructive to think of a much humbler creature that can perform quite comparable feats: the jumping spider (already mentioned in chapter 1). These are highly visual spiders that have several pairs of eyes, including a pair of large, frontally positioned eyes. Jumping spiders do

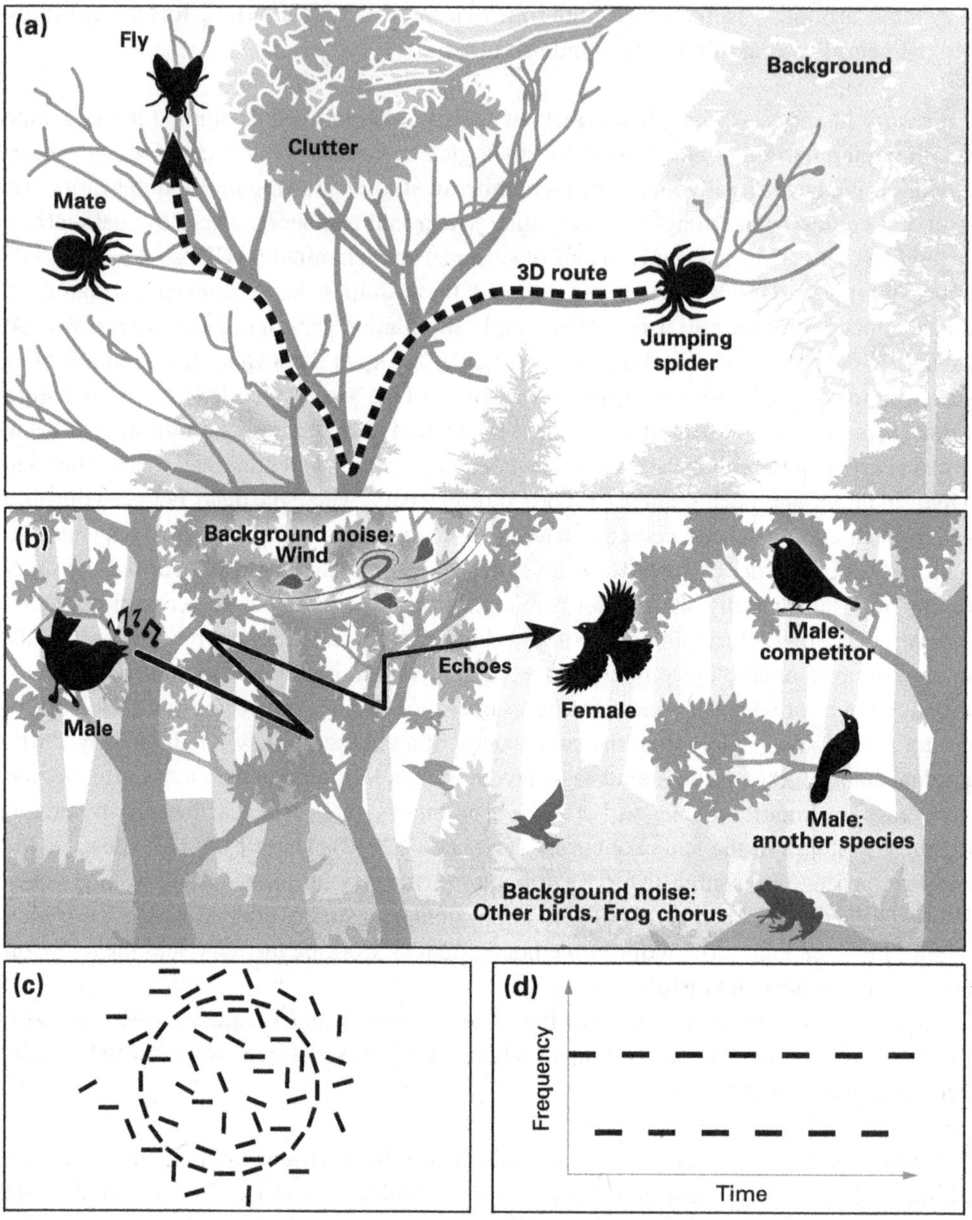

Figure 6.6
Scene analysis in the real world versus the laboratory. (a) The challenges of real-world visual scene analysis facing a jumping spider. (b) The challenges of real-world auditory scene analysis for a songbird. (c) Example of visual scene analysis as studied in the laboratory. (d) Example of auditory scene analysis as studied in the laboratory.

not build webs; rather, they stalk their insect prey and then rapidly attack it from a close distance. They have been shown to be able to plan a complex, 3D route to their prey (figure 6.6a). For example, a jumping spider may climb down from its perch, climb up another branch, and then make several correct choices at branch-bifurcations—even for branches that meander in 3D, one behind the other. Moreover, it does all of this in the face of visual occlusions and against a background of complex visual clutter that includes moving leaves that can easily obscure the little fly that the spider is hunting (Tarsitano & Jackson 1994). Remarkably, while the spider is on the move, it seems to know the 3D direction to its target at all times, even when the target is occluded, as is evident when it stops and turns its body to face the 3D direction of the target (Hill 1979).

Such a feat of 3D spatial perception, spatial memory, and sophisticated 3D route planning would baffle many humans, but the little jumping spider, with its tiny brain—the size of a poppy seed—is able to do this confidently and successfully. In other cases, the spider can spot a potential mate and needs to identify that it belongs to the same species. If it is a female that approaches a male, for instance, she will carefully inspect the courtship dance of the male and integrate the visual inputs with the vibrations created by his dance in order to decide if he is a suitable mate (Elias et al. 2003).

To achieve all of this, the spider must solve a set of difficult problems using a series of complex visual scans with its eyes. It needs to construct an internal representation of the 3D layout of the environment; it must separate foreground from background and clutter; it must identify a target; it must decide whether it wants to attack the target or mate with it; it needs to plan and execute a complex route in 3D; and it needs to integrate over time the changing visual information about the 3D scene and about the target location, all while on the move. Very few of these astonishing behaviors—or their analogs in mammals—have been studied in their full complexity in neuroscience laboratories.

Analogous problems occur in audition. Consider a female songbird in a forest (figure 6.6b). In the breeding season, she hears the songs of many male birds—some of them males of her own species and some from other species—and she needs to identify a potentially suitable mate based on the qualities of their songs (Gentner & Hulse 2000; Ballentine et al. 2004). She needs to do so despite many background noises, such as those of a frog chorus, insects, or wind. She then needs to fly toward the potential mate. Remarkably, the female is able to identify and track the song of the specific male that captured her interest, even in the presence of many other very similar songs that occur at the same time (Lewicki et al. 2014). Flight is a noisy affair, and the bird has to listen over her self-generated flight noises—and nevertheless she can locate the male over distances of dozens and sometimes hundreds of meters in dense vegetation. The song is also distorted by echoes—and in fact, the amount of echoes has been shown to provide useful information about the distance to the singer (Morton 1998; Lewicki et al. 2014): thus, as the female approaches the singer, the song becomes crisper and crisper. To complicate things further, the singer often stops singing. But this does not baffle the little female bird—songbirds have been shown to remember the 3D goal location over long periods of silence (Morton et al. 2006).

To do all of this, the bird needs to construct an internal representation of the 3D layout of the environment—just like the jumping spider—and similar to the spider, which integrates information from several senses (vision and somatosensory), the bird needs to integrate information across her senses (vision and audition). She needs to identify and localize a specific acoustic signature over background noise and self-generated noise and many competing similar sounds—a challenging example of the "cocktail-party problem." And, just like the spider, the bird needs to plan and execute a complex route in 3D space toward the singer. To do so, she must integrate over time, during flight, the rapidly changing visual information about the 3D scene and the target location. The latter behavior is a highly nontrivial feat: it is like trying to fly a drone through dense forest at high speed—something that is impossible for most humans. Yet birds do it all the time.

These are the kinds of real-world scene analysis problems that animals must solve on a daily basis (for more examples in other species, see Lewicki et al. 2014). But how do we typically study scene analysis in the laboratory? In the visual domain, simple 2D drawings are often used, where subjects need to do segregation and grouping, in the spirit of the Gestalt school of psychology (figure 6.6c). These are studied both psychophysically and at the level of neuronal processing (Gilad et al. 2017). Even if real photographs are used, these are nearly always static 2D images, requiring simple tasks such as 2D segmentation of objects from the background—a far cry from the much more complex task of reconstructing the 3D structure of the environment. Moreover, these static images are very different from the rapidly changing visual world of a jumping spider or flying bird.

In the auditory domain, streams of pure tones at different frequencies are often presented, and the subjects are tested on their ability to perform segmentation and grouping of these simple sounds into auditory streams (figure 6.6d). Here, again, a wealth of psychophysical results have been obtained over the years (Bregman 1994), and interesting neural correlates have been found in animals, which may reflect neural processes of segregation and streaming (Fishman et al. 2001, 2004). However, no background noises are presented in such studies, nor any self-generated movement noises, and the subject is never challenged with the need to plan a complex 3D route in the forest toward any of these sounds. Here again, the artificial, highly simplified laboratory tasks are a far cry from the real-world problems of 3D auditory scene analysis facing many animals in the wild, including songbirds, owls, bats, primates, and many vocalizing rodents. All these diverse animals must solve similarly complex problems of auditory scene analysis.

So what's wrong with using simple stimuli to try to dissect components of the scene-analysis problem? Visual and auditory segregation and grouping of simple stimuli (e.g., figures 6.6c and 6.6d), which have often been studied as proxies for scene analysis, indeed might be components of real-world scene analysis—although even that is far from being certain. However, as pointed out by Lewicki et al. (2014), the main problem with limiting ourselves to such "toy problems" (figures 6.6c and 6.6d) is that this leads us to underestimate the complexity of the problem, which is far, far more complicated than segregating streams of pure tones presented in complete silence. The subproblem of identifying targets against complex natural backgrounds has been given relatively little attention in sensory neuroscience. And even less attention has been paid to the bigger problem of reconstructing the 3D layout of the environment from 2D arrays of receptors (retinae). This is fundamentally a very ill-posed computational problem—namely, there simply isn't enough sensory information in the 2D retinae to uniquely solve the 3D structure of all the surfaces and occlusions and backgrounds in a complex 3D scene (Ulusoy et al. 2016). Consequently, the brains of animals and humans rely on an immense amount of top-down prior knowledge about possible 3D structures in the world, such as the 3D shapes of objects, the likely 3D shapes of occluded object parts, the structures of common visual backgrounds, and the 3D slant of the terrain (Lewicki et al. 2014). However, we know very little about the neural representation of this top-down knowledge, nor about the neural mechanisms of how this knowledge is incorporated into sensory perception.

To overcome the ill-defined problem of reconstructing the 3D world based on 2D retinal information, animals usually move through 3D space while scanning the world with their sensors. Such active-sensing strategies help disambiguate the 3D structure of the world, simply by looking at it from a variety of angles. Therefore, there is a tight link between the need to reconstruct the 3D structure of the world and the need to perform active sensing of the environment—and both of these problems are understudied in neuroscience. Even less studied is the link between the two.

Addressing the highly simplified problems of static 2D scene analysis—such as how to find a cat in a 2D image, as is done very effectively nowadays by deep neural networks—is likely only partially relevant for the real-world challenge of analyzing 3D scenes, which is a completely different computational problem. Moreover, object recognition (e.g., putting the label "cat" on an object in a 2D image) is not enough to drive the real-life behavior of a mouse that may be watching this cat. In order to survive, the mouse must rapidly analyze the 3D pose of the cat (is it resting or stalking?), its position within the 3D scene (is it close to me?), and its 3D movement trajectory (is it approaching me?), as well as more subtle properties, such as the eye gaze of the cat (is it looking at me?). If the mouse is to escape the cat, it also needs to have a good 3D representation of its environment, including possible shelters.

The problem of 3D scene analysis is therefore a key life-or-death problem for many species across the animal kingdom—predators and prey alike—from spiders and insects to rodents, carnivores, bats, and primates. Some headway on the problem of 3D scene analysis has been made in recent years in the computer-vision community, such as in projects of autonomous cars, but here, the problem is strictly limited to roads and their surroundings, which is a very specific and restricted subclass of 3D environments. By contrast, the wonder of biological sensory systems of humans and animals is that most creatures—even the humble spider, fly or honeybee—can cope with an astonishingly wide variety of different 3D environments. The engineers of Waymo, Toyota, Aurora, or Tesla are still quite far from achieving this level of robustness, generality, and flexibility. The problem of general real-life 3D scene analysis is therefore far from being solved in computer vision—and at the same time, we are far from understanding how the brain does it.

Finally, it is noteworthy that the 3D structure of the world, and all its 3D surfaces, are not necessarily represented in 3D Euclidean coordinates. In fact, the brain possibly employs some sort of dimensionality reduction to represent the 3D world efficiently; we really do not know. What *is* certain is that many animals perform 3D scene analysis very effectively, very robustly, and very rapidly—and based on what we know about the ill-posed nature of the computational problem, it is clear that biological sensory systems have to do this very differently (at the algorithmic level) from the artificial 2D toy problems that we typically study in the laboratory. To date, we simply have no idea how the brain of *any* animal is able to do 3D scene analysis. Therefore, it is time to approach scene analysis from a more complex, naturalistic, real-world perspective.

4. Sensory ecology. Another aspect of real-world sensory processing in the brain concerns the animal's knowledge about the physics of the world and how it affects the animal's sensory perception, a topic often dealt with in the field of sensory ecology. *Sensory ecology* is defined as the study of how organisms acquire and use environmental information, and it has been extensively covered in the beautiful books *Sensory Ecology* (Dusenbery 1992) and *Visual Ecology* (Cronin et al. 2014). Sensory ecology has been very influential in disciplines such as ethology, neuroethology, and behavioral ecology, but it has hardly penetrated mainstream neuroscience. But as I will argue, it *should* penetrate more, because sensory ecology has important implications for sensory neuroscience.

I will not attempt to discuss sensory ecology comprehensively here. Rather, I will just mention a few examples. One well-known example of a sensory ecology top-down influence in the visual system is illustrated by the so-called crater illusion. Take a look at figure 6.7a: the central shape on the left looks concave (like a crater), while the central shape on the right looks convex (like a hill), although these are in fact 180 degrees rotated versions of the same image. To convince yourself that this is an illusion, rotate the page by 180 degrees—the crater and hill will now flip. The common explanation for this illusion is that in our everyday lives, we are very strongly biased toward having a light source from above—the Sun—and therefore, our brain interprets the shadows in the left image as a "crater," or dip, that is illuminated from the top; while the shadows in the right image are interpreted as a "hill," or a protrusion that is also being illuminated from above by the Sun.

This example illustrates that our prior knowledge and experience with the physics of the world (in this case: the position of the Sun and shadows) is strongly integrated into our visual perception. However, it is unknown how this integration occurs neurophysiologically. Some clues to this enigma include an electrophysiological study that discovered in area V4 of monkeys a population of neurons that are sensitive to the direction of illumination (Hanazawa & Komatsu 2001); and an fMRI study showing that light direction is processed earlier than 3D shape in the human brain (Gerardin et al. 2010). But future work is required to truly elucidate the neural mechanisms of extracting shape from shading.

The animal's sensory ecology can also dramatically affect the physical structure of the visual system, adapting it to suit the ecological niche of the species. This can be illustrated by comparing the visual systems of predator versus prey animals, where the differences in visual ecology and their different behaviors strongly affect their visual systems (Walls 1942; Cronin et al. 2014). Many predators, such as cats, foxes, and owls, typically have their eyes placed frontally in the head, with both eyes sharing a large common portion of the visual field—a large "binocular overlap"—which in cats reaches a staggering 98 degrees (figure 6.7b, top left). This gives the cat good 3D perception, which is important for pouncing on prey, but it comes at the price of a large "dead zone" of 174 degrees behind the cat's head. By contrast, a terrestrial prey animal, such as the rabbit, does not care much about 3D perception at close distance. Instead, it is more important for the rabbit to have a surround vision of nearly 360 degrees, in order to look out for predators. And indeed, the eyes of the rabbit are placed much more laterally, leaving only 14 degrees of

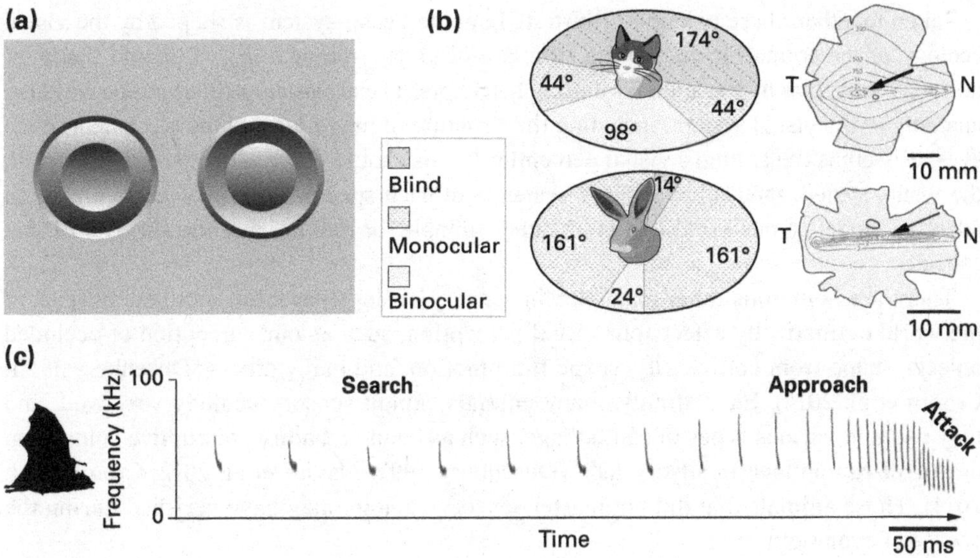

Figure 6.7
Sensory ecology. (a) The crater illusion: our perception of the central circle on the left as a "crater" and the central circle on the right as a "hill" is caused by our strong ecological prior knowledge that the shading is created by a light source shining from above—the Sun. This knowledge is acquired through our extensive experience with the statistics of the natural world. (b) The evolutionary history of each species has shaped its sensory systems, including the placement of the eyes, which determines the size of the binocular field and the blind zone behind the head (left). It also shaped the structure of the retina (right): Arrows on the right denote the area centralis in the retina of the cat (top) and the visual streak in the retina of the rabbit (bottom). Scale bars for the retinae, 10 mm. (c) Sensory ecology has shaped bat echolocation, including the acoustic structure of the search-approach-attack sequence of an insect-eating bat, which is shown here. (b) Adapted from Cronin et al. (2014); sizes of binocular and monocular fields were taken from Hughes (1971, 1977). (c) Adapted from Schnitzler et al. (2003); this is the same sequence shown in figure 5.14a, and it was replotted here for clarity.

blind zone behind the animal—but at the expense of having a very small binocular overlap of only 24 degrees (figure 6.7b, bottom left).

The different ecological needs of these two mammals also manifest in very different structures of their retinae. The cat has a focal area of high-acuity vision in its retina, with a particularly high density of ganglion cells—akin to the primate fovea, albeit less developed. This high-acuity region is called the "area centralis" (figure 6.7b, top right, see arrow)—and it is pointing forward, perfectly matching the 3D needs of a predator. By contrast, the rabbit has a horizontal "visual streak" across its retina—a linear area of high-acuity vision that points precisely toward the horizon (figure 6.7b, bottom right, see arrow); this visual streak allows detecting predators at the typical direction from which predators approach: that is, the horizon. Interestingly, the cat also has a horizontal visual streak, albeit a weak one (Fig. 6.7b, top right)—after all, most of the cat's potential prey is concentrated around the horizon. Even two closely related species may have very different visual systems if they live in different habitats. The red kangaroo, for example, has a very prominent horizontal visual streak in its retina (similar to the rabbit), matching its grassland habitat, which has a very prominent horizon. By contrast, the tree kangaroo has a circular area centralis in its retina, matching the complex structure of its forest habitat (Cronin et al. 2014).

Taken together, these examples illustrate how the visual system is shaped by the visual ecology of the animal, including the structure of its environment (e.g., Sun and shadows, horizon), as well as by the animal's natural behaviors. These can shape in dramatic ways the anatomy of the visual system, including the structure of the retina and the placement of the eyes, as well as the animal's visual perception.[6] This should come as no surprise: after all, the visual system (and other sensory systems) of each species have evolved to create an adaptive model of the world that is uniquely suitable for maximizing the survival of that species.

There are numerous other examples in sensory ecology revealing how the physics of the world dramatically affects our visual perception, such as our perception of occluded objects, shape from collinearity, shape from motion, and many others (Dusenbery 1992; Cronin et al. 2014). Interestingly, many animals exploit sensory ecology very well, and they use it in various types of camouflage, such as countershading, disruptive coloration, or mirror camouflage in silvery fish[7] (Dusenbery 1992; Davies et al. 2012; Cronin et al. 2014). Those animals that did not master sensory ecology may have perished during the course of evolution.

Sensory ecology is crucial not only in vision, but also in the other senses. Thus, the very reason why mammals use active sniffing in olfaction comes from the physics of diffusion: solving the diffusion equation shows that airborne odorants would simply take too long to enter our nose via diffusion—hence the need to actively inhale odors using sniffing (Dusenbery 1992; also see chapter 5).

Another example for the importance of sensory ecology comes from bat echolocation (Schnitzler et al. 2003; Ulanovsky & Moss 2008; Geva-Sagiv et al. 2015). Most bats use ultrasonic sonar calls. The use of ultrasound offers a distinct advantage over audible sound, because the short wavelength of ultrasound allows the bat to detect smaller objects, such as insects. But it also has the disadvantage of reduced detection range (due to the enhanced atmospheric attenuation of ultrasonic frequencies). For instance, when an insect-eating bat is attacking an insect, it rapidly changes its sonar pulse design in order to optimize sensory acquisition (figure 6.7c). As mentioned already in chapter 5, three phases of bat echolocation have been characterized: the "search phase," in the absence of targets, optimizes detection; the "approach phase" localizes a target, enabling approach; and the "attack phase" occurs just before reaching the target. The transitions between these phases are characterized by dramatic increases in the pulse emission rate (which provides a higher temporal resolution when maneuvering toward a target), an increase in the pulse bandwidth (improving the resolution of range measurement), and a decrease in pulse duration (which avoids temporal overlap between emitted pulse and received echo, thus preventing auditory masking) (Schnitzler et al. 2003; Geva-Sagiv et al. 2015). In addition, during the attack phase, many bat species lower their pulse emission frequency (see figure 6.7c, "Attack"), which, according to the classical piston model from physical acoustics, was predicted to widen their sonar beam, thus preventing them from losing sight of a rapidly maneuvering insect during the final attack (Mogensen & Møhl 1979). Indeed, recent studies confirmed experimentally the predicted beam-widening, by measuring the 3D sonar beam shape in freely flying bats (Jakobsen & Surlykke 2010; Jakobsen et al. 2013). Insects, of course, did not lag behind; they developed a suite of spectacular adaptations in their evolutionary arms race against bats (Ter Hofstede & Ratcliffe 2016). Mantises, for example, evolved ultrasonic hearing tuned to bats' sonar calls and have specialized neuronal circuits that serve the sole purpose of triggering evading maneuvers when the mantis hears the sounds of an echolocating attack sequence emitted by an approaching bat (Triblehorn et al. 2008). Tiger moths take acoustic countermeasures and produce special clicking sounds during the bat's attack phase—which, amazingly, are able to jam the bat's echolocation (Corcoran et al. 2009). Some bats, in turn, developed "stealth echolocation," to counter the insects' countermeasures (Ter Hofstede & Ratcliffe 2016).

Overall, our rich understanding of the sensory ecology of echolocation has been crucial for elucidating the bat's behaviors and has generated predictions for neurobiological experiments. These predictions, in turn, led to Nobuo Suga's discovery in the bat auditory cortex of delay-tuned neurons, and also of tracking neurons that adaptively track the shortening distance to the target (see chapters 2 and 5). Suga discovered these amazing sonar-analyzing neurons only because he had a good grasp of the sensory ecology of echolocation and of the bat's behavior. Likewise, understanding the sensory ecology of the bat-insect arms race has guided the neurophysiological study of the fascinating auditory systems of many insects (Pollack et al. 2016). These and many other examples demonstrate how important it is for students of sensory neuroscience to understand the sensory ecology of their animal model, in order to elucidate the animal's behaviors, as well as the neural basis of real-world sensory perception.

5. Ecological psychology. Finally, a discipline that is related to visual ecology is that of *ecological optics,* which forms a part of a larger approach of *ecological psychology.* Ecological optics and ecological psychology were founded by the psychologist James Gibson, who described them in many writings, most notably in his book *The Ecological Approach to Visual Perception* (Gibson 1979). I will discuss it here only briefly and then go into some more details on ecological psychology in chapter 7, which will focus on human studies. In his early work, in the 1940s and 1950s, Gibson focused on the sensory information available in the environment and on sensory invariants that exist in the environment. (Note that the focus on the information available in the environment is a common thread between ecological psychology and sensory ecology, although the former approach originated from psychology and the latter originated from ecology, they seem to have converged on some common themes.) I will mention here only three types of such sensory information that Gibson emphasized, and which are heavily utilized by many animals and humans: optic flow, surface texture, and the horizon.

Optic flow is created by the rotation or translation of an animal within the environment, whereby all the visual elements surrounding the animal are moving and creating an optic flow field around the animal. When we rotate while standing in one place, we experience a rotational optic flow: a unidirectional rotation of the entire scene (figure 6.8a: see arrows on the left). Forward translation is characterized by translational optic flow: an expanding optic-flow field in front of the animal (figure 6.8b: see arrows on the left).

Optic flow provides very rich information about the environment and the animal's movement within the environment. When moving forward, the magnitude of the translational optic flow (the length of arrows) conveys information about the *distance to surfaces.* Distal objects create weak optic flow, which is why the arrows in figure 6.8b,c (left) become shorter farther away. Nearby objects create strong optic flow: note the rock in figure 6.8c, which

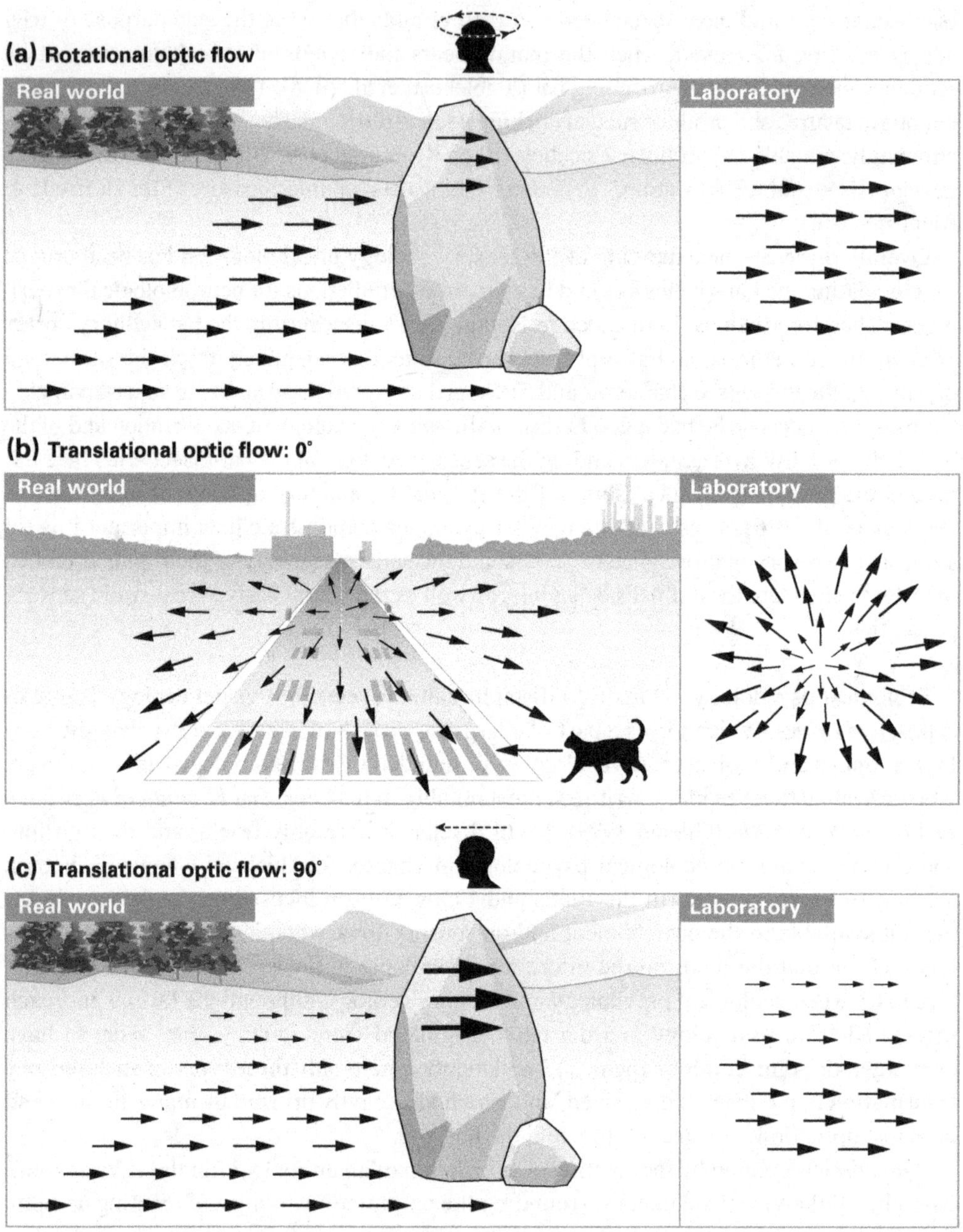

Figure 6.8
Optic flow provides animals and humans with surprisingly rich information about the structure of the environment, and about their own movements. (a)–(c) Left: optic flow fields (arrows) in a realistic scene. The optic flow is shown from the perspective of the *viewer* (e.g., the reader), when: (a) the viewer rotates the head, (b) walks into the scene ("into the page"), or (c) walks parallel to the scene (parallel to the page). Right: Simplified optic-flow stimuli as used in the laboratory.

elicits a particularly large optic flow that signals a nearby object. By contrast, rotational optic flow creates a uniform flow field because it depends only on the turning rate of the head. Hence, rotational optic flow is independent of distance and is therefore less informative about the environment (figure 6.8a). If a flying animal knows its flight speed, then the magnitude of translational optic flow beneath the animal provides precise information about the altitude above ground. Further, the epicenter of the focus of expansion of translational optic flow indicates the heading direction of the animal, as well as its expected point of impact. A pilot, for example, is guaranteed to hit the runway so long as he centers the optic-flow expansion epicenter on the runway (figure 6.8b, left). Importantly, the optic-flow epicenter shows the *true heading direction of the animal*, regardless of side winds or side currents, which are often very difficult to estimate. Thus, optic flow provides the most reliable information source on true heading direction (Gibson 1979).

Finally, if one considers the rate of expansion of translational optic flow (looming), one can estimate the *time to collision*, which is important for obstacle avoidance. Estimating the time to collision does *not* require knowing the animal's movement speed or the distance to the target—the time to collision can be derived *directly* from the optic flow (Lee & Reddish 1981; Dusenbery 1992). Remarkably, some neurons in the midbrain nucleus rotundus of birds were shown to encode the time to collision of a looming stimulus precisely, in a speed-invariant manner (Wang & Frost 1992). Moreover, the information derived from optic flow about time to collision was shown to explain extremely well the precise timing of the wing-closing behavior of sea birds (gannets) as they perform a plunge dive into water (Lee & Reddish 1981). Thus, optic flow seems to be important both for the visual neurons of birds and for their visually guided behaviors (Nakayama 1994).

Notably, the movement of individual objects (e.g., animals) can be easily detected even on top of the optic-flow field produced by self-motion. Consider, for example, the cat walking across the scene in figure 6.8b (left)—the local optic flow vector produced by the cat is very different from all its surroundings, making it easy for the observer to detect the cat. Thus, by separately analyzing wide-field motion and the motion of point targets, the visual system can disentangle the optic flow generated by self-motion from the localized motions of objects.

The honeybee is another expert user of optic flow, as was shown in a series of beautiful experiments by Mandyam Srinivasan and colleagues, where they manipulated optic flow in a variety of ways (Srinivasan 2011; but see Luebbert and Pachter 2024). They showed that honeybees fly exactly through the middle of narrow passages, and the researchers hypothesized that they do this by equalizing the optic flow on the left and the right. To test this hypothesis, they moved the visual pattern on one of the walls with or against the bees' flight direction, and the findings showed that the bees moved accordingly closer to or farther from the moving pattern, demonstrating that honeybees indeed use translational optic flow in order to center their flight through narrow passages (Srinivasan et al. 1991). This is a wonderful real-world strategy for negotiating multiple obstacles: The bee does not need to compute distances to objects, or even to identify them. All it needs to do is simply equalize the optic flow on both sides, which will ensure that it neatly passes between the obstacles. Interestingly, the same optic-flow-based strategy was also demonstrated in birds flying through narrow passages (Bhagavatula et al. 2011), suggesting that convergent computational principles may guide behavior across very distant species.

Honeybees also use the optic flow to land, by keeping the horizontal component of the optic flow constant as they approach the landing surface. They even know the expected time to landing from the optic flow, which they signal by extending their legs to land with precisely the correct timing. Remarkably, this optic-flow-based strategy ensures a smooth landing without any explicit knowledge about flight speed or height above ground (Srinivasan et al. 2000b).

Further, Srinivasan and colleagues were able to elegantly show that the bee's distance meter (odometer) is based purely on optic flow. The bee uses the integral of the optic flow during a foraging trip to assess the distance that it has flown, and this is the distance that it later signals to its hive-mates in the famous waggle dance (Srinivasan et al. 2000a; Esch et al. 2001). Not surprisingly, honeybees—and in fact, many other insects—possess special visual neurons that signal wide-field rotational optic flow and translational optic flow (Krapp & Hengstenberg 1996; Borst & Helmstaedter 2015).

But optic flow is used not only by humans, birds, and insects—the study subjects of Gibson and Srinivasan. In fact, a wide variety of animals rely heavily on optic flow in their behavior and possess neural circuits that encode optic flow. In vertebrates, the analysis of visual motion starts already in retinal circuits (Gollisch & Meister 2010; Borst & Helmstaedter 2015), and there is evidence that direction-selective ganglion cells in the mouse retina may be organized to optimally extract optic-flow information (Sabbah et al. 2017). Monkeys have two cortical visual areas—area MT and area MSTd—where neurons encode rotational and translational optic flow, respectively (Albright 1984; Tanaka & Saito 1989; MT is the medial temporal area, and MSTd is the dorsal region of the medial superior temporal area). Experiments that study the coding of optic flow in areas MT and MSTd of monkeys are typically conducted using reduced versions of optic flow, with no scenery or background landmarks. Instead, researchers present a set of random dots that are either moving uniformly in a parallel manner in a certain direction (figure 6.8a, right), mimicking rotational optic flow; or radially expanding dots (figure 6.8b, right), mimicking translational optic flow.

Tom Albright and Bill Newsome demonstrated that MT neurons are tuned to the direction of motion of rotational optic flow—that is, encode whether the dots (or other motion stimuli) are moving left, right, up, or down. Further, electrical microstimulation of MT has biased the judgment of monkeys toward the direction of motion encoded by the stimulated MT neurons, which suggests that the neurons' directional tuning is causally related to the monkey's subjective perception of the direction of motion (Albright 1984; Salzman et al. 1990). Area MSTd contains neurons that respond to either rotational or translational optic flow, with most neurons preferring a combination of both types of optic flow (Tanaka & Saito 1989; Duffy & Wurtz 1991, 1995). Further, MSTd neurons encode the direction of the epicenter of expansion of translational optic flow (namely, the location of the epicenter within the field of view), with different neurons having different preferred epicenter directions (Duffy & Wurtz 1995). Because real-life optic flow is created primarily by self-movement, we might expect that MSTd neurons will encode both optic-flow and vestibular signals—as both types of signals are generated by self-movement. Indeed, Dora Angelaki and colleagues have shown that MSTd neurons perform such multisensory integration and encode a weighted combination of visual translational optic flow and vestibular translational information (Gu et al. 2008; Angelaki 2014). This result supports the notion that MSTd neurons analyze the animal's heading direction.

Later studies showed that in addition to MSTd, several other areas in the parietal cortex of monkeys encode heading direction by combining optic flow and vestibular information (Angelaki 2014). Notably, the vestibular stimulation in Angelaki's experiments was delivered to the head-fixed monkey via large hexapodal acceleration machines on which the monkey and the visual screen were placed—the monkey was not *actually* locomoting. It would be crucial, therefore, to record from neurons in MSTd and related areas in head-fixed monkeys, and then compare these responses to the activity of the same neurons as the monkey moves around and performs real-world ambulatory movements, in order to definitively test the idea that these neurons encode the direction of ambulatory movement. Such an experiment would be a difficult technical feat, but it is becoming possible to achieve, given the technological advances in recent years (chapter 3). This experiment will thus combine the power of traditional, controlled approaches in neuroscience, with the power of the natural neuroscience approach.

The first step in this direction was recently done by Angelaki's group by recording in the MSTd of monkeys that were head-fixed but allowed to move their eyes in a target-catching task (Noel et al. 2022). Under these free-viewing conditions, they found a strong representation of eye gaze by MSTd neurons—a representation that is abolished in classical experiments with eye fixation. This shows, yet again, that unexpected neural signals are uncovered when the researchers allow the animal even a simple degree of freedom of naturalistic behavior (in this case, free-viewing), and it illustrates the importance of going to more naturalistic and less constrained experiments. Even more surprising results will surely be found when animals will be allowed to locomote completely freely.

Surface texture is another very rich source of information that is available in the environment. Gibson emphasized that surface texture allows perception of both the size of objects placed on the surface and the slant of the surface (Gibson 1979; Nakayama 1994). Let's start with *size*. An object of a given size covers *N* texture elements, regardless of the object's distance from the observer. One does not need to compute the distance to the object and then compute the object size based on the distance, as suggested by classical theories. Instead, argued Gibson, one can directly perceive the object size if the observer knows the size of the surface texture elements, such as the size of each flower in a field of flowers, or the size of waves out at sea. Likewise, Gibson showed that one can estimate the *slant* of a surface from the gradient in the size of texture elements—again, regardless of distance. Hence, both the absolute size of objects (in meters) and the absolute slant of surfaces (in degrees) can be directly perceived, in a *distance-invariant* fashion, by utilizing the textures of real-world surfaces such as fields, deserts, or seas. Indeed, both object size and surface slant are important *invariants* of ecological optics (Gibson 1979; Nakayama 1994). Studying them requires, however, that the experiment is done in a real-life situation: namely, that a real object is placed on a real, textured surface.

The *horizon* is yet another important invariant emphasized by Gibson (1979). Our vision is referenced to the horizon. Even when we cannot see the horizon, we know where it is—both because it is perpendicular to our sense of gravity (encoded by the vestibular system) and because as we move forward on a horizontal surface, the point of expansion of the optic flow in front of us specifies precisely where the horizon is. Gibson emphasized that the horizon provides surprisingly rich visual information about the world. As one example, all terrestrial objects of the same height are cut by the horizon at the same height ratio, regardless

Figure 6.9
The horizon provides animals and humans with surprisingly rich visual information about the structure of the environment and the sizes of objects.

of the object's distance (and thus regardless of its angular size): that is, all these objects have the same *horizon-ratio relation*. Importantly, the height at which the horizon cuts the object (e.g., figure 6.9, arrow) exactly equals the observer's height (more precisely, the height of the observer's eyes). Consider, for example, figure 6.9. Try to estimate the precise height of trees 1 through 4, in meters, based on this image alone, with no extra information. This may seem like an impossible job, but try it anyway. What is your guess? Most likely, you have estimated the heights of trees 1 and 2 at roughly 5 m and of trees 3 and 4 at roughly 15 m. The reason for this is that the horizon cuts trees 1 and 2 to 1/3 of their height, which means that these trees are 3 times higher than the observer's eyes (i.e., 3 times higher than *you*), which comes to about 5 m. By contrast, the horizon cuts trees 3 and 4 to 1/9 of their height, indicating a height of roughly 15 m for these trees (i.e., 9 times higher than the observer). In other words, our visual system contains prior information about our own height, which strongly biases the visual perception of naturalistic scenes. This is a remarkable observation, and yet it is completely unknown where and how in the brain the horizon-ratio relation is computed. The same is true for many other invariants of ecological optics, which were discovered by Gibson and his colleagues.

In his later work, in the 1960s and 1970s, Gibson developed a radical new approach to visual perception, called *direct perception* (e.g., Gibson 1979), which I will discuss in chapter 7. However, three key aspects of this approach are important to mention briefly now. One of these is the importance of *surfaces* and understanding their 3D layout. A second aspect is the importance of *movement through the world* and looking at the visual scene from multiple angles of observation, which Gibson emphasized as absolutely essential for disambiguating visual occlusions, and for understanding the 3D layout of surfaces in the environment. He therefore vehemently opposed the typical laboratory setups for visual psychophysics, with the head restraint, the bite-bar, and the visual fixation point. He strongly argued that these artificial methodologies are irrelevant for understanding vision in the real world—an extreme stance that I personally think goes a bit too far. But I definitely agree that we need a combination of classical laboratory methods and more naturalistic experiments where subjects can move around.[8] A third key aspect of Gibson's approach was the theory of *affordances.* He argued that we treat surfaces and objects in the world in terms of what they afford us—that is, what we can *do* with them. This point will be elaborated upon in chapter 7. Here again, it is clear that in order to test Gibson's ideas—and similar theories—we must conduct 3D visual-perception experiments in which subjects are free to move around and manipulate objects. In other words, we must perform experiments that are quite different from the dominant paradigm in visual neuroscience.

Finally, allowing animals to move freely is even more crucial if we wish to understand the interaction between sensory perception and other cognitive functions—especially in complex behaviors such as navigation, foraging, social interactions, and many other natural behaviors that require the integration of sensory information over space and time, as well as planning and attention. Dramatic effects will likely be found in sensory brain areas of animals performing such naturalistic complex behaviors. I will return to this issue in chapter 8.

6.5 Summary

In this chapter, I reviewed some failures and some successes. On the one hand, there are systematic major failures of the reductionist attempt to use simple classical artificial stimuli in order to explain neuronal responses to complex natural sensory stimuli. On the other hand, I reviewed some markedly successful adaptations of sensory neurons for processing natural stimuli in the auditory, visual, somatosensory, and olfactory systems. Further, I have reviewed evidence and arguments coming from a broad range of approaches—including *sensory coding of natural stimuli, natural stimulus statistics, scene analysis, sensory ecology*, and *ecological psychology*—all of which support the importance of studying sensory processing in the brain using more natural stimuli and more natural behaviors than typically used in present-day brain research. The founders of these various approaches came from very different backgrounds, including physiology, computational neuroscience, ecology, and cognitive psychology. Nevertheless, they were all inspired by the richness of the sensory information that is available in the real world: a richness that stands in stark contrast to the highly simplified, artificial stimuli used in most studies of sensory neuroscience. I think that researchers from these various disciplines and approaches should talk to each other more than they typically do today; this will be very beneficial, as they are essentially looking at the same problems from different angles. Importantly, all these approaches point to a common action plan for the future—the need to diversify the neurophysiological and mechanistic studies of sensory perception into a more naturalistic style by using natural stimuli and natural behaviors.

7

Human Cognitive Neuroscience and Natural Behaviors

Two of the most longstanding controversies in human cognitive neuroscience have focused on real-life versus laboratory studies of perception and real-life versus laboratory studies of memory. It is difficult to understate the ferocity of these debates. For example, Ulric Neisser, a major proponent of the real-life study of memory, who is often considered the founder of cognitive psychology, wrote that "the naturalistic study of memory is an idea whose time has come. . . . If X is an interesting or socially significant aspect of memory, then psychologists have hardly ever studied X. . . . the results of a hundred years of the psychological study of memory are somewhat discouraging. We have established firm empirical generalizations, but most of them are so obvious that every ten-year-old knows them anyway" (Neisser 1978). In his influential book *Cognition and Reality* (Neisser 1976), he criticized more broadly the classical tradition of psychological research, including research on human memory, and called for studying cognition under real-life, naturalistic conditions. In turn, proponents of the laboratory study of memory responded, for example in a paper titled bluntly "The Bankruptcy of Everyday Memory" (Banaji & Crowder 1989), which stated that "an obsession with ecological validity of method can compromise genuine accomplishments . . . we have not been able to see any new principles of memory emerging from the everyday memory studies."

Some of the examples that will be discussed in this chapter argue against this latter statement. Notably, the furious back-and-forth exchange between the two camps, accusing each other of the futility of the opposing approach, serves to illustrate the heated nature of the debate. A similar debate about real-world versus laboratory studies took place in the field of perception—particularly visual perception. Thus, many of the arguments on controlled-versus-naturalistic studies, which unfolded in tranquility within the animal-neuroscience community, have exploded in parallel in a much more agitated manner within the community of human cognitive neuroscience. But while the heat levels of the arguments were rather different, the content of the disputes was in fact quite similar in the two communities. It is thus worthwhile, in a book that focuses primarily on the animal literature, to discuss in brief the parallel debates in the human literature. This survey will necessarily be very concise, but hopefully instructive.

I will start with the fiercest controversy—that of real-life versus laboratory studies of human memory. I will then continue with the ecological approach to studying human perception, and will finish with a future outlook for naturalistic experiments in humans.

7.1 Real-Life Human Memory Research

The classical approach to memory research in the laboratory is based on using random lists of words, syllables, or pictures. These random lists are presented to a subject, who is asked to study them and subsequently is tested on cued memory recall of these items (Dudai 2002). As in all the preceding chapters, there are two main arguments for criticizing this methodology and advocating the adoption of a more naturalistic, real-life approach to the study of memory: an empirical argument and a theoretical argument.

The empirical argument focuses on experimental evidence showing that classical memory research using cued recall of word lists is only weakly related to real-life memory. For example, amnesic patients can perform surprisingly well in some laboratory tests of cued recall but be profoundly impaired in real-life memory—specifically, in their ability to acquire and recall new **declarative memories** (Buschke 1984; Shimamura & Squire 1988). Conversely, people with extraordinary real-life memory abilities are often mediocre on laboratory-based memory tests (Patihis et al. 2013). This suggests that different brain circuits may be engaged by laboratory-based versus real-life memories. And indeed, a meta-analysis of thirty-two neuroimaging studies has shown that very different brain regions were activated during memory retrieval in memory tests that involved cued recall of word lists, as opposed to recall of autobiographical memories (McDermott et al. 2009). In fact, these brain areas were almost completely nonoverlapping[1] (figure 7.1, note that the regions of overlap are tiny). This overwhelming difference suggests that fundamentally different brain circuits are underlying laboratory-based versus real-life memories. Moreover, the set of brain regions activated by real-life autobiographical memory recall—most notably the hippocampus—was in line with the neuropsychological literature on brain regions whose lesions underlie real-life amnesia. But importantly, the hippocampus was *not* activated in standard laboratory memory tests (McDermott et al. 2009). These empirical findings thus cast doubt on the classical rationale for using word lists to study memory processes. The assumption underlying the classical approach is that findings from such artificial laboratory tests would generalize and provide insights into real-life memory, as formulated by Endel Tulving (1983): "I know of no compelling reasons why the general principles that apply to remembering of mini-events in the laboratory should be greatly different from those governing the remembering of real-life experiences. Rememberers do not leave their brains and minds behind, or switch them off, when they enter the memory laboratory." Well, it turns out that there *is* a compelling reason—if Tulving were right, then we should have expected at least *some* overlap between brain regions activated by memory recall in laboratory memory tests and by recalling autobiographical memories, which would have pointed to a core set of memory centers and processes that generalize across the two approaches. However, the activated brain regions were so strikingly separate and nonoverlapping (figure 7.1) that it seems reasonable to conclude that the classical assumption of lab-based memory research simply does not hold.[2]

The theoretical argument against classical studies of human memory is that, at their very core, they lack *ecological validity* (Neisser 1978). There are, in fact, a number of different arguments here (table 7.1). First, the approaches differ dramatically in their timescales: months/years/decades for autobiographical memories versus minutes/hours/days for most laboratory tests—and this large difference in timescales means that different neuronal

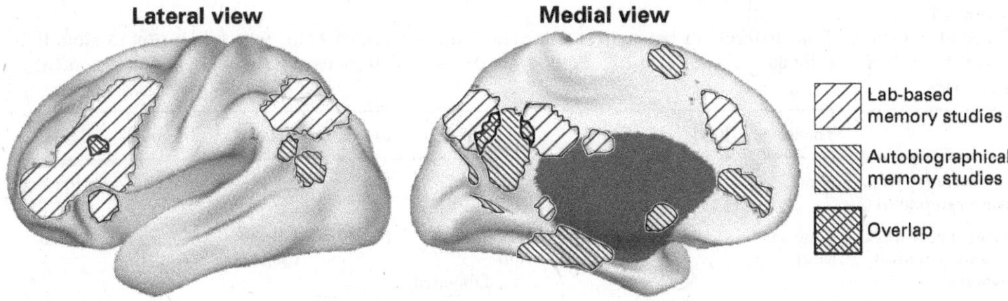

Figure 7.1
There is almost no overlap between brain areas activated in lab-based studies of cued memory recall of random word lists versus more naturalistic studies of autobiographical memory recall. Adapted from McDermott et al. (2009). The hatched areas summarize a meta-analysis of thirty-two fMRI studies, showing brain voxels that were consistently activated across studies during memory recall. Note the near-absence of overlap (very small cross-hatched areas), suggesting that different brain regions are activated by lab-based as opposed to real-life memory recall.

mechanisms are likely involved in real-life versus laboratory memories. Furthermore, memory over long timescales (i.e., years) usually cannot be studied in the laboratory, and hence nonclassical methods have to be applied in order to study real-life autobiographical memories. Second, life events are complex and spatially and contextually rich—none of which is captured by testing random word lists or random syllables.[3] Third, autobiographical memories are emotionally laden, and the emotional aspect is missing in tests that use random word lists. Fourth, autobiographical memories are self-focused, while the recall of a word in a laboratory test has no intrinsic self-relevance. Fifth, autobiographical memories are often confabulated, containing false memories (Bartlett 1932; Loftus 1979; Schacter & Loftus 2013). By contrast, subjects are not very likely to confabulate words or nonsense syllables in a laboratory test. Sixth and finally, autobiographical memories are often highly social in both content and context, while laboratory tests traditionally examine nonsocial memories. All these factors may have contributed to the fundamental differences in brain regions activated by laboratory memory tests versus autobiographical memories—as seen in figure 7.1 (McDermott et al. 2009; see also Roediger & McDermott 2013). Therefore, I would argue that memory research should not focus exclusively on using random lists; instead, it should diversify its approaches and experimental paradigms in order to bridge the gaps that are summarized in table 7.1.

The natural real-world types of memories in table 7.1 could be extended beyond autobiographical memories, to include additional rows about false memories, social effects on memory, expert memory performance, and more (all of these types of real-life memory are discussed later in this section). However, because the most-studied of these were autobiographical memories, this table focused on those. Two notes are due here: First, *all* of these natural real-life facets of memory differ fundamentally from memories of random lists. Second, the gaps listed in this table can be studied either in the lab or outdoors—there are merits to both options—but the important point is that they *should* be studied in order to get a fuller picture of real-life memory mechanisms.

Another interesting example of the differences between real-life and natural memories comes from working memory. *Working memory* is a limited-capacity part of memory that

Table 7.1
Summary of some of the differences between standard laboratory memory tests, which focus on random lists of words, syllables or images, versus real-life memory research, which focuses largely on autobiographical memories.

	Timescale	Items	Order	Recall	Emotion	Relevance	Context	Social
Lab memory tests using random lists	Short	Simple	Random	Binary (yes/no)	Low	None	Simple	Nonsocial
Natural real-world autobiographical memories	Long and short	Complex	Sequential, structured	Graded, often confabulated	High	Self-focused	Complex, real-life	Social

allows for storing information temporarily and manipulating it to support ongoing behavior and cognition (Baddeley & Hitch 1974, 2010). Two of the hallmark properties of working memory, that have been characterized by using random lists of words or images, are as follows. First, there is the fact that working memory has a *very limited capacity*: In his famous article "The Magical Number Seven, Plus or Minus Two," George Miller argued that we can remember 7 ± 2 random items, such as random words or random numbers—and in some cases, the capacity is even strikingly less than 7 (Miller 1956). Second, there are the *primacy effect* and *recency effect*: the fact that we tend to remember better the first and last items in a random list (Hurlstone et al. 2014). However, when it comes to real life, none of these fundamental findings on working memory seems to be particularly relevant. During a long conversation, or when reading a long text such as this chapter, you clearly remember a vastly larger amount of information than 7 ± 2 items. If you remembered only 7 items in a conversation, you wouldn't be able to talk intelligently about anything. The key difference is that real-life working memory always comes in context—unlike isolated random lists that have no context—and we use this context during conversation to remember much larger numbers of items than 7 ± 2. And what are the relevant items for working memory during a conversation: words, sentences, ideas, stories? Real-life conversations and behaviors occur on multiple timescales, from milliseconds to minutes and hours, unlike random lists that have a single timescale, which again argues against the ecological relevance of random-list laboratory experiments. The same applies to the primacy and recency effects: Do these effects mean that in a real-life conversation, we will best remember the last word or the first word of the conversation? This is clearly not the case: we best remember those parts of the conversation that were most relevant for us, or emotionally laden, or conferred an interesting idea. Random lists do not address relevance, emotions, or interest, so the random-list approach misses the most basic facets of what *is* best remembered in working memory during real-life behaviors and conversations. This is not to say that these classical findings on working memory are wrong—they are certainly correct, albeit specifically for random lists and in the confines of the laboratory. Rather, the point is that these classical notions are somewhat irrelevant for real-life behavior.

Together, these two examples—of autobiographical memory and working memory—suggest that focusing purely on highly simplified laboratory experiments with random lists can lead us astray. Such focus can lead us to overemphasize findings that have little ecological relevance. And even worse, it can lead us to develop concepts and theories that have rather limited relevance to natural, real life behavior and cognition, such as the notion that working memory has a very low capacity. This means that we must bring into the lab much more of real life in order to make the research more relevant.[4]

This point was made very nicely by Ulric Neisser. In a famous speech (Neisser 1978), which is considered the manifesto of the real-life movement in memory research, he described a whole host of interesting phenomena of real-life memory that had hardly been studied until then—and are still understudied today:

> You need only tell any friend, not himself a psychologist, that you study memory. Given even a little encouragement, your friend will describe all kinds of interesting phenomena: the limitations of his memory for early childhood, his inability to remember appointments, his aunt who could recite poems from memory by the hour, the regrettable recent decline in his ability to recall people's names, how well he could find his way around his home town after a thirty years' absence, the differences between his memory and someone else's. . . . One does not remember only events that one has personally experienced, but also those known at second-hand: things that have happened to other people. We learn from the experiences of our friends and acquaintances, and also from historical figures whose lives are somehow relevant to our own.

How can this rich array of memory phenomena be encapsulated by studying memory for random word lists or nonsense syllables? Neisser attacked forcefully the classical laboratory approaches and highlighted the successes of ethology in explaining natural behaviors of animals in the wild—and then, inspired by the ethologists, he called for a real-world approach to studying memory. He wrote:

> "Natural conditions" does not mean in the jungle or in the desert, unless that happens to be where our subjects live. It means in school and at home, on the job and in the course of thought, as carefree children and as reflective old men and women. . . . The psychological laboratory is the easiest of these settings in which to work, but it is also among the least interesting; we ourselves are the only people who spend much time there voluntarily. . . . I am not recommending an aimless accumulation of ecological minutiae. We will surely be guided by our general conception of human nature and human social life, as well as by more particular hypotheses about the phenomena we study. Without such conceptions and hypotheses, we can make little progress. The challenge will be to shift from testing hypotheses for their own sake to using them as tools for the exploration of reality. (Neisser 1978)

Neisser's ideas have been further elaborated by Bruce, who coined the term *ecological memory* and called for studying the real-life functions of memory, as well as its evolutionary and comparative aspects (Bruce 1985).

Neisser was perhaps a bit too extreme, in my opinion, in rejecting *altogether* all classical reductionist memory experiments. Reductionism is a graded affair—there are various levels of reductionism. On the one hand, there are highly reductionist laboratory assays, which allow detailed mechanistic studies at the expense of ecological validity. However, while allowing studying the mechanisms of some aspects of memory, the classical reductionist laboratory assays likely miss some key mechanisms altogether because, as summarized in table 7.1, many basic mechanisms are simply not at play in highly reductionist experiments.

On the other hand, there are very naturalistic nonreduced behaviors that feature high ecological validity, perhaps at the price of less detailed mechanistic study. However, importantly, natural behaviors do engage many more key mechanisms and thus can reveal new phenomena that are simply not at play in reduced classical experiments. So as usual, it's a trade-off. I think there is something to be learned from all these levels of reductionism. We need a variety of approaches. Nevertheless, I agree with Neisser's strong criticism of the state of affairs in his time, when memory research was almost *only* highly reductionist. Much more diversity, and in particular much more real-life memory research, were needed. And indeed, following Neisser, the study of real-life memory—that is, memory acquired, used and investigated in natural settings—has truly taken off since the 1980s (Dudai 2002). This approach has focused on investigating memory phenomena that are relevant to daily life using familiar and/or meaningful stimuli, such as autobiographical memories, narratives, and movies.

In one such study, Hasson, Davachi and colleagues (2008a) studied memory encoding in subjects that watched a novel movie stimulus—an episode of a TV sitcom—while their brain activity was scanned using fMRI. Three weeks later, they tested the subjects for their memory of the narrative content of the movie. They then computed, voxel by voxel, intersubject temporal correlations of brain activity between pairs of subjects, using the fMRI scans taken when they watched the movie. These correlations were computed separately for time epochs in which both subjects correctly recalled the movie details and time epochs where they did not remember the content. They found a number of brain regions that were strongly correlated between subjects during remembered periods, as compared to nonremembered ones. Some of these regions are typically found in traditional studies using random image sequences as memoranda, such as the parahippocampal gyrus. Surprisingly, however, some brain regions that are not traditionally associated with episodic memory encoding have emerged from this analysis, including the medial prefrontal cortex, temporo-parietal junction, and superior temporal gyrus—regions strongly implicated in social cognition.

Because the naturalistic movie stimulus contained many social interactions, this study suggested that the formation of memory for social situations and social interactions also involves the social brain network—a network of regions that extends far beyond the classical brain regions found in traditional memory research using random word lists or random image sequences (Hasson et al. 2008a; Adolphs 2009; Yeshurun et al. 2021). The authors concluded, "Given that much of what we encounter and value on a daily basis is fundamentally social, these findings highlight the important role that social cognitive processes play in episodic memory formation, which to date has been relatively underappreciated in the cognitive neuroscience of memory" (Hasson et al. 2008a).

Social processes affect memory also in a rather different way: social pressure and interpersonal influences may permanently affect long-term memory (Bartlett 1932), a process known as "memory conformity." This process has far-reaching societal implications, such as for political propaganda, commercial advertising, or the reliability of eyewitness testimony in court (Schacter & Loftus 2013). The neuronal correlates of long-term memory conformity were investigated by Yadin Dudai and colleagues (Edelson et al. 2011). They presented groups of five subjects with a movie, and after three days, they tested them individually on their memory of movie details. Four days later, subjects underwent an fMRI scan while their memory was manipulated by leading each person to believe that the other four participants remembered movie details differently. This social manipulation was extremely efficient, inducing subjects to erroneously conform to the majority opinion in 68 percent of the trials, as opposed to only 16 percent without social manipulation. Dictators over millennia have known how efficient social manipulation is, of course—but nevertheless, it is remarkable that such high efficiency was reached in a laboratory setting. The authors then found that fMRI blood-oxygen-level-dependent (BOLD) activation of the amygdala, as well as amygdala-hippocampus functional connectivity, were strong predictors of whether memory conformity would be long-lasting after two weeks or not, and they showed that these activations were specific to social manipulations (conformity), as opposed to nonsocial manipulations of memory. These results suggest that the amygdala—an important region for emotional processing in the brain—is also a key hub of the interface between social cognition and memory formation. Regrettably, classical memory research using random items—syllables, word lists, or images—has little to say about such interesting natural facets of human memory.

Another example of real-life memories that are difficult to capture by traditional memory assays are *flashbulb memories*: the exceptionally detailed and vivid "snapshot" of a surprising and emotionally arousing moment (e.g., a moment when shocking news was first heard or directly experienced). Flashbulb memories are thus a particularly salient subtype of autobiographical memories. Tali Sharot, Elizabeth Phelps, and colleagues studied the flashbulb memories of subjects who were in New York City during the September 11, 2001 (9/11) terrorist attacks (Sharot et al. 2007). They conducted their study three years after the 9/11 events, and found that subjects who were in downtown Manhattan on 9/11, close to the World Trade Center—who experienced the sights, sounds, and smells of the attack up close—remembered the 9/11 events more vividly and in more detail than they remembered other personally important events from the summer of 2001 (such as a wedding or a summer vacation). By contrast, subjects who were in midtown Manhattan on 9/11—and thus had much weaker firsthand experiences of the attack—did not exhibit a more vivid memory than for other personally important autobiographical memories from summer 2001. Concomitantly, the downtown group, but not the midtown group, showed enhanced fMRI activation of the amygdala when recollecting the events of 9/11. This study thus supported the notion that flashbulb memories are formed by separate mechanisms that involve the amygdala, but only if the experience was exceptionally emotional. As some of the downtown participants said in this study: "I saw with my own eyes: The towers burning in red flames, noises and cries of people . . . The explosion caused everyone in the area to automatically duck for cover" (Sharot et al. 2007). Random word lists are unlikely to evoke such vivid emotional feelings or to tap into the neural networks that are activated by emotional memories.

Another direction taken by researchers of real-life memory is to study the *memory performance of experts* and its neural substrates. One example of this is the study of navigation in London taxi drivers, who are experts in navigation and spatial memory (Maguire et al. 1997). I will come back to this example in section 7.3.

These examples illustrate some of the directions taken by real-life memory researchers, who study autobiographical memories, flashbulb memories, false memories, social effects on memory, expert performance, and much more. Note that these studies did not venture physically into the real world: they were not conducted on city streets or in the forest; rather, they were carefully conducted in the laboratory. However, their subject matter involved aspects of real-life memory that were hardly ever tested in the first 100 years of memory research. Other studies did venture into the real world. For example, one case study video-documented two days in the life of a subject to obtain ground truth objective information about every single event during that time, and then tested the subject months and years later to assess the accuracy and confidence of her autobiographical memories, while performing neuroimaging in an fMRI scanner (Mendelsohn et al. 2009). They found that over long time periods, of months to years, the subject came to accept an increasing number of false memory details as true—and concomitantly, growing portions of the brain exhibited activity that was correlated with confidence rather than with accuracy of the memory. Another study (Nielson et al. 2015) used a smartphone "life-logging" system that took thousands of photographs from the subjects' daily lives over a timescale of 1 month and spatial scale of 30 km. This study found that during subsequent memory recall that was cued by these photographs, fMRI multivoxel activation patterns in the hippocampus were more similar for events that occurred closer together in space and time. Importantly, all these studies maintained good levels of experimental control even while studying real-life aspects of memory.

Notably, even the biggest opponents of the real-life memory research movement (Banaji & Crowder 1989) endorsed such an approach, saying: "Now, nobody would deny that, other things being equal, the [condition] in which ecologically valid methods are used to achieve generalizable results is the best situation in which to find oneself." And then, when discussing some classical experiments in social psychology, they commented: "Several of these studies incorporated laboratory manipulations that mimicked the real world and concurrently maintained the control necessary to infer a causal relationship." Many other authors have likewise underscored the need for experiments that would examine real-life aspects of memory while offering good control, calling for coexistence of the naturalistic approach and the laboratory-based controlled methods (Klatzky 1991; Tulving 1991; Cohen & Conway 2007). Indeed, as I have emphasized in previous chapters of this book, the best approach for understanding the neural bases of behavior is to study a natural or naturalistic behavior while compromising as little as possible on the level of control (i.e., to be at the top-right corner of figure 1.2). Suitable behaviors for such studies can be identified by performing careful observations of natural behaviors outside the laboratory (i.e., in the real world). The key challenge for natural neuroscience—and for real-life memory research—is to identify those behaviors and situations that maximize naturalness while minimizing the loss of control: a min-max principle of scientific research.

7.2 The Ecological Approach to Human Perception

Sensory perception is another major area of human cognitive psychology, in which debates raged about real-life versus laboratory approaches. The debate focused primarily on visual perception. In this section, I will mostly discuss ideas and debates surrounding James Gibson's *ecological optics,* which I already started discussing in section 6.4. Beyond their historical interest, these ideas remain crucial, to this day, for questions of real-life vision.

Ecological optics is part of a broader approach of *ecological psychology*, which also encompasses Ulric Neisser's real-life memory research. Ecological optics was described most comprehensively in Gibson's book *The Ecological Approach to Visual Perception*

(Gibson 1979), where he highlighted the information available in the environment—such as optic flow, surface textures, and the horizon—as discussed in detail in section 6.4. It is noteworthy that Gibson discovered optic flow and the rich information that it carries for visually guided behaviors because he very seriously considered the natural behaviors of humans and animals within their environment.

Gibson emphasized perception as an active process, based on a *direct pickup of available information* (Gibson 1979). Three key aspects of his approach were as follows: first, the importance of *surfaces* and understanding their 3D layout; second, the importance of active *movement* through the world and looking at a visual scene from multiple directions; and third, a key aspect of Gibson's approach was the theory of *affordances*. He argued that we treat surfaces and objects in the world in terms of what they afford us (i.e., what we can *do* with them). Some examples of affordances include "portable," "graspable," "wearable," "seatable," "hide-able" (affords hiding), or "navigable." Gibson suggested that when we see an object, we do not care so much about its physical properties, such as its reflectance or color; what we do care about is its affordances. Is it a food utensil? A tool? Is it something I can write with, like a pen or a pencil? Is it a weapon, such as a knife? The difference between a knife used for spreading butter and a knife that can kill you is very subtle in terms of reflected photons—but the affordances of these knives are very different, and the ability to rapidly determine the affordance of a knife could mean life or death. Similarly, determining whether a place affords hiding or whether it is navigable may be crucial for survival.

How fast do we identify such affordances? An interesting study suggested that we perceive affordances such as *navigability* extremely rapidly (Greene & Oliva 2009). Human subjects were presented with photographs for a very brief time, followed by maskers. Surprisingly, the subjects were able to categorize global properties of the scene (such as navigability, concealment, or naturalness) much faster than lower-level properties (such as whether the scene contained a mountain or a lake). Amazingly, subjects were able to categorize the navigability of the scene (i.e., whether it contained a clear navigable path that is free of obstacles) after an average presentation time of only 36 ms—much faster than the 50 ms that it took them to categorize the semantic content or the low-level features of the scene. This result is in line with many studies showing that humans perceive the gist of a complex scene extremely rapidly (Thorpe et al. 1996; Fei-Fei et al. 2007). It is also in line with the "reverse hierarchy theory" of Shaul Hochstein and Merav Ahissar (Ahissar & Hochstein 1997; Hochstein & Ahissar 2002): This theory suggested that we first perceive high-level information about the scene ("seeing the forest before the trees")—a vision at a glance, which is done rapidly and automatically—and only later do we perceive low-level visual details through focused attention and slow, conscious scrutiny.

It is thus clear that a prominent aspect of our visual system is that it can rapidly perceive the affordances of things in our environment. However, nothing is known about how this is done in the brain. To understand whether and how affordances are represented in neural circuits, it is crucial to conduct real-life-like experiments in the laboratory, where the subject can actively manipulate the world (i.e., *do* things). This can be accomplished using one of two approaches, both of which became technically possible only in recent years (chapter 3). One approach is to record neural activity in the visual system of animals or humans that are moving around freely in the environment and manipulating objects—doing different things with them. Another approach is to use high-quality VR systems—which are becoming available for both animals and humans—where realistic-looking scenes and objects can be created and manipulated by the subject's avatar, while the content of the three-dimensional (3D) scene can be controlled precisely by the experimenter. At least in humans, advanced immersive-VR systems can deliver very believable 3D experiences. The combination of these approaches may open the door to a future systematic study of the neural basis of ecological optics.

While many researchers shared Gibson's enthusiasm for the importance of real-life vision and the ecological information available in the environment (as discussed in detail in section 6.4), there was another suggestion by Gibson that has sparked a huge debate. He suggested a "direct perception" theory, whereby the world is perceived directly, with no internal representations or computations (Gibson 1979). Thus, Gibson's ideas consisted of two key pillars. The first pillar was his direct-perception approach and denial of the need for internal representations and computations in the brain. This notion was strongly opposed by many researchers. For example, Shimon Ullman, in his famous rebuttal article "Against Direct Perception" (Ullman 1980), provided a detailed theoretical analysis against direct perception and argued for the importance of internal representations—and computations performed over these internal representations—to 3D visual perception. He gave a number of examples where our brain generates percepts that are not really there in the real world—visual illusions—so they must reflect internally constructed representations (e.g., consider illusory figures such as the Kanizsa triangle or shape from shading, as shown in figure 6.7a). One can think of many other phenomena that would be very difficult to explain without internal representations, such as dreams, hallucinations, or even bistable perception, in which the brain randomly switches between perceiving an image in one of two possible ways (e.g., the Necker cube). In fact, Gibson's idea that there are no internal representations and no computations in the brain seems so unlikely that I tend to think that he did not mean this in the literal sense. More likely, he meant to argue against the idea that the brain is performing computations akin to the classical computer-science algorithms of the time, so it's not *any* computations that he argued against, but specifically computerlike computations.

The second pillar of Gibson's ideas was the importance of natural stimuli and ecology, as was discussed previously, which many researchers, including Ullman, supported wholeheartedly (Ullman 1980; and S. Ullman, personal communication). In fact, many opponents of the direct-perception idea were sympathetic to Gibson's second pillar—the ecological pillar—and called for merging the real-world insights of ecological optics with the computational/representational approach and information processing (Ullman 1980). As a possible example for this merger, we may consider Gibson's own idea of optic flow, which he discovered and studied extensively. Although optic flow is reflected by neuronal activity at the periphery—in the retina—one would need to perform *computations* in order to extract anything useful from this raw input. For example, we discussed in section 6.4 that there are neurons in the MSTd cortical area of monkeys that represent the direction of the focus of expansion of translational optic flow (Duffy & Wurtz 1995), thus signaling the upcoming point of collision, which may be crucial for collision avoidance. Such high-order selectivity does not seem to exist in the primate retina (Heukamp et al. 2020), and hence it must be extracted from the retinal input via some computation. It is very difficult to imagine that achieving such selectivity to the imaginary focus of expansion does not require *any* computations, even if they are not performed by classical

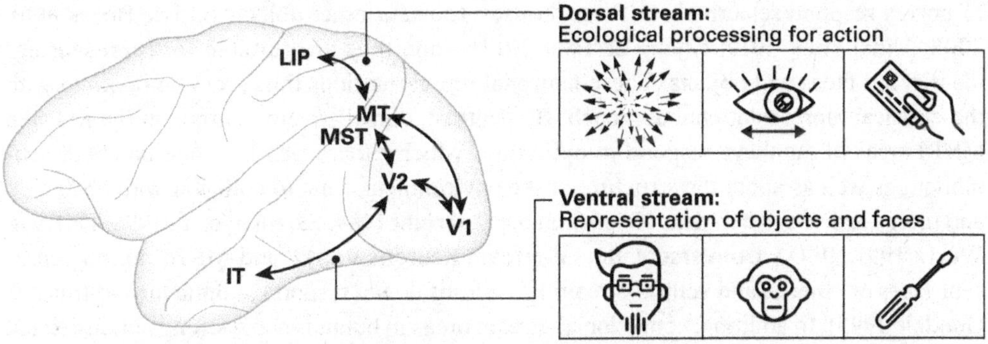

Figure 7.2
The ventral and dorsal streams of visual processing have very different functions: the representation of objects and faces in the ventral stream versus the sensory processing of action in the dorsal stream. The drawing of the monkey brain is based on Lehky & Sereno (2007). IT, inferior temporal cortex; LIP, lateral intraparietal cortex; MT, medial temporal area; MST, medial superior temporal area; V1, primary visual cortex; V2, secondary visual cortex.

computerlike algorithms, but rather by neural networks. Indeed, there are multiple layers of neural networks between the retina and MSTd, allowing rather complex computations—and the challenge is to understand which computations are done, and how.[5]

Ullman's approach, which is the dominant one in cognitive psychology, is often called the *computational/representational approach* or the *constructivist approach* because the key premise of cognitive psychology is that brain function in general relies on the construction of internal representations and on computations performed over these internal representations (Neisser 1967; Ullman 1980; Marr 1982). The basic idea of the computational/representational approach is that the brain takes simple elements, such as the oriented bars of Hubel and Wiesel, and uses bottom-up computations to construct internal representations of complex objects, such as the ones seen in the inferior temporal (IT) cortex. These internal representations are then used heavily in vision: thus, you do not see an apple; rather, you perceive it through your filter of an internal representation of an apple. This notion is in stark contrast to Gibson's ideas, whereby you directly perceive visual scenery and objects in a manner that is directly compatible with guiding action.

The debate between the direct-perception ecological approach and the computational/representational approach has raged over the decades, ever since the publication of Gibson's last book in 1979. An interesting possible resolution to this debate was suggested by Joel Norman (2001, 2002), who pointed out the possible link between the dorsal stream of the visual system—which is concerned primarily with visual guidance of *action*—and Gibson's ecological approach; and likewise the possible link between the ventral stream—which is concerned primarily with the representation and recognition of shapes and objects—and the computational/representational approach. The dorsal stream is often called the "how/where stream" or "action stream," mirroring the key emphasis of the ecological approach on the link between action and perception, while the ventral stream is often colloquially called the "what stream," mirroring the emphasis of the computational/representational approach on representing visual objects (figure 7.2).

Two main lines of evidence support this notion. First, neurophysiological findings reveal very different neuronal selectivities in the two visual streams. Ventral-stream neurons in the

IT cortex respond selectively to faces, houses, tools, or other objects (Op de Beeck et al. 2008; Vidal et al. 2010; Chang & Tsao 2017)—and thus are suitable for representing/identifying faces and objects. These neuronal representations thus seem compatible with the classical representational approach. By contrast, dorsal-stream neurons in the MT and MSTd areas of monkeys respond to optic flow, which carries rich information about self-motion, as well as about the structure of the environment, time to collision with obstacles, and more, as discussed in detail in section 6.4 (Albright 1984; Salzman et al. 1990; Duffy & Wurtz 1991, 1995). Conversely, dorsal-stream neurons in MT and MSTd do *not* represent faces or objects, and ventral-stream IT neurons do *not* respond to optic flow (Milner & Goodale 1995). In addition, some dorsal-stream areas in humans show strong action-related fMRI BOLD activation to movie clips depicting action (object manipulation), but such action-related activity is not seen in the ventral stream (Shmuelof & Zohary 2005).

Interestingly, in another dorsal-stream region, the lateral intraparietal (LIP) area of monkeys, neurons are involved in encoding and generating saccades toward an attended location (Bisley & Goldberg 2010)—which seems very different from the function of MT or MSTd. However, these dorsal-stream areas in fact do have something in common: Both saccades (encoded in LIP) and optic flow due to self-motion (encoded in MT and MSTd) are related to the *active pickup of information*—the hallmark of Gibson's theory (which is very closely related to the modern notion of active sensing, as discussed in chapter 5). Thus, the dorsal stream may be important for a Gibsonian pickup of visual information by a mobile agent that is moving through the environment and is actively looking at things.

Second, neuropsychological findings on some patients with a damaged ventral stream, suggest that they are totally capable of inserting a card into a rotating slit, matching the hand's grasp to the object to be grabbed, or avoiding obstacles while walking—but they do this with no conscious awareness of the slit, the object, or the obstacles: an intriguing example of perception for action without apparent object representation (Milner & Goodale 1995; Norman 2002). In Gibson's terms, this may reflect the pickup of the affordances of an object such as the slit ("card-insertable") or the cup of coffee (graspable, drinkable) without encoding an explicit representation of the object itself. Conversely, some patients with lesions in the dorsal stream exhibit severe deficits in moving toward a visual target or acting upon it, but they have absolutely no problem with the visual identification and discrimination of objects (Milner & Goodale 1995). Thus, according to Norman (2002), both the action-oriented ecological approach and the computational/representational approach are (at least partially) correct—which is why it is so difficult to settle the debate: in fact, they coexist in different parts of the brain (figure 7.2). Further, Norman proposed a dual-process approach to visual perception, whereby "much of our day-to-day pickup of visual information is carried out by the dorsal-ecological system without involving much conscious awareness . . . [it] allows the organism to function within its environment, that is, Gibson's affordances. The ventral-constructivist system, on the other hand, is a "higher" system that deals with the interface between the visual input and cognition, and we are normally conscious of its output" (Norman 2002). While Norman's suggestion has received substantial criticism (see the commentary on Norman 2002),[6] it may still provide a useful reconciliation of some of the debates surrounding Gibson's ideas: Gibson perhaps was correct about the dorsal processing stream, while his opponents were correct about the ventral stream.

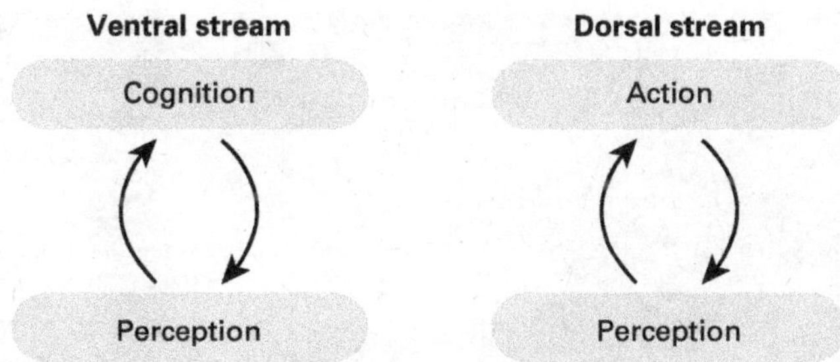

Figure 7.3
The ventral and dorsal streams of visual processing may have different interactions with higher brain functions: the ventral stream interacts heavily with cognitive processes, while the dorsal stream interacts with motor action (Ullman 1996). Two points are noteworthy here. First, there are direct anatomical connections between some dorsal-stream visual regions and motor/premotor regions (see figure 5.2a). Second, by "action," I mean non-eye-related movements, such as locomotion of the body or reaching with the hand—the types of actions that seem to interact primarily with the dorsal visual stream. However, there is another type of action—eye movements—that is important for vision more broadly, for both the dorsal and ventral streams (see chapter 5).

One may expand Norman's suggestion to other sensory systems. Dual sensory streams for action and for object perception seem to exist not just in the visual system—where the evidence for dual streams is extensive in both monkeys and humans (Ungerleider & Mishkin 1982; Milner & Goodale 1995)—but possibly also in the auditory system (Romanski et al. 1999; Bizley & Cohen 2013). If so, then dual brain processing, with one system processing information for action (possibly subserving the ecological approach), and another system for object-representation (subserving the representational approach) might be a general feature across sensory modalities.

A related but somewhat different view of the dorsal versus ventral visual streams suggests that they are not "Gibsonian" as opposed to "computational/representational," but that their functions differ in more complex ways. In particular, it is not clear whether the dorsal stream indeed requires simpler or fewer computations than the ventral stream, as would be expected from Gibson's direct perception. For example, computing the focus of expansion of optic flow, which signals the point of collision, may require neural circuitry that is rather complex—perhaps even as complex as the circuitry needed for categorizing objects. (Notably, artificial deep neural networks that successfully compute optic flow are as complex as deep networks that can categorize objects (Ilg et al. 2017), suggesting that optic-flow computations in the brain may require networks in the dorsal stream that have similar complexity to the object-related networks in the ventral stream.) Nevertheless, the functions of the two visual streams are quite distinct: the dorsal stream does *perception for action*, while the ventral stream does object perception, or more generally, *perception for scene understanding* (Milner & Goodale 1995). Accordingly, a key difference between the two streams is the involvement of cognition (Ullman 1996): The ventral stream requires constant interaction of perception with *cognition*, while the dorsal stream does not need to interact as much with cognition—at least not high-level cognition—but it does require constant interaction of perception with *action* (figure 7.3).

To illustrate the interaction between vision and cognition for the purpose of scene understanding, we may consider the scene depicted in the painting *The Raft of the Medusa* by Théodore Géricault (figure 7.4). To understand this scene, it is not enough to simply identify

Figure 7.4
The Raft of the Medusa by Théodore Géricault. Scene understanding of this painting requires elaborate interaction between visual perception and cognition.

the different people using the face-selective neurons in the ventral visual stream. True understanding of the scene requires identifying the corpses and processing the facial expression and postures of the people—which, through a ping-pong between visual perception and cognitive processing, reveals a story of shipwreck, despair, and death. Then you shift your gaze to the two people waving pieces of cloth: To whom are they waving? Their frantic motion suggests that there might be a glimpse of hope in sight. This realization immediately leads you to do a visual search, which reveals the tiny mast of a ship on the horizon—and this percept instantly turns upside-down your entire scene understanding: from a story of total despair to a story of hope and imminent salvation.[7]

By contrast, the dorsal visual stream, according to the view shown in figure 7.3, is related to action, and is less in need of such complex ping-pong between perception and cognition. Instead, the dorsal stream requires mostly a complex ping-pong between perception and action, as well as the cognition that enables action. In particular, the two most important types of actions that are mediated by the visual system are *navigation* in space and *manipulation* of objects. Both of these functions may thus be key to understanding the dorsal visual stream—and in order to elucidate their neurophysiological basis, it would require conducting electrophysiological recordings of neurons in the dorsal visual stream during navigation (as already discussed in section 6.4) and during object manipulation.

Finally, as noted earlier in this chapter and in section 6.4, Gibson emphasized the importance of self-movement—the *actions* of the subject—as part of visual perception. He argued that the function of perception is to guide action, but also conversely, action and movement are necessary for perception. Thus, Gibson (1979) wrote: "So we must perceive in order to move, but we must also move in order to perceive."[8] He therefore vehemently opposed head-fixed experiments as being almost irrelevant to real-life perception. Gibson's ideas subsequently generated a whole host of other theories that emphasized the importance of action as part of perception—so-called *dynamic theories of perception*—which are tightly related to notions of active sensing (see chapter 5). These theories include, in addition to Gibson's own theory of *direct perception*, O'Regan and Noë's theory of *sensorimotor contingencies* (O'Regan & Noë 2001); van Gelder and Port's theory of *dynamic perception* (van Gelder & Port 1995); and Ahissar's theory of *closed-loop perception* (Ahissar & Assa 2016). A full discussion of all these (and related) theories of perception is beyond the scope of this section (some of these ideas were discussed in chapter 5). However, it is important to stress that *all* these theories emphasize that a crucial factor for normal visual perception is the human's (or animal's) own movements: eye movements, whole-body locomotion, and the ability to inspect 3D objects and surfaces from multiple viewing angles.

The best way to test these ideas mechanistically would be to conduct electrophysiological recordings from the visual system of freely moving animals that are locomoting and inspecting real 3D objects from various angles, and then compare the neuronal responses to those obtained under a passive, head-fixed presentation of photographs of the same objects taken from various angles.[9] Can visual neurons tell this difference between active looking versus passive presentation? We may perhaps expect such a difference, for two reasons. First, active self-movement was shown to strongly influence the activity of V1 neurons (in the mouse), as discussed in detail in chapter 5, and hence active visual exploration and locomotion may evoke very different neuronal responses in V1 than passive stationary visual presentation. Second, a likely way to piece together the 3D structure of a 3D object would be to use information about viewing directions, as encoded by 3D head-direction cells (Finkelstein et al. 2015; Laurens et al. 2016; Angelaki et al. 2020), and goal-direction cells (Sarel et al. 2017), and combine it with the visual snapshot information seen from the various angles. But the activity of the head-direction system is quite different between a condition of free-locomotion and active looking from different head directions, versus a condition of head fixation when looking at a screen from just one head direction. Therefore, and because the head-direction system sends strong polysynaptic inputs to the visual system (Furtak et al. 2007; Taube 2007), we might expect important differences in the activity of neurons in the dorsal visual stream between active looking versus passive, head-fixed presentation of images.

Thus, it would be very important to conduct freely moving electrophysiological recordings systematically along the dorsal visual stream in monkeys—especially in the MT and MSTd areas—as well as in lower regions, such as V1. Such experiments have not been done to date. Of equal importance would be to conduct similar experiments also in humans (epileptic patients): to record from dorsal-stream neurons, while the subjects are freely locomoting and actively inspecting objects from various angles. Again, such experiments have not been done. Thus, we are in a sorry situation, whereby one of the major debates in the history of sensory perception—the ecological approach of Gibson and his followers versus the mainstream approach of constructivist/representational perception—has yet to be tested properly at the level of brain circuits and electrophysiological responses of visual neurons. To do so would require taking the natural neuroscience approach—conducting visual-neuroscience experiments under conditions that are close to the animal's natural behavior.

7.3 Human Natural Neuroscience: From Movies to Experiments Outside the Lab

Both Neisser, in his real-world memory research, and Gibson, in his ecological psychology, emphasized the critical importance of conducting naturalistic behavioral experiments; however, neither of them had any real interest in the nuts and bolts of brain mechanisms. They inspired generations of real-world memory researchers and vision scientists, but they were both practically silent about the possible neural circuits that may underlie memory or visual perception. Recent years, however, have seen a growing trickle of real-world experiments in human neuroscience that *do* look into the brain, driven by approaches such as Gibson's and Neisser's, as well as by recent technological advances that enable studying the human brain under more naturalistic conditions. In chapter 3, I reviewed some of the technological revolutions in *animal research*, which enabled the development of natural neuroscience; here, I will focus on research in humans and discuss both conceptual and methodological advances that now enable more real-world brain studies in humans. I will *not* claim here that we need to conduct all our experiments outside the lab: this is indeed desirable in some cases, but this is the end of the spectrum. There are many other ways to study the brain in a naturalistic manner, *inside* the lab, using more real-life approaches than classically done, in order to engage the brain fully.

One such conceptual advance is to use more naturalistic sensory stimuli, such as movies. The basic rationale is that in standard fMRI experiments, subjects are lying passively in the scanner and are shown static pictures—neither of which is very engaging. A Zoom meeting is most likely much more interesting and engaging than a standard experiment in human cognitive neuroscience—and as we all know from the COVID-19 pandemic, Zoom meetings are rather boring and do not engage the brain very well. One clear way to better engage the brain is to present interesting movies. The use of movies was already mentioned in section 7.1 with regard to studying memory, but rich movie stimuli also have been used for studying vision. This approach, which requires nonconventional fMRI experimental designs and analysis methods, was pioneered by Uri Hasson and Rafael Malach. In their first study (Hasson et al. 2004), they presented subjects with 30 minutes of a highly engaging movie—Sergio Leone's *The Good, the Bad, and the Ugly*—and then compared the temporal correlations between subjects' brain activations. They searched for matching voxels in all the pairs of brains, which were significantly correlated over time (synchronized), and they found strong intersubject correlations that were widespread over large portions of the cortex. These correlated regions included the classical sensory areas (visual and auditory), but they extended far into high-order regions such as the superior temporal sulcus, retrosplenial cortex, and cingulate cortex—well beyond the classical areas that are typically activated in standard fMRI experiments, which use discrete, passively presented images or sounds. In another analysis, Hasson and Malach identified events in the movie that evoked particularly strong intersubject correlations. Some of these events, consisting primarily of emotionally arousing actions, such as gunshots and explosions, or surprising turns in the movie's plot, activated large swaths of the cortex. Other events elicited intersubject correlations that were restricted to category-selective portions of the cortex. For example, close-ups of the faces of the characters elicited strong intersubject correlations in cortical face-selective areas. Likewise, buildings and outdoor scenes elicited strong intersubject correlations in building-selective areas.

This approach also allowed the authors to find unexpected selectivities. For example, in the middle postcentral sulcus, they found strong intersubject correlations in response to scenes depicting motor manipulations of objects with the hands—such as loading a gun (Hasson et al. 2004). In later studies, Hasson and coworkers used movie stimuli that were scrambled at different timescales and found a hierarchy of temporal receptive fields across the human cortex—from rapid responses in early sensory areas, to a selectivity for timescales of dozens of seconds in higher areas (Hasson et al. 2008b). Similar findings were obtained by using the soundtrack of narrated stories, which were scrambled at different timescales (Lerner et al. 2011). Here again, early auditory areas responded to the momentary acoustics of the stimulus, while higher areas parsed the story into longer and longer units of meaning, such as sentences, or even entire paragraphs. The longest timescales were found in the medial prefrontal cortex and precuneus.

The authors also directly compared fMRI responses to audiovisual movie clips versus responses to spoken narrated stories, which represented one of two schemas (eating in a restaurant versus going through an airport). They found many cortical areas that represented the multiple timescales of the schema, regardless of the details of the story or the modality of its presentation (Baldassano et al. 2018). Overall, the studies of Hasson and his colleagues demonstrated that using naturalistic stimuli for fMRI studies (e.g., movies or spoken narratives) reveals insights into brain representation of complex events and relationships that are not revealed by simple, well-controlled classical stimuli.[10]

A different benefit of movie stimuli was demonstrated by Malach and colleagues, who used an 8-minute segment of an engaging movie to study fMRI-based functional connectivity patterns in the visual cortex of human subjects (Wilf et al. 2017; Strappini et al. 2019). "Functional connectivity" refers to the correlation patterns between different brain areas based on the fMRI BOLD signal, capturing which brain areas tend to activate together. The authors found that spontaneous functional connectivity patterns measured during rest (resting-state connectivity) resembled the functional connectivity patterns elicited when subjects viewed the naturalistic movie—but they did *not* resemble the patterns elicited when subjects viewed a scrambled version of the movie, static image of faces and objects, or classical artificial visual stimuli (such as flickering checkerboard wedges and expanding rings). These results were found both in early retinotopic areas of the visual cortex and in high-order visual areas. These findings suggested that the spontaneous connectivity patterns across the human visual cortex reflect the statistics of cortical coactivation during natural vision. In other words, the adult human cortex seems to be wired to respond most effectively to real-life natural stimuli. Notably, conceptually similar results also were found in animal studies using ensemble recordings of many neurons (Berkes et al. 2011).

Movie-viewing experiments are much more naturalistic than traditional fMRI experiments, but they are still not very natural and are very different from ecological vision, in that they require the subject to lie motionlessly inside a scanner and avoid any movements of the head.[11] Moreover, these experiments decouple the visual input from the subject's head and eye movements—a highly unnatural *open-loop* situation, which is very different from the *closed-loop* condition of natural vision, wherein the visual input is changing upon movements of the eyes, head, or body (chapter 5). Therefore, in order to test Gibson's ideas about the crucial role of self-movement and inspecting objects visually from a variety of angles, one would have to record human brain activity from subjects that are allowed to move around freely—or at least are free to move their heads and eyes (Stangl et al. 2023). Very few such experiments have been done. However, important technological steps toward this goal were taken recently by Nanthia Suthana and the neurotechnology company NeuroPace, who have developed methods for recording intracranial brain activity from freely walking, freely behaving human patients (Aghajan et al. 2017; Topalovic et al. 2020; Stangl et al. 2021, 2023). This system, however, can only record LFP, not spikes. Clearly, it would be important in upcoming years to develop methods that allow recording both spikes and LFP in freely walking human patients.

Another recent technological breakthrough came from the group of Matthew Brookes, who developed a miniaturized, wearable magnetoencephalography (MEG) machine that allows neuroimaging while subjects are freely moving their heads (Boto et al. 2018, 2019). By developing novel quantum sensors, they eschewed the need for liquid-helium cooling, which is the chief reason why conventional MEG scanners are very large and cumbersome (figure 7.5a). By contrast, the new, wearable MEG device allowed high-quality neuroimaging even as the subject was moving its head over a range of up to about 20 cm. This enabled, for example, an unobstructed brain measurement in a subject playing a ball game (figure 7.5b), which revealed the anticipated brain activation in the contralateral motor-cortex wrist area (figure 7.5c). In subsequent work, they expanded the scope to standing and walking scenarios and demonstrated the ability to perform neuroimaging while subjects walked naturally over distances of >1 m (figure 7.5d) (Seymour et al. 2021).

Like any MEG system, it has an important advantage over electroencephalography (EEG), in that it allows imaging subcortical structures—even those as deep as the hippocampus (Barry et al. 2019)—which is not possible with EEG (table 7.2). This wearable-MEG system may allow neuroimaging studies to be conducted in the near future in freely moving subjects, at millisecond-level resolution, addressing some of the key questions raised by Gibson. For example, how do we visually construct 3D surfaces and object shapes based on multiple views that the subject acquires while moving around and inspecting an object from different head directions? In which brain areas does this construction process take place? While this wearable MEG has the major drawback that it has very low spatial resolution of a few centimeters, averaging over millions of neurons—unlike intracranial recordings in epileptic patients, which allow single-neuron resolution—the wearable-MEG *does* have the great advantage of being noninvasive, and hence usable in wide human populations.

The wearable MEG, with all its advances, is still just a laboratory device: it requires a magnetically shielded room, and it allows relatively limited movement because large movements create large motion artifacts in the MEG signal. The most recent experiments showed that it is possible to walk distances of a little more than 1 m during recordings with the wearable MEG (Seymour et al. 2021). Despite the limitations in walking distance, the subjects *can* freely move and rotate their heads, which is a big step forward. This will allow, for example, for conducting immersive-VR neuroimaging experiments in subjects walking on a 2D omnidirectional treadmill—a large step toward performing more realistic VR experiments in

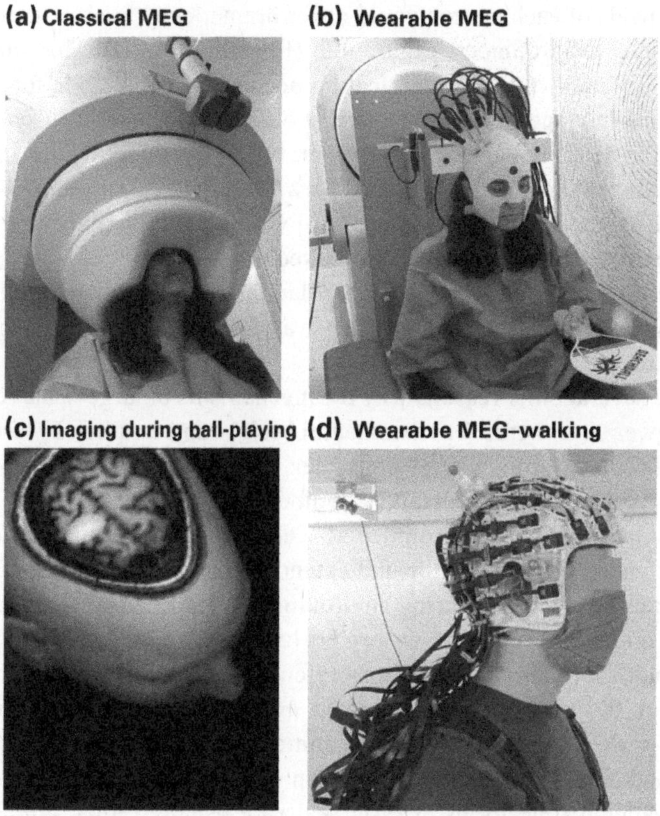

Figure 7.5
Classical versus wearable MEG systems. (a) Classical MEG does not allow any head movements. (b) The new, wearable MEG allows free head movements up to about 20 cm, and object manipulations with the hands. (c) The right motor cortex was activated during naturalistic ball-playing using the left hand (as seen in (b)). The center of the activation was in the arm and wrist area of the right motor cortex. The broad peak of the activation underscores the very low spatial resolution—a major caveat of MEG and most other neuroimaging methods (although not all methods: see table 7.2). (d) A newer version of the wearable MEG, which allowed expanding the scope of the MEG experiments to naturalistic walking over distances of >1 m. Note also the marked miniaturization of the MEG cap within a few years; compare (d) to (b). (a)–(c) reproduced from Boto et al. (2018); (d) reproduced from Seymour et al. (2021).

cognitive neuroscience. However, an even more extreme approach seeks to record human brain activity in people who are totally untethered and unrestrained in their movements and can walk and behave entirely naturally, inside or even outside the lab. Such an approach may open new directions in multiple fields of research, including active vision, complex motor coordination, natural social interactions, and the neurobiology of navigation.

As already discussed in chapter 4, it is crucial to study the neural basis of navigation in freely moving subjects navigating in large environments because natural navigation in large-scale, complex environments may activate very different brain mechanisms than small-scale navigation in a small box. And indeed, neural recordings that we conducted in the hippocampus of bats that were flying in a 200-m-long tunnel have revealed a completely different kind of neural code than found in small environments. Specifically, individual place cells in the dorsal hippocampal area CA1 exhibited many place fields, and different place fields of each neuron could differ dramatically in size: a multifield, multiscale code for large environments (figure 4.2) (Eliav et al. 2021). This surprising neural code was very different from the classical hippocampal spatial code for small environments, where single place fields are typically observed for each neuron. These results were discussed in detail in chapter 4. However, we do not know if such dramatic differences also exist in the *human* hippocampus between large-scale and small-scale environments. Some differences have been hinted at by neuroimaging experiments in London taxi drivers—expert navigators—who were asked to recall large-scale navigational routes in London, as opposed to recalling individual landmarks. When the taxi drivers recalled large-scale navigation, a stronger activation was found specifically in the right hippocampus (Maguire et al. 1997).

But there are several other reasons why the mechanisms of large-scale realistic navigation in complex environments (environmental scale) can be rather different from navigation in small open environments (vista scale): these reasons were discussed already in chapter 4, and were reviewed in detail by Wolbers and Wiener (2014), so I will mention them here only briefly. First, environmental-scale navigation requires integrating information over extended periods of time and extended spaces, while small-scale navigation at vista scale does not. Second, during environmental-scale navigation, the goal to which we navigate is initially beyond the sensory horizon (i.e., not accessible to the navigator); and the same holds for many relevant sensory cues and landmarks. By contrast, in vista-scale navigation, the goal and all landmarks are directly available throughout. Third, environmental-scale navigation requires planning and executing complex navigational trajectories—and moreover, the plans are often refined on the go during navigation. By contrast, navigational trajectories at vista scale are usually rather simple and straight. Fourth, environmental-scale navigation requires relying on complex cognitive mechanisms: memory, planning, decision-making, integration of information over time, monitoring of the planned path, error-correction, and so on. By contrast, small-scale navigation at vista scale requires very little of those because the goal is directly visible. All these reasons suggest that studying navigation at vista scale substantially limits what we can learn about real-world navigation at environmental scale—the two are simply too different. Therefore, neuroimaging experiments on human navigation must also be conducted in subjects navigating through complex, large-scale environments.

Such complex navigation experiments can be carried out either using VR systems, or by actually navigating outdoors. Arguably, VR may not be satisfactory—at least not with fMRI—because the VR displays used in fMRI experiments require the subjects to lie motionlessly inside the fMRI scanner, thus abolishing the natural proprioceptive, motor, and vestibular signals that occur during real-world navigation. In turn, the lack of normal vestibular inputs is known to dramatically affect head-direction cells and the entire brain's navigation system (Taube et al. 2013); also see chapter 3 for a more complete discussion of the limitations and advantages of VR. Furthermore, there is mounting evidence that real ambulatory navigation differs a great deal from navigating in VR. Several studies have shown that spatial learning of a real environment is substantially better than spatial learning in VR for a digital rendering of the *same* environment—most likely due to the difficulty of subjects to update the perceived heading direction after a virtual rotation in VR (Richardson et al. 1999; Montello et al. 2004; Weisberg et al. 2014). This calls, therefore, for more naturalistic experiments. As previously mentioned, one possibility is to conduct experiments in highly immersive VR, where subjects walk on a 2D omnidirectional treadmill while wearing high-end VR goggles and are imaged (e.g., with the wearable-MEG system). An even more naturalistic possibility is to conduct neuroimaging in humans who are freely walking and navigating outdoors.

First steps in the latter direction have been taken by a number of researchers who are using wearable EEG (mobile EEG) to record EEG signals in humans navigating outdoors on foot or by car, using a wearable, multielectrode EEG cap that connects to a laptop or tablet on the subject's back (Mavros et al. 2016; Park et al. 2018; Ladouce et al. 2019; Park & Donaldson 2019; Rito Lima et al. 2020). Nevertheless, despite these encouraging first steps, wearable EEG does not seem to be the most suitable methodology for mobile-neuroimaging during navigation outdoors. There are three major limitations of wearable EEG. First, like any EEG technology, the spatial resolution of wearable-EEG is very poor (about 5 cm), which limits the scope of the investigation. Second, EEG suffers from large motion artifacts from eye and head movements, which pose a major challenge to mobile EEG experiments. While these artifacts can be somewhat reduced by using printed flexible electrode arrays that hug the skin (Inzelberg & Hanein 2018), even these improved electrodes cannot eliminate the largest artifacts, which come from eye movements, because these major artifacts stem from physiological electromyogram signals coming from the eye muscles. Third, EEG does not allow imaging deep inside the brain, which is crucial for imaging the hippocampus and other navigation-related medial-lobe structures in humans (table 7.2). These challenges may be better addressed by other wearable imaging technologies, which have higher spatial resolution, are less prone to motion artifacts than EEG (or MEG), and/or might allow imaging deep inside the brain. The two most promising alternative technologies are functional near-infrared spectroscopy (fNIRS) (Pinti et al. 2018, 2020) and functional ultrasound (fUS) (Deffieux et al. 2018).

Like fMRI, fNIRS utilizes the fact that neural activity elicits a local hemodynamic response and enhanced blood oxygenation. The fNIRS imaging method relies on measuring the differences in absorption of near-infrared light between oxygenated and deoxygenated hemoglobin, which can be measured by noninvasively shining near-infrared light through the skull with a light source and measuring the backscattered light spectrum with an appropriate sensor. Importantly, fNIRS and fMRI neuroimaging methods were shown to yield very similar results when compared directly in laboratory settings (Huppert et al. 2006; Cui et al. 2011; Liu et al. 2017), attesting to the robustness of the fNIRS method.

Wearable fNIRS has become a mature technology in recent years, allowing noninvasive neuroimaging during completely free locomotion in humans—in the lab or outdoors—with imaging data being transmitted wirelessly or via a cable to a portable acquisition device on the subject's back (figure 7.6) (Park et al. 2018; Pinti et al. 2018, 2020). Wearable fNIRS does not show artifacts due to head or eye movements, and thus solves most of the problems of motion artifacts that plague wearable EEG. In addition, fNIRS has a somewhat better spatial resolution than EEG (about 2 cm versus about 5 cm: table 7.2). However, fNIRS only allows imaging the cortex; it cannot image deep brain structures such as the hippocampal formation. In addition, fNIRS measures blood oxygenation rather than direct neural

Figure 7.6
A portable, wearable fNIRS system that allows neuroimaging in freely moving subjects outdoors. Adapted from (Pinti et al. 2020).

activity, and thus has a considerably worse temporal resolution than EEG (table 7.2). Notably, fNIRS has a somewhat better temporal resolution than fMRI (Pinti et al. 2020). Despite its limited imaging depth, fNIRS allows robust neuroimaging from a variety of cortical regions in human subjects as they freely navigate the outdoors.

For example, Hugo Spiers and colleagues have been conducting an fNIRS study where they imaged the activity of the prefrontal cortex in subjects navigating outdoors through their university campus. They found increased activity in the right dorsolateral prefrontal cortex when the route demanded solving detour problems actively, compared to the same route experienced when the participant was instructed by others on which route to follow, and thus no decisions or detours were required (Hugo Spiers, personal communication). Also, fNIRS could potentially be used to image the parietal cortex, which was previously shown in rodent and human studies to be involved in navigation (Nitz 2006, 2012; Whitlock et al. 2008; Ekstrom et al. 2017).

More generally, fNIRS also could be used to study a host of other natural behaviors, including social behaviors, complex motor control, and foraging. The robustness of fNIRS to motion artifacts makes it a very appealing technology for the neuroimaging of multiple humans simultaneously ("hyperscanning"), during real-life social interactions, for studies of social cognitive neuroscience. Indeed, fNIRS has already been used to image between two and four subjects simultaneously during verbal communication, storytelling, and cooperative singing (Czeszumski et al. 2020). Recent development efforts have advanced the fNIRS technology substantially by building miniaturized time-domain fNIRS systems, which offer a much-improved signal-to-noise ratio (SNR) and deeper imaging depth inside the brain, compared to standard fNIRS systems (Ban et al. 2022; Ortega-Martinez et al. 2023). The fNIRS technology continues to improve rapidly in terms of SNR and imaging

Table 7.2
Comparison of various wearable imaging technologies for human mobile neuroimaging outdoors and indoors.

	Wearable MEG	Wearable EEG	Wearable fNIRS	Wearable fUS	Wearable ECoG
What is measured	Direct neural activity	Direct neural activity	Hemodynamic response to neural activity	Hemodynamic response to neural activity	Direct neural activity
Principle of operation	Magnetic fields generated by electrical currents flowing in axons	Summed electrical activity of a local neuronal population	Absorption difference of near-infrared light between oxygenated and deoxygenated hemoglobin	Ultrafast ultrasound imaging of changes in cerebral blood flow	Summed electrical activity of a local neuronal population
Invasiveness	Noninvasive	Noninvasive	Noninvasive	Invasive (in adult humans)	Invasive
Allows recording while walking outdoors?	No. Requires a magnetically shielded room, but subjects can walk inside the room	Yes	Yes	No. However, may allow whole-brain neuroimaging during ambulatory walking in the hospital	No. However, may allow whole-brain neuroimaging during ambulatory walking in the hospital
Motion artifacts during locomotion and eye movements	Large	Large	Small—medium	Small—medium	Large
Imaging depth	Cortex, as well as some subcortical structures	Cortex gyri only	Cortex gyri only	Whole brain, including deep brain structures	Cortex gyri only
Spatial resolution	~2 cm	~5 cm	~2 cm	~100 μm	~0.5 cm
Temporal resolution	Several milliseconds	Several milliseconds	Hundreds of milliseconds—several seconds	Hundreds of milliseconds—several seconds	Several milliseconds
Number of sensors available in wearable systems	Tens	Hundreds	Hundreds	Multiplane imaging through a single transducer	Hundreds

This table is based on Fukushima et al. (2015), Sieu et al. (2015), Mavros et al. (2016), Krishnaswamy et al. (2017), Boto et al. (2018), Deffieux et al. (2018), Pinti et al. (2018), Boto et al. (2019), Pinti et al. (2020), and Boto et al. (2021). Not listed here are intracranial electrophysiological recordings of single neurons in freely walking, freely behaving humans: these are in their infancy, but once available, they will address many of the caveats listed in this table—and in particular they will improve tremendously over ECoG in terms of motion artifacts (almost none), recording depth (record in deep structures), and spatial resolution (single cells).

depth. Thus, fNIRS seems to be a very promising neuroimaging method for natural neuroscience experiments in real-world environments.

The fUS technology relies on ultrasound imaging of changes in cerebral blood flow that occur following neural activity (Deffieux et al. 2018). Unlike fNIRS, where the adult human skull is almost transparent to near-infrared light—and thus allows noninvasive imaging through an intact skull—this is not the case with ultrasound, which is strongly attenuated by the skull. To date, fUS imaging through intact skull was achieved only in very small animals such as mice and young rats, which have thin skulls (Brunner et al. 2021), and, using acoustic contrast agents (microbubbles), also in adult rats (Errico et al. 2016). But in humans, fUS imaging cannot work through the skull and requires imaging through a craniotomy (Deffieux et al. 2018), which is a major drawback. On the other hand, fUS is portable (although it still requires cables attached to an ultrasound machine); and most important, it allows imaging deep into the brain, in contrast to portable fNIRS systems and unlike traditional invasive imaging methods such as electrocorticography (ECoG), all of which can only image the cortex. In addition, fUS also provides extremely high spatial resolution of about 100 μm—much better than fNIRS or ECoG (table 7.2).

Portable fUS devices could potentially be used for high-resolution, whole-brain imaging in subjects who have undergone a surgical procedure in the hospital: For example, one can envision navigation experiments in subjects walking within the hospital ward, concomitant with fUS imaging of their hippocampal formation; or fUS imaging during real-world social interactions. Portable fUS neuroimaging for humans is currently less mature than portable fNIRS: problems still need to be solved on the clinical side due to the invasiveness of the method; and in addition, fUS systems must be further miniaturized to allow neuroimaging in completely freely locomoting subjects. Nevertheless, fUS imaging holds great future promise for human neuroscience due to its potential ability to perform whole-brain deep imaging in freely moving human subjects: the only neuroimaging method that offers this ability.

A word of caution is needed here. All these neuroimaging methods have the fundamental limitation that they average the activity of large groups of neurons, and therefore they cannot yield a fine-grained understanding of the mechanisms of natural behaviors. For this, it will be crucial to use single-neuron electrophysiology in order to record the activity of individual neurons (or even better, ensembles of individual neurons) from freely walking, freely behaving humans—for instance, epilepsy patients walking around the corridors of the hospital ward. This has not been achieved yet. Further, it would be very valuable to compare single-neuron data to neuroimaging data in freely moving humans, where both sets of data have been collected during similar naturalistic behaviors.

Another cautionary note is that naturalistic neuroimaging experiments, indoors or outdoors, also require designing tasks that maximize naturalness while maintaining a reasonable amount of control (the top-right quadrant of figure 1.2). In the field of navigation, such complex-yet-controlled navigation tasks and analytical tools have been used in fMRI experiments on subjects who virtually navigated in London (Spiers & Maguire 2006; Howard et al. 2014; Javadi et al. 2017)—some of these approaches will be described next—and similar experiments can be conducted on subjects performing real-world navigation outdoors. In fact, it would be of particular interest to compare neuroimaging results of human subjects navigating in the exact same environment in VR versus actual reality. Importantly, naturalistic outdoor experiments, as complex as they are, offer experimental designs and analysis methods that permit robust conclusions.

The various approaches for analyzing brain signals during continuous natural experiences have been reviewed by Spiers and Maguire (2007); here, I will mention only a few of the approaches. First, even in a complex-navigation experiment, it is possible to employ an event-related design—and thus to obtain multiple repetitions of a similar event, which allows reaching reliable statistical conclusions on the relations between brain activity and behavior. For example, when conducting neuroimaging in subjects that navigate on city streets and are performing repeated navigational decisions at street junctions, we can treat each navigational-decision as a repeating event. Even though this event occurs at different locations of junctions, the effect of location can be averaged out, allowing us to focus only on the decision events.

Second, immediately at the end of the experiment, the subjects can be shown their own navigational trajectories (or even better, a video taken through an on-head camera), and be asked to verbally describe their moment-by-moment navigational decisions—and those decisions can then be correlated to the brain activity recorded previously during the actual navigation. This approach has been used in fMRI experiments using VR navigation (Spiers & Maguire 2007), and it can also be used for neuroimaging real-world navigation.

Third, one can analyze brain activity with respect to known navigationally relevant variables, such as the subject's position and direction, and the distance and direction to the goal (Howard et al. 2014). Here, the slow hemodynamic response of the brain (if using fNIRS neuroimaging) may potentially pose a problem; however, the slow pace of walking during real-world navigation, whereby the subject's location and direction change rather slowly compared to the scale of the environment, are actually advantageous for overcoming the slowness of the hemodynamic response. For example, if one performs fNIRS neuroimaging in humans who are navigating outdoors in a city neighborhood sized 500×500 m, then assuming a walking speed of 4 km/h, the subject would walk about 1 m during the about 1-s temporal resolution of fNIRS: this distance is negligible compared to the size of the environment.

Fourth, intersubject correlation analysis (as in Uri Hasson's movie experiments described earlier in this chapter) can be used to compare the brain responses of subjects that navigate along similar navigational trajectories. However, this should be done cautiously because different people may employ different navigational strategies: this problem can be alleviated by combining this approach with a verbal self-report about navigational decisions, and then computing the intersubject correlation separately for groups of subjects that employed similar strategies (e.g., subjects that used wayfinding versus subjects that relied on route following).

All in all, it is clear that there is still work to be done. In chapter 8, we will further discuss some of the challenges and solutions for analyzing the complex, multidimensional neuronal data generated by natural behaviors. However, it is also clear that there is no option but to tackle these challenges head on. If we wish to understand the neural mechanisms of real-life navigation in humans—or understand human real-life behaviors more generally—we must ultimately measure human brain activity outdoors, during natural behaviors, locomotion, and navigation.

In summary, these novel emerging methodologies for measuring neural activity in freely moving humans, including wearable neuroimaging devices (MEG, EEG, fNIRS, fUS) and mobile intracranial electrophysiological recordings, open the door to unprecedented new research directions on the neurobiology of human behavior and cognition. These will include novel studies on the topics discussed in this chapter: (1) real-life memory; (2) navigation; and (3) sensory processing—but also studies on a multitude of other topics: (4) sleep research; (5) the neural bases of real-life motor coordination, going beyond the classical 2D manipulandum to study realistic 3D complex movements; (6) decision-making—that is, going beyond binary forced decisions to study the neural mechanisms of decision-making in real-world scenarios where choices are continuous, multioption, and temporally dynamic; (7) emotional processing; (8) language; and (9) social interactions.[12]

Further, fNIRS in particular can be used to image brain activity in children, and even infants, because fNIRS provides a safe, quiet, and convenient neuroimaging method (as compared to the noisy and stressful fMRI scanner). Therefore, it allows for conducting neurodevelopmental studies in children on the neural mechanisms of real-world navigation, perception, naturalistic movement, social learning, and other topics (Pinti et al. 2020). New, more miniaturized devices will be developed in the coming years, and this will allow for studying the human brain—in both adults and children—under evermore naturalistic conditions: in the lab, as well as outdoors. Exciting years lie ahead for human natural neuroscience.

8

Future Outlook

In the preceding chapters, I argued that, parallel to pursuing classical controlled experiments, neuroscience should move toward studying the brain under more complex, naturalistic conditions and focus on naturalistic behaviors. I called this the "natural neuroscience" approach. Here, I will briefly summarize some of the main arguments made in the previous chapters for why we should be going in the direction of natural neuroscience.

First, *neural circuits in the brain were shown to be activated very differently for natural versus artificial behaviors*. Previous chapters of this book were dedicated to describing in detail the evidence supporting this statement (in particular, chapters 4, 5, 6, and 7). The brain clearly operates differently under natural conditions. We therefore must study the brain in more natural situations.

Second, *there are many fascinating complex behaviors that animals do not exhibit in standard laboratory conditions—or exhibit them in impoverished ways*. These behaviors include:

1. Long-distance navigation in mammals—over the scale of kilometers or thousands of kilometers—which is difficult to study in a standard laboratory space.[1]
2. Short-distance navigation—as standard laboratory environments lack many of the natural cues that animals use for real-life navigation, such as the Sun and Moon, polarization cues, and wind.
3. Complex hunting behaviors and antipredator behaviors, such as group hunting and communal defense—which are not expressed at all in small laboratory spaces.
4. Social behaviors—many of which are impoverished in confined laboratory spaces.
5. Active-sensing behaviors—many of which are impoverished in confined laboratory spaces.
6. Tool use.
7. Camouflage behaviors.
8. Long-term memory over years.
9. Animal cultures (Whiten 2021).

Many other examples abound. Today's mainstream behavioral neuroscience does not study almost any animal behavior that comes even remotely close in complexity to the kinds of behavior seen outdoors, as documented, for instance, in some of the wonderful nature documentaries produced by the BBC or National Geographic. Think about the incredible cooperative hunting shown by killer whales, wolves, and lions. Think about how spiders spin their webs and the amazing hunting behaviors of net-casting spiders. Think about bowerbirds decorating their bowers, and about tool use in primates or in New Caledonian crows. (Just Google some of these behaviors, if you haven't seen them!) These fascinating natural behaviors are what prompted many of us to study the neural basis of behavior in the first place—but then we settled for studying highly simplified, tightly controlled, impoverished behaviors in the lab. Now, I am not necessarily claiming that we should study the neural basis of the most complex and most spectacular behaviors in existence—there are plenty of simpler natural behaviors, which are still much more complex and rich than the typical behaviors studied in the laboratory. And, importantly, technological advances allow us to record for the first time neural activity and document in detail the behavior of freely moving animals roaming in large, naturalistic enclosures—or even in the wild. This will allow us, at last, to study the neural basis of truly natural behaviors, which were neglected for far too long.

Third, *several general arguments reinforce the need to study the brain during complex, naturalistic behaviors*. These theoretical arguments come in several flavors, as follows:

1. *The evolutionary argument*: The brain has evolved under strong evolutionary pressures to perceive and parse natural visual scenes, natural sounds, and natural odors rapidly; to move, forage, and navigate in the natural world; to remember real-life events; to make real-world decisions; and to carry out natural social interactions. These natural pressures must be reflected in the neurophysiology of brain circuits. Therefore, we should study the brain under natural or naturalistic conditions rather than being content with classical, highly simplified laboratory experiments.
2. *Complexity of brain networks*: The brain is a highly complex system, exhibiting high degree of nonlinearity and strong collective behaviors with emergent properties that cannot be elucidated by trying to control parts of the system one by one (chapter 1). Therefore, the reductionist paradigm originating from Bacon, Descartes, and Galileo—which promoted the exclusive use of highly controlled, simplified experiments for studying isolated phenomena (in this case, studying isolated brain functions or isolated simple behaviors)—is severely limiting our understanding of a complex system like the brain.
3. *Mixed selectivity of neurons*: Multiple studies have shown that neurons in higher brain areas, such as the prefrontal cortex, parietal cortex, and hippocampus, exhibit mixed selectivity—that is, they are tuned to multiple behavioral parameters of a complex task or a complex setting (Rigotti et al. 2013; Eichenbaum 2015; Fusi et al. 2016; Hardcastle et al. 2017; Sarel et al. 2017; Noel et al. 2022; Ray et al. 2024; Testard et al. 2024). Further, these behavioral parameters may interact in complex nonlinear ways, which *cannot* be uncovered by studying one behavioral parameter at a time (Sarel et al. 2022; also see figure 4.5b). Therefore, in order to reveal the complex mixed selectivities of neurons and to uncover their true dimensionality, we must study the brain in complex behavioral scenarios, with multidimensional behavioral variables, rather than studying simple, one-dimensional (1D) behavior in isolation.

There are many other reasons why we should move toward studying the brain in the context of more naturalistic behaviors: these reasons were discussed in detail in chapter 1 (see table 1.1) and in chapter 2, and I will not repeat them here. However, simply running experiments under more naturalistic conditions is *not* enough. Such experiments are necessary but not sufficient. What must accompany them, in parallel, are new ways of thinking about brain research—as well as new research questions. In what follows, I will attempt to portray some possible future directions for this research.

8.1 Synthesis: Linking Natural Neuroscience with Behavioral Ecology

More than half a century ago, Niko Tinbergen—one of the founders of ethology—published a very influential paper, in which he proposed that the study of behavior must follow four basic questions (Tinbergen 1963), given in table 8.1. These questions, which followed in the footsteps of Aristotle's "four causes," address different aspects of behavior: (1) development (ontogeny), (2) mechanisms, (3) evolution, and (4) utility (or function). (Note that although Tinbergen framed his questions explicitly in the context of behavior, the same questions can be applied very fruitfully across a wide range of biological problems—not just those pertaining to behavior.) Tinbergen argued that in order to gain a full, multilevel understanding of a particular behavior, all of these four questions should be applied to this behavior. He stressed that these questions are not mutually exclusive, but are *complementary* ways of looking at behavior. Tinbergen's four questions can be conveniently summarized in a 2 × 2 matrix, as shown in table 8.1 (Bateson & Laland 2013; and modified from Nesse 2013).

The rows in table 8.1 are separated according to "how?" versus "why?" questions—also known as *proximate* versus *ultimate* questions, following the classification by Ernst Mayr (1961). The columns of table 8.1 are separated according to questions concerning dynamics (past history) versus statics (present situation). Notably, David Marr's famous three levels of analysis (Marr 1982) can also be mapped onto this classification: Tinbergen's proximate question of mechanisms embodies Marr's two lowest levels of analysis—the algorithmic and

Table 8.1
Tinbergen's four questions (Tinbergen 1963), organized in a 2 × 2 matrix

	Dynamics (Past)	Statics (Present)
How? (Proximate questions)	**Development (ontogeny)** How do this behavior and its mechanisms develop during the ontogeny of the individual? ***Developmental neuroscience*** ***Developmental ethology***	**Mechanisms** How does it work? What are the mechanisms (neural circuits, cell types, hormonal effects, and genes) that control this behavior? ***Systems neuroscience*** ***Behavioral neuroscience*** ***Cellular and molecular neuroscience*** ***Behavioral genetics*** → *Marr's algorithmic level and implementation level*
Why? (Ultimate questions)	**Evolution (phylogeny)** Why did this behavior evolve? What is the evolutionary history of the behavior and of its underlying mechanisms? ***Evolutionary neurobiology*** ***Evolutionary psychology***	**Utility (function)** Why is this behavior needed—what is it good for? What is the current utility/function/adaptive significance of this behavior for the individual in terms of survival value or reproduction? ***Behavioral ecology*** → *Marr's computational (functional) level*

implementation levels, while Tinbergen's ultimate question of utility (function) equates to Marr's highest level of analysis—the computational or functional level (Mobbs et al. 2018).

To understand the complementary views provided by Tinbergen's questions, consider the example of bird migration—specifically, why does a migratory bird take off and fly north on a particular date in the spring? The *mechanism* of this behavior is likely related to the lengthening of the daily photoperiod, which triggers a decision-making circuit in the bird's brain, and the bird then retrieves the long-term memory of the migratory direction or route—and off it goes. In birds migrating in flocks, social brain circuits are likely also engaged. But even the most detailed understanding of the brain mechanisms of this behavior would not replace the need to understand the *utility (function)* of this behavior—the "why?" question—for which the answer is that the bird needs to arrive on time to the summer breeding grounds in order to reproduce. If the bird starts its migration too late, the best nesting locations would already be taken, and its newborn chicks thus would be raised on the outskirts of the bird colony, where predation risk is much higher. If, on the other hand, the bird starts migration too early, it may perish in the last winter storms that hit the northern part of its migratory route or its breeding grounds. This leads to the question of *evolution*, where we can now understand that, across many generations, those birds with a genetic predisposition to leave at the right time—neither too early nor too late—have their genes take over the population of their species. This is not to say that there is a single migration-timing gene, or even a group of migration-specialist genes; in fact, this may be a complex set of genes related to the neural mechanisms of photoperiodicity, decision-making, memory, and sociality, including genes controlling the development of these neural circuits. Finally, the *development (ontogeny)* of this migratory behavior is crucial for us to understand as well, because in most bird species, young birds in their first migratory season learn from experienced adult birds both the precise timing of migration and their migratory route. Likewise, Tinbergen's four questions provide complementary angles for many other behaviors—such as birdsong (i.e., we can ask how and why birds sing: the mechanisms, utility, evolution, and development of this behavior) (Bateson & Laland 2013); or we can ask these questions about active-sensing behaviors, such as whisking in mice or echolocation in bats, or pretty much any behavior. The general notion is that considering all four questions enriches our understanding of the behavior.

However, for most behaviors, only some of these questions are being studied—and furthermore, they are usually studied separately by different researchers, without any attempt at synergizing. In fact, Tinbergen's questions have been typically addressed by several very different research disciplines, exhibiting relatively few interactions. At the risk of oversimplifying, I listed these disciplines in table 8.1. The "how?" question of *development* has been addressed by the disciplines of developmental neuroscience and developmental ethology. The "how?" question of *mechanisms* has been addressed by systems neuroscience, behavioral neuroscience, cellular and molecular neuroscience, and behavioral genetics. The "why?" question of *evolution* has been addressed by evolutionary neurobiology and evolutionary psychology. And the "why?" question of *utility* (or function) has been addressed mostly by behavioral ecology (Davies et al. 2012; Bateson & Laland 2013).

Historically, there has been some horizontal cross-talk within this table, such as within the "how?" row, between researchers interested in mechanisms and those interested in development. Sometimes the same people studied both questions: It suffices to mention Hubel and Wiesel, who studied the mechanisms of vision in V1 of adult cats, but also the development of these mechanisms in kittens (Hubel & Wiesel 1963a; Wiesel & Hubel 1963), as well as O'Keefe and the Mosers, who studied the mechanisms of hippocampal place cells and entorhinal grid cells in adult rats, but also their development in young rat pups (Langston et al. 2010; Wills et al. 2010). Likewise, there has been horizontal cross-talk within the "why?" row between people interested in utility (function) and those interested in evolution (Davies et al. 2012). Here again, the same people sometimes study both subjects. For instance, to take an example outside of neuroscience, research by Hopi Hoekstra and colleagues demonstrated a direct link between experimentally measured survival rates of wild deer mice that differed in fur coloration, and the evolution of fur-color-related genes within the deer mouse population (Barrett et al. 2019). However, there has been very little *vertical cross-talk* in neuroscience between people interested in "how?" questions and people interested in "why?" questions. It just seems that very different people are interested in "how?" versus "why?" questions of behavior (a chasm noted already by Mayr 1961), with the notable exception of some neuroethologists and comparative neurobiologists, who are interested in both kinds of questions. Here, I wish to argue that a major challenge for natural neuroscience would be to further build vertical links—in particular, strong links are needed between behavioral neuroscience and systems neuroscience on one hand, and behavioral ecology on the other hand (see the right column of table 8.1). I will argue next that such linkage could lead to a whole host of new questions, novel research directions, and fruitful experiments concerning the neural basis of natural behaviors.[2]

I will now give some concrete examples of the importance of establishing a vertical link in table 8.1, between the neural mechanisms of a behavior and the utility (function) of that behavior—that is, linking neuroscience with behavioral ecology.

* * *

Example 1: Foraging and decision-making. When animals make decisions about foraging—where and when to go, which food items to take, and more—they trade off a variety of considerations, which have been examined extensively in the behavioral-ecology literature, both experimentally and theoretically (Stephens & Krebs 1987; Stephens et al. 2007; Davies et al. 2012; Mobbs et al. 2018; Prat & Yovel 2020). The animal is usually trying to maximize the energy gain from food while minimizing many other factors, including energy loss from travel (or the travel time), handling time of prey, risk of injury from prey, predation risk, and load size to be carried. This could be written, in its simplest form, as a linear optimization problem—where the animal is trying to maximize the following function:

$$\begin{aligned}\text{Max}\,(&\beta_1 \times \textit{Energy gain from food}—\beta_2 \times \textit{Energy loss from travel}—\\ &\beta_3 \times \textit{Travel time}\text{ or }\textit{Search time}—\beta_4 \times \textit{Handling time}—\\ &\beta_5 \times \textit{Load size to be carried}—\beta_6 \times \textit{Risk of injury from prey}—\\ &\beta_7 \times \textit{Predation risk}—\beta_8 \times \textit{Competition}—\beta_9 \times \textit{Hunger level})\end{aligned} \tag{8.1}$$

This equation is certainly too simplistic. First, it is a linear equation, while in reality, there could be nonlinear interactions between the various factors. Second, these factors may dynamically change over time.[3] For the sake of simplicity, these important complexities are ignored in the following discussion. Third, additional factors are likely also important—both external (e.g., seasons, weather) and internal (motivational state, health). Equation (8.1) is definitely not comprehensive; it encompasses only a *partial* list of all the factors that may be important for decision-making during foraging.

On the other hand, studies in behavioral ecology have shown that, for a given species in a given behavioral context, only a subset of these factors are actually important. For example, some animals are trying to minimize energy loss from travel but do not care about travel time ($\beta_3 = 0$), while for others the opposite is true ($\beta_2 = 0$) (Davies et al. 2012). Or consider another example: if we set the weights $\beta_3 = \ldots = \beta_9 = 0$, and $\beta_1 = \beta_2$, then we get the simplest form of optimal foraging theory: Max(*Energy gain from food—Energy loss from travel*), whereby animals try to maximize energy gain from food minus the energy loss from travel. Equation (8.1) highlights the advantages of having good sensory systems, which allow for minimizing *Search time* for food and minimizing *Predation risk*. It also highlights the importance of relying on learning and memory. Thus, relying on spatial learning and spatial memory in foraging (i.e., being able to navigate as straight and as rapidly as possible to known food locations) minimizes the variables *Energy loss from travel*, *Travel time*, and *Predation risk*. This may underlie the straight navigational trajectories of Egyptian fruit bats (figure 8.2a) (Tsoar et al. 2011) or the food-storing behaviors of birds (figure 4.6a) (Clayton & Krebs 1995). Likewise, social learning—such as in meerkats who teach their pups how to safely handle dangerous prey such as scorpions (Thornton & McAuliffe 2006)—minimizes the variables *Handling time*, *Risk of injury from prey*, and *Predation risk*. Thus, while equation (8.1) is somewhat simplistic, it can at least qualitatively explain the results of many experiments in behavioral ecology.

Another interesting aspect of equation (8.1) is that it may be linked directly to neuroscience. Specifically, neurons might potentially compute it since the linear combination inside the parenthesis looks very much like a weighted sum of synaptic inputs. Could it be that some decision-making neuronal circuit is receiving inputs about these various factors and literally computes equation (8.1), with the $\beta_1 \ldots \beta_9$ values being synaptic weights (β_1 = excitatory synapse, $\beta_2 \ldots \beta_9$ = inhibitory synapses)? This computation could potentially be done by single neurons at the subthreshold level, with evidence accumulation implemented via synaptic integration culminating in a decision spike. Such an evidence-accumulation mechanism in single neurons was shown in a study by Miesenböck and colleagues, who studied olfactory neurons in *Drosophila melanogaster* flies and showed evidence accumulation leading to a decision spike in individual neurons (Groschner et al. 2018).

Alternatively, this computation could be done by larger neural circuits, which could compute the weighted sum of benefits and risks for decision-making during foraging, akin to equation (8.1). These circuits could sum up the firing rates of different inputs, according to equation (8.1), until a decision threshold is reached—with the computation done at the level of the neuronal population rather than in a single neuron. Such a thresholding mechanism was demonstrated for escape behaviors in mice and was localized to the synapses between the superior colliculus and the periaqueductal gray (Evans et al. 2018). This study showed that neurons in the medial superior colliculus encode the saliency of a threat stimulus, and these neurons project monosynaptically to the periaqueductal gray, where neurons encode the decision to escape—and this decision is made by integrating the evidence for threat, coming from the superior colliculus, and then reaching a decision threshold at the population level. Thus, while a direct neural implementation of equation (8.1) seems naive, it is certainly feasible and could be investigated.

Further, regardless of the naiveté of equation (8.1), my main point here is that the classical approaches for studying the neurobiology of decision-making usually focus on only two of the factors from equation (8.1)—the energy gain from food (reward size) and the search time (delay, or probability of reward)—but equation (8.1) illustrates that many more factors are crucial in real-life decision-making during foraging. Importantly, behavioral ecologists have come up with experimental methods to quantify many of the factors listed in equation (8.1) (Davies et al. 2012). We can therefore imagine future experiments in natural neuroscience, where an animal will be freely moving and foraging, while these behavioral variables are manipulated in the tradition of behavioral ecology. For example, in order to increase the variable *Predation risk*, one can present predator odors or predator vocalizations, a lifelike stuffed predator, or the silhouette of a predator, as was done behaviorally by Milinski and Heller (1978). Or one can add extra weight to the animal to increase the variable *Load size*. Or one can introduce competitors (conspecifics) to increase the variable *Competition*. All these manipulations are fully feasible behaviorally and have been used in studies of behavioral ecology (Davies et al. 2012)—but not in conjunction with neuronal recordings. It would therefore be interesting to manipulate these behavioral variables and to search for individual neurons or distributed neural circuits that linearly (or nonlinearly!) combine all these variables. This can be done by using wireless neural loggers to record the activity of neurons in freely moving animals, and then fit the model using equation (8.1) to the measured responses in order to estimate the weights $\beta_1 \ldots \beta_9$. These behavioral manipulations would be very useful for elucidating neuronal mechanisms, so long as the underlying behavior which they probe is a naturalistic behavior. Such experiments would require using sophisticated data-analysis methods to determine the contribution of each of the behavioral variables to the neuronal activity, as this is a multidimensional behavioral space: such analysis methods will be discussed in section 8.2.

In addition, one may use the arsenal of modern optogenetic and chemogenetic tools to perform brain manipulations during these behaviors: to excite or inhibit the relevant inputs or to directly alter the synaptic weights that may correspond to the weights $\beta_1 \ldots \beta_9$ in equation (8.1). Such future experiments may reveal the neural circuitry that computes the weighted sum of benefits and risks for real-life decision-making during actual foraging, and will thus help elucidate the neural mechanisms of the naturalistic decision-making processes that have been studied by behavioral ecologists for the past half century. As an important consequence, such experiments would also highlight the complex set of factors that affect real-life decision-making more generally—most of which are rarely (or never) considered in neuroscience experiments (Mobbs et al. 2018). This would go a long way toward a more natural approach to decision-making.

Example 2: Social behaviors and the case of reciprocal cooperation. Social behaviors in animals come in a spectacular variety of forms and contents (Davies et al. 2012). They include coordinated group hunting, as in wolves, lions, and killer whales; communal defense, as in many herbivores and group-nesting birds; individuals serving as altruistic sentinels for the entire group, as in meerkats; group foraging and "information centers," as in ravens, bats, and honeybees; complex examples of competition over food, as in bats deliberately jamming the sonar of other bats (Corcoran & Conner 2014); and group decision-making, as in social insects, fish, birds, and many mammals. Group decision-making can be done by local decision rules—as in fish (Couzin et al. 2005); or it can be done by explicit voting, as in buffalo and baboons (Prins 1996; Strandburg-Peshkin et al. 2015). For example, baboons decide where to go via a remarkable voting behavior, whereby the entire group follows the direction of the vector average heading of the first individuals who initiate movement; but if the initiators exhibit a large amount of disagreement (i.e., large angular separation), the group moves in the direction of the majority—truly democratic decision-making. Social behaviors include acts of aggression, but also many examples of cooperation, such as collective pup-rearing and even acts of empathy, consolation, and altruism (Burkett et al. 2016; de Waal & Preston 2017).

Over the past half century, behavioral ecologists and ethologists have been studying in detail these and many other naturally occurring social behaviors, as shown in figures 8.1a and 8.1b (Chang et al. 2013; Clutton-Brock 2016; de Waal & Preston 2017). In stark contrast, neuroscientists are studying social behaviors mostly in inbred mice, in impoverished laboratory environments, using highly artificial setups such as the "tube test," or the "three-chamber test" (figures 8.1c and 8.1d). The problem with these tests are not only that they are artificial, but that in most cases, the analyses focus on *time-averaged behavior* and quantify *one-dimensional (1D) behavioral variables* such as the time spent in one of the chambers of the three-chamber test (although some recent studies have performed richer behavioral analyses of the tube test: see Zhou et al. 2017; Kingsbury et al. 2019). This static, 1D view ignores the rich dynamics of behavior and ignores the multidimensionality of rodent behavior, which includes social vocalizations, social sniffing, social whisking, and many other social aspects that are typically not measured by these standardized tests. In fact, it was shown that dozens of behavioral parameters are needed to characterize complex social behaviors (Forkosh et al. 2019). One cannot summarize behavior with one number, but this is precisely what standardized social tests attempt to do. Thus, mainstream neuroscience research rarely considers the natural complexities, subtleties, and dynamics of social behaviors—but it should.

Reciprocal help among genetic relatives is well understood theoretically in behavioral ecology through the theory of kin selection, which describes quantitatively why relatives should help each other to promote their common genes (the "selfish gene" idea promulgated in Dawkins 1976). This theory can even explain extreme cases of altruism, where helping your relatives comes at the expense of your own survival or reproduction—but nevertheless promotes your genes (Davies et al. 2012). A more difficult-to-explain situation is that of cooperation and reciprocal help in nonrelatives—a phenomenon sometimes called "reciprocal altruism," which occurs most prominently in some primates (Schweinfurth & Call 2019), including humans.

One of the famous examples of cooperation and reciprocal help in nonrelatives is that of *food sharing in vampire bats*—highly social bats inhabiting neotropical forests in Central and South America (figure 8.1a). These bats feed exclusively on blood, and because of their high metabolism, skipping a blood meal for more than one night may be fatal. In a series of studies, Gerald Wilkinson and Gerald Carter showed that vampire bats regurgitate blood to feed their group members—both relatives and nonrelatives (Wilkinson 1984; Carter & Wilkinson 2013, 2015). Field observations and laboratory experiments have shown that blood donations are given only to very familiar bats. The feeding is usually initiated by the donor, who identifies the hungry bat and voluntarily approaches it, which

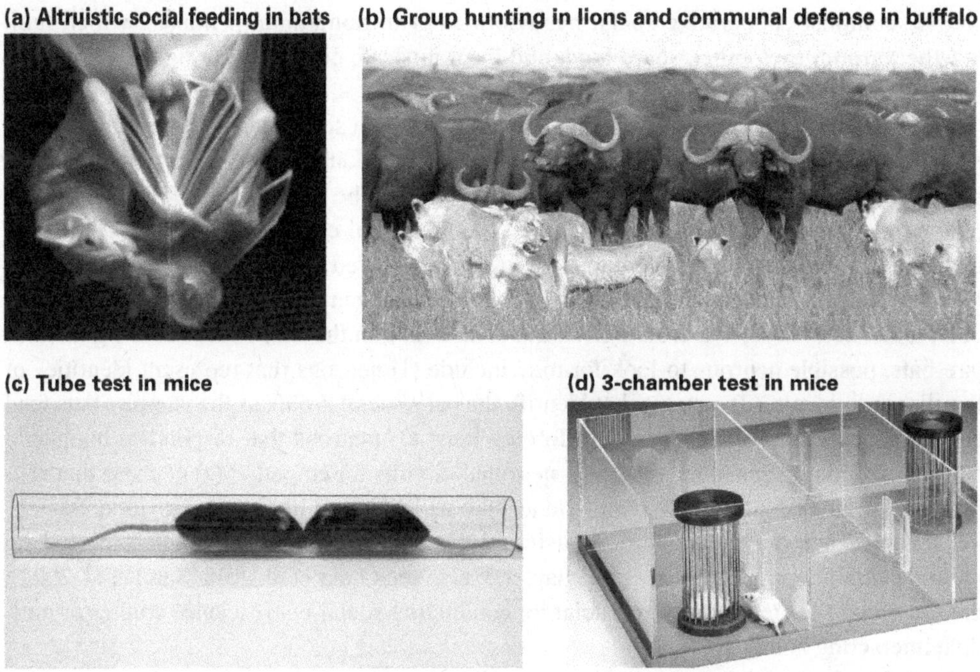

Figure 8.1
Natural social behaviors versus laboratory social assays. (a)–(b) Natural social behaviors. (a) Altruistic social feeding in vampire bats (Wilkinson 1984; Carter & Wilkinson 2013, 2015). Shown here is a well-fed, white-winged vampire bat feeding a hungry conspecific by regurgitating blood—a spectacular example of reciprocal altruism in animals. Photo courtesy of Gerald Carter. (b) Group hunting in lions and communal defense in buffalo. Lions exhibit sophisticated group-hunting tactics (Stander 1992) and are rivaled by the equally sophisticated group-defense tactics of African buffalo (Prins 1996). Photo: Brian Sedgbeer / Dreamstime. (c)–(d) Laboratory social assays. (c) Tube test for social dominance in mice. Photo taken from a movie in Zhou et al. (2017). (d) Three-chamber sociability test in mice. Photo: Catalog of Stoelting Co.; photo credit: Ugo Basile, Italy.

cannot be explained through harassment by the receiver bat. Donations are preferentially given to specific individuals; and, most important, blood donations are reciprocal, meaning that if bat A often donates to bat B, then bat B often donates to bat A. In fact, the amount of food received is by far the best predictor of how much food is donated—nearly nine times more predictive (in terms of explained variance) than genetic relatedness. Moreover, bat pairs that tend to feed each other also tend to engage in social grooming (Carter & Wilkinson 2013). These social relations are not short-term "tit-for-tat," but rather seem to accumulate information over extremely long timescales: The accumulated integral of blood donations and social grooming over the past few years (!) is a strong predictor of the likelihood of reciprocation (Carter & Wilkinson 2015; Carter et al. 2020). This suggests that these long-lived bats—which were shown to live together in the same social group sometimes for up to twenty years—form long-term "social support networks," with highly stable, long-lasting social bonds. Reciprocal cooperation in these social networks can operate through multiple interchangeable "currencies," including food donations, social grooming, and social thermoregulation. Complex, long-term cooperative social networks also occur in other bat species, such as greater spear-nosed bats—a highly social bat species exhibiting collective pup-rearing and reciprocal "babysitting" (Wilkinson et al. 2016). And of course, complex cooperative social behaviors exist in nonhuman primates, which hunt together, groom each other, share food, and form intricate coalitions for fighting and mating (de Waal & Suchak 2010).

What are the neural mechanisms that may underlie such sophisticated cooperative social behaviors in bats, primates, and other mammals? This is completely unknown, but one can outline a list of candidate types of neurons that we should be searching for. The likely brain regions to search for such neurons would be the prefrontal cortex, cingulate cortex, hypothalamus, amygdala, and hippocampus—regions implicated in social behaviors and social memory (Hitti & Siegelbaum 2014; Li et al. 2017; Murugan et al. 2017; Sliwa & Freiwald 2017; Omer et al. 2018; Wu et al. 2021; Yang et al. 2022). In the case of food sharing in vampire bats, possible neurons to look for may include (1) neurons that represent identities of familiar individuals; (2) neurons that identify the bat's social group, as the vampire bats feed only other bats from their group, not foreign bats; (3) neurons that respond to hunger in other bats, and perhaps more generally, neuronal circuits for empathy; (4) neurons that represent the preferred individuals (i.e., "friends") to which the bat likes to donate food. Searching for such neurons could be done using miniature wireless neural loggers placed on multiple animals simultaneously (see chapter 3; also see Omer et al. 2018; Sarel et al. 2022); the use of wireless technology is crucial for conducting social-neuroscience studies of multiple interacting animals.

First steps in this direction were done recently in multianimal social experiments in fruit bats, where we discovered that hippocampal neurons track the identities and locations of *specific* individuals (social-spatial identity coding)—and also encode affiliation (i.e., who are my friends) (Ray et al. 2024; see also section 4.3).

Another intriguing question is the mechanisms by which neural circuits in these bats integrate information over timescales of several years—in particular, how they integrate the variable "he donated to me." Such long-term memory over years, or long-term "neural integrator" mechanisms, may explain the behavioral accumulation of "long-term donation history" observed in vampire bats. This ultra-long-term memory may manifest both in synaptic plasticity mechanisms and in long-term changes in neuronal responses. It may be difficult in practice to study such mechanisms of accumulation over years, so we may need to settle for studying them over timescales of weeks or months, instead of years. This may be technically feasible, with the recent advent of head-mounted, miniature, two-photon microscopes, which can track over many weeks the same neurons and possibly even the same dendritic spines, in freely behaving animals (chapter 3) (Zong et al. 2017, 2021). It would be highly interesting indeed to understand the mechanisms of such long-term accumulation of the history of cooperative reciprocity, which is crucial for understanding this fascinating behavior in bats.

Likewise, similar types of neurons and circuit mechanisms should be looked for in nonhuman primates, and in rodents, in search for the neural mechanisms underlying their complex social behaviors. Very few of the abovementioned types of neurons have been found so far in any animal species—and were not even looked for to date in studies of social neuroscience. As previously noted, we found hippocampal neurons that represented social identity and social affiliation in bats during free social interactions. Very important steps in this direction were also done by Ziv Williams and colleagues. In primates, they found prefrontal-cortex neurons that encoded the identity of other individuals[4] (Báez-Mendoza et al. 2021), and cingulate-cortex neurons, which predicted the behavior of other individuals

(Haroush & Williams 2015). In the cingulate cortex of freely moving mice, they found neurons that represented multiple social variables, including social rank (Li et al. 2022). Much more work is needed in order to understand the neural basis of cooperative reciprocity in mammals. We should first find suitable natural behaviors (e.g., in bats or primates or rodents), which could then serve as models for neuroscientific studies. The same goes for all the other fascinating natural social behaviors that were listed earlier in this section—as well as any other complex social behaviors. The first and foremost challenge is to identify a suitable behavior; and the next challenge is to establish a naturalistic variant that is close to the original, but is amenable for laboratory studies.

* * *

In summary, behavioral ecology has provided a wealth of wonderful explanations for "why?" questions about animal behavior (Davies et al. 2012). In addition, it has received its share of criticism (Gould & Lewontin 1979; Bradbury & Vehrencamp 2014). One of the points of criticism was that behavioral ecology is quite reductionist. Bradbury and Vehrencamp (2014), for example, made the point that in behavioral-ecology studies, only pairs (dyads) of animals are usually tested, rather than full complex animal groups—and they emphasized that pairwise interactions cannot fully explain the social dynamics of the group. Further, many experiments in behavioral ecology are done in laboratory conditions.[5] Nevertheless, behavioral ecology has two key features that should serve as a source of inspiration and new ideas for neuroscience. First, *behavioral ecology deals with understanding natural behaviors*. Many observational and experimental (interventional) studies in behavioral ecology are performed in the wild, and even those studies done indoors are clearly addressing natural behaviors. Second, *behavioral ecology focuses on "why?" questions of behavior (function),* unlike neuroscience, which mostly focuses on "how?" questions (mechanisms) (see table 8.1). Therefore, although behavioral ecology is often overly reductionist, it can nevertheless serve as a source of inspiration for new questions, new ideas, and new experiments in behavioral neuroscience, as I tried to outline in the two examples given here.

There are many more such examples. In fact, I would argue that *natural neuroscience should blend three distinct influences: neuroscience, ethology, and behavioral ecology.* Specifically, we ought to apply the mechanistic "how?" approach of neuroscience in order to understand natural behaviors as studied by ethology, which are illuminated by functional "why?" explanations of behavioral ecology. Such a triple combination holds promise to generate novel questions and new research directions on the neural basis of behavior.

8.2 Natural Neuroscience: Future Vision and Challenges

It should be clear from the previous chapters that the natural neuroscience approach does not aim to replace the classical, highly controlled approach to brain research, but rather to complement, expand, and diversify it. The goals of this diversification are to conduct experiments under evermore naturalistic settings and to record neuronal activity in animals performing more natural behaviors in order to elucidate the neural basis of natural behavior.[6] In what follows, I will outline some of these future directions for natural neuroscience, as well as some of the main challenges it faces and how they might be overcome.

So how should we approach future experiments in natural neuroscience? The best approach would ideally be to record from the same individual neurons (or the same ensembles of neurons), both under classical controlled conditions, which allow a better mechanistic handle, and under more naturalistic conditions, which allow direct investigation of the neural basis of natural behaviors. Thus, I do not propose here an "either/or" approach, but rather an "and" approach, which will allow direct comparison and gaining richer insights.

How would the more naturalistic experiments look like? This depends, of course, on the questions being asked. In social neuroscience, for example, I argue for a stronger focus on natural, complex social behaviors (figures 8.1a and 8.1b), rather than on using only laboratory-based artificial social assays (figures 8.1c and 8.1d). Such behaviors could include reciprocal cooperation, as already discussed for bats and primates; observational learning; teaching; play; and many other dyadic social behaviors that have long been studied by ethologists (Dugatkin 2020). Furthermore, there should be more emphasis on social interactions in large natural social groups, rather than studying only dyadic interactions between isolated pairs of animals—as is currently the norm in both behavioral neuroscience and behavioral ecology. As summarized by Bradbury and Vehrencamp (2014), the use of dyadic interactions cannot explain the behavior of the group as a whole because it misses the importance of complexity, the network structure of animal groups, nonlinear interactions between animals, complicated group dynamics, and emergent collective behaviors. It is thus necessary to identify some interesting natural social behavior, and then to study the neurobiology of this natural social behavior in whole animal groups. Examples of such multianimal social behaviors include collective decision-making, group hunting, communal defense, and the formation of coalitions and alliances (Prins 1996; Couzin et al. 2005; Davies et al. 2012; Strandburg-Peshkin et al. 2015; Dugatkin 2020). These behaviors, which were documented in many animal species, arguably require *neural mechanisms for simultaneously tracking the positions and behaviors of multiple animals*—both friends and foes. We may expect, for example, to find neurons that track animals, and moreover, track specific individuals. Indeed, several studies have found neurons that track the positions of other animals during dyadic behaviors (Danjo et al. 2018; Omer et al. 2018), as well as during multianimal group behaviors (Ray et al. 2024; also see section 4.3). Many more multianimal experiments need to be done, including experiments searching for neurons that track groups of animals as a whole (i.e., track a group entity). This can be achieved by using arrays of cameras and other sensors to document the detailed behaviors of *all* the group members, while recording neural activity from the brains of *multiple* animals in the group, simultaneously, using the wireless electrophysiology and imaging methods discussed in chapter 3. Further, the neurobiology of complex social behavior should best be studied across a *range* of animal species that vary in the level of their sociality (Zilkha et al. 2021), ranging from the antisocial (solitary) blind mole rat to highly social species such as naked mole rats, meerkats, primates, and bats.

In the neurobiology of navigation, what we would eventually like to understand is the neural basis of natural navigation. To cite some examples: how a Norway rat can navigate up to about 1 km every night and find its way back to its burrow (Taylor 1978; Russell et al. 2010); how an Egyptian fruit bat navigates tens of kilometers from its cave to its favorite fruit tree, night after night (figure 8.2a); or how a stork navigates from its nest atop a chimney in Germany, all the way to South Africa, and then back to the same chimney, year after year (figure 8.2b). Laboratory studies on the various types of spatial neurons, as recorded in small laboratory setups, give us important clues about the possible mechanisms that may underlie these navigational feats. However, the spatial scale of natural navigation is so much

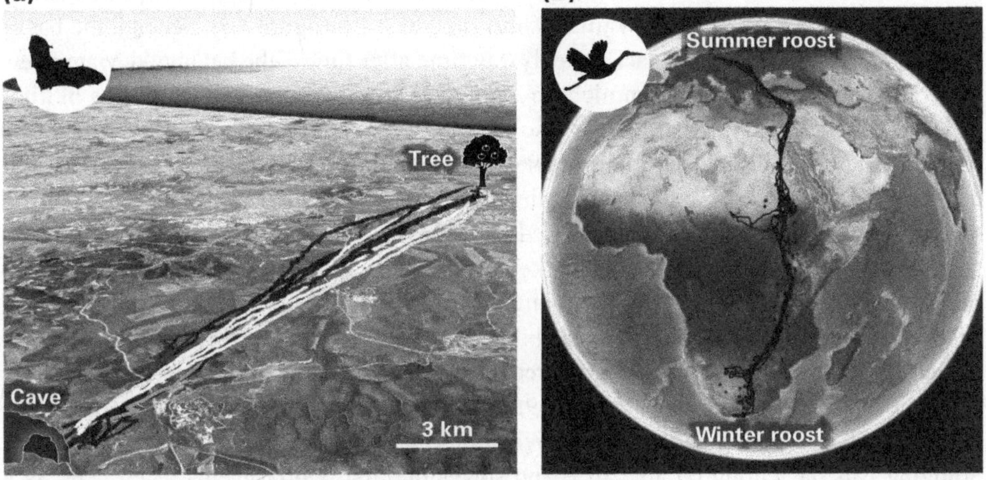

Figure 8.2
Natural navigation. (a) GPS tracking of a single Egyptian fruit bat in Israel. This bat flew from its cave to forage in the same fruit tree, night after night, for seven consecutive nights. Each line shows data from a different night; trajectories are superimposed on a satellite image from Google Earth. Adapted from Geva-Sagiv et al. (2015). (b) Annual migrations of a single stork, named Princess, from the town of Loburg in Germany to near the town of Clarkson in South Africa (Berthold et al. 2004). This stork holds the current world record for the number of years that a single animal has been tracked: she was tracked over a span of twelve years, from 1994 to 2006. Black dots denote the stork's migration flights in the years 1994–2004, while the black lines show the stork's migration flights in the years 2004–2006. Trajectories were taken from Movebank, courtesy of Martin Wikelski, and were superimposed on a satellite image from Google Earth.

larger than the scale of standard laboratory setups that we are still far from elucidating the neurobiology of natural navigation (figure 8.2). So should we implant tetrodes in the hippocampuses of rats, bats, and storks and release them into the wild in order to address these questions? On the one hand, this would be an important advance, as it would give us direct data on "the real thing"—that is, neural recordings in animals navigating outdoors. With today's wireless neural loggers and GPS loggers, such experiments are becoming feasible, as was demonstrated by our recent recording from place cells and head-direction cells in bats that were navigating outdoors on a remote oceanic island (figure 4.3).

On the other hand, such experiments outdoors also suffer from nontrivial difficulties. First, in order to obtain reliable estimates of the neural correlates of navigation, it is important to acquire *multiple* repeated navigation trajectories while recording from the neurons—but this means that one would need to record from the *same* neurons over multiple years in storks (figure 8.2b) or over multiple nights in bats (figure 8.2a). This is not yet feasible in terms of the microSD memory required to store such enormous amounts of data—although this problem will likely be solved within a few years because memory capacity increases exponentially over time. More important, such long-term neuronal recordings are not yet feasible in terms of recording stability. However, miniaturized one-photon microscopes, or "miniscopes," now allow long-term stability over days and weeks (Ziv et al. 2013), and with the recent advent of wireless miniscopes, it is becoming possible to conduct wireless calcium imaging of the same neurons over many days (chapter 3) (Aharoni & Hoogland 2019; Shuman et al. 2020). An alternative approach is to conduct neural recordings in a restricted area outdoors—such as an isolated island—which *does* allow for obtaining repeated coverage of the same region of space within a short time of < 1 h (figure 4.3). Second, the bat or the stork does not follow the exact same flyway time after time—the bat may deviate a few hundred meters left or right from night to night, and the stork may deviate dozens of kilometers in its flyway from year to year—so we shall never get multiple repeats of the *exact* same track. This is problematic for reliable estimation of the neuronal responses. Third, it may be challenging to recapture a wild animal outdoors in order to retrieve the data. Fourth, there are always unknown and unmeasured factors when conducting experiments outdoors, such as possible encounters with nontracked conspecifics, predators, or unexpected meteorological conditions. Thus, performing single-neuron recording in the wild would not be a trivial enterprise. Nevertheless, the feasibility of this approach was recently supported by successful wireless electrophysiological recordings from song nuclei in dueting songbirds, conducted outdoors (Hoffmann et al. 2019), as well as by our recent recordings from place cells and head-direction cells in bats navigating outdoors on a small island (figure 4.3). Studying bats navigating on a small island solves the first two problems listed here, as it allows for obtaining dense behavioral coverage, with multiple flights through one location, all within less than 1 h (figure 4.3)—thus solving the repeatability, stability, and spatial-coverage problems. Studying bats on a small remote island also makes it much easier to recapture the animal. Thus, despite the challenges, wireless neural recordings outdoors are feasible to make and hold great future promise.

Another approach would be to take a middle way between completely free outdoors experiments and the traditional small-scale laboratory experiments indoors—namely, to conduct experiments in large, complex, seminatural laboratory environments. For example, one could use wireless neural loggers to make recordings from the brains of rats navigating through complex natural burrow systems: given the dramatic reorganization of the barrel cortex, which was elicited when rats were allowed to dig natural burrows in soil (figure 5.11) (Polley et al. 2004), we may perhaps expect similarly dramatic effects on hippocampal spatial maps. One also could use neural-logging technology to record from the brains of rodents, monkeys, or other study species, as they move around and navigate in huge outdoor pens—for example, a piece of forest that is fenced off to recreate a portion of the natural environment—or in a very large greenhouse or a barn. Examples for such large setups are the 200-m and 700-m-long tunnels that we built for recording hippocampal place cells in bats flying in ultra-large spatial scales (figures 4.2 and 4.5). These experiments in bats have revealed very surprising results: individual place cells fired in multiple locations (multiple place fields), with different place fields for the same neuron having very different sizes—a multifield, multi-scale code that is very different from place-cell firing is small laboratory setups (see figure 4.2). Hippocampal recordings in large outdoor pens are now being performed in rodents by the group of Azahara Oliva and Antonio Fernández-Ruiz (Paudel et al. 2023). It would also be crucial to study hippocampal coding in animals navigating through complex, large, multicompartment environments where the animal needs to navigate beyond its sensory horizon. This is crucial because navigation in such complex large environments requires *memory* and *integration of information over time*—neither of which is required when navigating in small-scale environments within the animal's sensory horizon (its "vista scale"; also see Wolbers & Wiener 2014). Large outdoor pens, barns, or greenhouses could also be used for studying *collective group navigation,* a highly understudied branch of animal navigation (Berdahl et al. 2018). Such facilities can also be used for studies of natural foraging, hunting, escape responses, and for investigating natural social behaviors in large natural social groups, where the subject animals could move around and interact with other animals.

As discussed already, many social behaviors require very large spaces, which are not available in small laboratory setups. That said, it also would be very important to increase the usage of medium-sized, laboratory-based seminatural arenas: the key point is to use a complex, seminatural environment, with multiple animals, because experiments in such environments can reveal fundamentally new results about brain mechanisms (Zilkha et al. 2016), as discussed in chapter 2 in more detail.[7] Further, as described in chapter 3, there is rapid progress in the miniaturization of relevant electronic components and circuits, and in making them more energy efficient. Thus, we are nearing the day when we could wirelessly record neuronal data from many animals simultaneously, and in each animal record from multiple multisite electrophysiological probes, such as tetrode arrays (Wilson & McNaughton 1993; Pfeiffer & Foster 2013; Widloski & Foster 2022), Neuropixels probes (Jun et al. 2017; Steinmetz et al. 2021; Gardner et al. 2022), polymer arrays (Chung et al. 2019), and their future derivatives—that is, from hundreds or even thousands of contacts in each animal. Thus, we can expect that in upcoming years, researchers will be able to wirelessly record the activity of thousands of single neurons per animal, in multiple animals simultaneously, during free natural behaviors in seminatural conditions—indoors and outdoors.

This ability to record neural activity in naturalistic conditions, from thousands of neurons in each animal—and also from entire groups of animals simultaneously—will be complemented by highly detailed measurements and quantifications of the animals' behavior[8] (chapter 3). Such outdoor pens, greenhouses, and seminatural arenas can be wired with dozens of cameras, microphone arrays, and various localization devices to track the location of the animal, as well as its 3D posture (pose), which is known to strongly affect neuronal firing in certain brain areas, such as the posterior parietal cortex (Mimica et al. 2018). In addition, it is already possible today to mount a whole slew of miniaturized behavioral sensors on animals that are freely moving in natural or naturalistic settings, including (figure 3.3):

1. Miniature cameras that track eye movements in freely moving animals, together with small, on-board cameras that document the external world from the animal's point of view (Wallace et al. 2013; Meyer et al. 2018; Michaiel et al. 2020);
2. Three-axis accelerometers, which provide valuable data on the animal's fine-scale body movements (Kano et al. 2018; Brandes et al. 2021);
3. On-board microphones, including ultrasonic microphones, which allow recording social vocalizations (Kays et al. 2015; Greif & Yovel 2019).

In addition, laser-scanning and multicamera methods can be used to acquire the detailed 3D structure of the natural environment, down to the last stone and last twig—which, together with miniature eye-trackers and on-head cameras, allows for reconstructing the full retinal image that was available to the freely moving animal, moment by moment, as well as reconstructing exactly what the animal looked at (Stürzl et al. 2015; Murray & Zeil 2017; Walter & Couzin 2021). These techniques enable seeing the world "from the animal's perspective," thus allowing for gaining insights into the neurobiology of animals' real-life decision-making processes.[9]

This last point raises yet another future direction for natural neuroscience: real-world decision-making. I already mentioned how behavioral ecology is approaching questions of decision-making by considering a multitude of complex, real-life factors that influence the animal's decisions, such as *competition* with conspecifics or *predation risk*—which is very different from today's neuroscience, where decision-making is usually studied in animals performing very simple two-alternative forced choice (2-AFC) tasks. These real-life factors, such as competition and predation risk, turn out to have a profound influence on decision-making during foraging or while making social choices (Mobbs et al. 2018). Real-life decision-making differs from 2-AFC in many other ways (Pearson et al. 2014; Mobbs et al. 2018; Prat & Yovel 2020; Yoo et al. 2021). First, it is not two-alternative: *in the real world, there are often many more than two binary alternatives.* In navigational decisions, for example, there are usually *multiple* relevant navigational goals—and many possible routes to get there. In social decisions, rodents, monkeys, and bats interact with many individuals (many options), and moreover, the interaction itself is usually complex, graded, and dynamically changing over time: not something that can be captured by a simple binary choice.

Second, *real-life decision-making is often a continuous, sequential process.* Thus, during foraging, animals continuously need to decide whether to stay in a food patch and exploit it, or search for a new food patch—the so-called exploration-exploitation tradeoff (Mobbs et al. 2018)—which is a continuous, sequential, and ongoing process. This sequential, continuous nature of decision-making during foraging has dramatic implications for computational models of optimal foraging, and it provides interesting predictions for the neurophysiology of decision-making (Davidson & El Hady 2019; Kilpatrick et al. 2021). Prey pursuit also requires continuous decision-making: think of a cheetah pursuing a gazelle or a dragonfly pursuing a fly—these predators need to make new decisions almost every millisecond, continuously deciding how to change their speed, direction, and pose. In navigation, too, an animal that moves around outdoors needs to continuously make small decisions: should I veer left or right, and by how much? Such decisions are usually continuous in time and continuous in the space of possible outcomes—certainly not binary like 2-AFC. Even active sensing can be thought of as a sequential decision process: for instance, a bat or a mouse need to continuously decide how to change its echolocation calls, whisking, or sniffing patterns[10] (chapter 5).

Third, *real-life decision-making is volitional.* The animal makes its decisions whenever and wherever it chooses to, rather than being *forced* by the experimenter to choose, as in 2-AFC. Of course, 2-AFC offers an experimental advantage: by forcing the animal to decide, it allows the researcher to easily determine the timing (i.e., *when* did the animal decide). The disadvantage is the absence of free will. This may seem like a philosophical argument about free choice and free will—but we truly do not know how this may affect neural activity in brain areas that are involved in decision-making. Thus, the 2-AFC paradigm is clearly rather different from real-life decision-making.

Furthermore, neuroeconomics—one of the prevailing frameworks in laboratory studies of the neurobiology of decision-making—has several additional key limitations, which are beyond the scope of this chapter: these were discussed in detail in critical reviews by Peter Dayan and Michael Platt and colleagues, who called for more naturalistic experiments in decision-making (Pearson et al. 2014; Mobbs et al. 2018). In particular, the theoretical frameworks of *control theory* and *reinforcement learning* may be better suited for modeling continuous decisions than the classical framework of economics and neuroeconomics, which better fits highly discretized decisions (Yoo et al. 2021).

It is quite likely that novel discoveries about the neurobiology of decision-making will emerge from natural neuroscience experiments in which manipulations and recordings of large neuronal populations would be done in freely moving animals under naturalistic conditions during *foraging*—including group foraging—as well as *social decision-making*. In foraging experiments, natural decision-related variables such as *Predation risk* and *Competition* could be manipulated (see section 8.1 and equation (8.1)), together with rich measurements of the animal's behavior (chapter 3). This should be combined with detailed measurements of the spatiotemporal distribution of food, as well as measurements of the sensory information available to the animal: the sensory information that is being used by the animal to make its decisions. These sensory measurements could be achieved via on-head cameras to record the visual information available to a rodent, as well as the animal's whisker movements; by using thermocouples for measuring olfactory sniffing; or by using on-head microphones and ground-based microphone arrays to measure the sonar image available to a bat (Kothari et al. 2018). Extremely rich and detailed information can now be acquired about decision-related variables, the animal's behaviors, and the sensory information that is available to individual animals during natural foraging. Further, this can be done in groups of freely foraging and socially interacting animals, while recording from the brains of the entire group. The wireless technology is already here—now it is up to our imagination to catch up.

* * *

The approach described so far holds great promise for uncovering fundamentally new insights about the neural bases of natural behaviors—but it also faces some difficult challenges. Perhaps the most central challenge has to do with the point that was already discussed in chapter 1; namely, by going to more natural behaviors, we are relaxing our experimental control and run into the *curse of dimensionality*, stemming from the complexity and high dimensionality of behavior. Specifically, the difficulty stems from a combination of factors:

1. During natural unconstrained behavior, the animal is varying multiple behavioral variables simultaneously—not just one variable, as in classical controlled experiments. This means that in complex, multidimensional behaviors, the data fills the behavioral space more sparsely than for 1D, simple laboratory experiments, and hence longer recordings are required to obtain enough behavioral data to fill well the multidimensional behavioral space. For a high-enough dimensionality, these long recordings may become impractical because of limitations on recording time (both technical limitations—how long neurons can be recorded, and behavioral limitations—how long the animal is willing to behave in a particular way). Moreover, behavior is not random, and therefore it will *never* fully fill up the entire N-dimensional behavioral space, even if we could record for very long times. On the other hand, this is also an *advantage*—because the behavior fills up only that part of the behavioral space that is *relevant* for the animal; any behaviors outside of this subspace are irrelevant to investigate in the first place.

2. During natural behaviors, the multiple behavioral variables are often correlated to each other over time, with the correlations being created by the animal's nonrandom dynamic behaviors.
3. The functional tuning of cortical and hippocampal neurons is usually multidimensional, exhibiting mixed selectivity—that is, tuning to multiple variables, including multiple behavioral variables (Rigotti et al. 2013; Raposo et al. 2014; Eichenbaum 2015; Hardcastle et al. 2017; Sarel et al. 2017, 2022; Mallory et al. 2021; Noel et al. 2022; Ray et al. 2024; Testard et al. 2024). This mixed selectivity was proposed to confer two benefits: it allows large coding capacity, and it enables downstream brain areas to read the information using a simple linear decoder (Fusi et al. 2016). At the same time, this high dimensionality means that it is more difficult to characterize the neurons.

Points 1, 2, and 3 (i.e., the combination of multidimensional behavioral correlations during natural behaviors and of multidimensional neuronal tuning) means that it becomes a real challenge to understand what the neuron is responding to.

Several approaches have been used to tackle this problem in the past, including generalized linear models (Truccolo et al. 2005; Acharya et al. 2016), data-based reconstruction methods (Rubin et al. 2014; Sarel et al. 2017), and partial correlations (Stark et al. 2006), as well as more sophisticated mathematical methods (Lepage et al. 2012; Wu et al. 2017; Williams et al. 2018; Stringer et al. 2019). However, all these methods have certain drawbacks. The generalized linear model, for example, relies on a linear model of the data, as its name implies—and is therefore limited in its ability to capture nonlinear data. This problem is solved in part by generalized additive models (GAMs) (Hardcastle et al. 2017; Mallory et al. 2021; Balzani et al. 2022; Noel et al. 2022). A GAM is an extension of the generalized linear model, which is more powerful for capturing complex tunings; but it retains the major advantage of GLM—namely, it allows for *quantifying* the contribution of each behavioral variable to the neuronal activity. We have recently made progress in using GAMs while also addressing the problem of correlated behavioral variables (Ray et al. 2024). However, there is still much work to be done.

Thus, a major challenge for theoretical neuroscientists in upcoming years is to develop better analysis methods for elucidating the tuning and selectivity of individual neurons during natural behaviors, given that many neurons exhibit multidimensional tuning to multiple behavioral variables (i.e., mixed selectivity), and given that complex natural behaviors are composed of many correlated behavioral variables. This challenge, in my view, constitutes a key theoretical challenge for the future of neuroscience.[11] The new computational methods that will be developed must then be applied to data from high brain areas such as the prefrontal cortex, parietal cortex, and hippocampus—where we know that mixed selectivity is very prominent—but also to lower brain areas, where these questions have not been examined (yet). It might turn out that multidimensional mixed selectivity is in fact widespread throughout the brain.

A second challenge is to develop better approaches for elucidating the population activity of *large ensembles of neurons* during natural behavior. This is both an experimental challenge and a theoretical challenge. Experimentally: since natural behavior involves many brain areas, the relevant neural activity is likely distributed across multiple areas. This requires developing better experimental methods for recording large populations of individual neurons across multiple brain regions—in freely moving, freely behaving animals. This is a nontrivial technical feat, but it is becoming evermore feasible, e.g. through the use of multiple Neuropixels probes. This is also a theoretical challenge: previous computational work on quantifying the dynamics of neuronal ensembles has focused on relatively simple, low-dimensional, and highly controlled behaviors (Vyas et al. 2020). Recent work has begun to elucidate relationships between neural ensembles and more naturalistic behaviors (Markowitz et al. 2018; Rubin et al. 2019; Bagi et al. 2022); however, these methods have mostly succeeded in investigating repetitive and stereotypical behaviors. Since most natural behavior is *not* very stereotypical, it would be very important to extend this work to complex, nonstereotypic naturalistic behaviors, and to develop computational methods that allow capturing the relation between multidimensional complex behaviors and multidimensional neuronal population dynamics (Williams & Linderman 2021). The difficulty is that population activity is quite complex, even during relatively simple behaviors (Mante et al. 2013; Vyas et al. 2020; Gardner et al. 2022)—so it might be even more complex during natural complex behaviors. A key challenge thus remains: to develop new data-analysis methods for elucidating the relation between multidimensional ensemble neuronal activity and multidimensional natural behavior.

A third challenge is quantitatively coping with the *complexity of behavior*. Behavior is often hierarchical, evolves over multiple timescales, and is context dependent. A related problem is that in less-controlled, naturalistic environments, it is often difficult to measure what the animal is paying attention to, and what it is responding to behaviorally. However, this set of problems is now becoming feasible to address via the combination of (1) the rich multisensor behavioral measurements described above and (2) the novel machine-learning methods for automatic analysis of behavior, which were discussed in chapter 3—specifically, methods for automatic tracking, segmentation, and pose estimation of animals performing natural behaviors (Datta et al. 2019; Mathis & Mathis 2020). This was not possible until a few years ago, but it is now facilitated by the rapid advances in sensor technology and in automated behavioral analyses (see chapter 3).

A fourth challenge is identifying the *algorithms of natural behavior*. If we consider Marr's three levels of analysis (Marr 1982) with respect to the neural basis of behavior, then a gap emerges. "The neural basis" usually refers to biological implementation—that is, Marr's bottom level (implementational level; e.g., the level of neurons and neural circuits). "Behavior" often refers to the functional level—that is, Marr's top level (function). What is missing is the algorithmic level—Marr's intermediate level. The algorithmic level forms a critical midlevel link between neural circuits and behavior. As already discussed, Matteo Carandini argued that a crucial step is to understand the *algorithms of neural computation* (Carandini 2012)—that is, he placed the emphasis on identifying algorithms in the brain. But equally important is to identify *behavioral algorithms* (i.e., the algorithms that generate behavior), which may be equally difficult to elucidate. Nevertheless, promising steps have been made. For example, substantial headway has been achieved in deciphering the behavioral algorithms of *pursuit* and *evasion* behaviors (as reviewed in Hein et al. 2020); and some progress has been made in deciphering the behavioral algorithms of path-integration *navigation* behaviors (Cheung & Vickerstaff 2010; Vickerstaff & Cheung 2010). However, it is clear that algorithmic understanding of natural behaviors remains a major open challenge for the future (Hein et al. 2020).

A fifth challenge is to take into account the animal's *behavioral state*, as well as its *motivational state*. These are fundamentally difficult to measure but may strongly affect neural activity. Regarding behavioral state, the problem is not just measuring it, but the basic difficulty is that behavior exhibits strong sensitivity to small differences in the input (i.e., it is highly nonlinear). Specifically, a small transient difference in sensory input can create a large behavioral switch—a long-lasting change in the behavioral state of the animal. For example, when a primate observes a conspecific displaying an aggressive facial expression, it will react very differently than if the conspecific displays a "play face" (an invitation to play; see Dugatkin 2020). In the first case, a long agonistic interaction may ensue, while in the second case, a long play bout will take place. Thus, very short sensory input (facial expression), which lasts only a second or two and differs only subtly at the level of sensory input, can make a huge behavioral difference, propelling a behavioral switch of the animal into an agonistic state or into a play state, which may last for many minutes. The neural mechanisms that underlie such dramatic switches or bifurcations in behavioral state remain enigmatic—and are fundamentally difficult to study due to the long timescales involved—yet such bifurcations in fact are rather common in animals' natural behavior and should be incorporated more regularly into our thinking. Arguably, the neural mechanisms of such switches should be best studied under natural conditions, which is yet another example of the need to pursue natural neuroscience research.

Regarding motivational state, we may hypothesize that for any behavior, the activity of neural circuits should reflect not only the overt behavior of the animal, but also the *reason* for this behavior (i.e., the animal's internal motivational state). For instance, when an animal navigates outdoors from location A to location B, it can do so for a number of very different reasons (i.e., internal motivations such as foraging, predator escape, homing, migration, or navigation for reproduction), and these internal factors themselves may potentially be represented by navigation-related neurons, such as hippocampal place cells.[12] This possibility implies that we need to study how neural activity may be affected by various real-life motivational factors, which are crucial for shaping the natural behaviors of animals—but are usually neglected in mainstream neuroscience.

A sixth and final challenge, which is related to some of the other challenges, is that *in natural behaviors, we never obtain exactly the same behavior twice*—so how can we understand the neuronal activity elicited by a behavior, if the behavior itself is not precisely reproducible? The answer to this has several facets:

1. We need to find and study specific natural behaviors that are as reproducible as possible, naturally—that is, where the animal chooses to perform the behavior multiple times in a similar manner. In other words, we should seek natural behaviors where the animal "is doing the control for us"—in the sense that it naturally repeats the same behavior over and over in a reproducible manner. Specifically, as discussed in chapter 1, it is important that the animal naturally performs *several* distinct behaviors, each of them being repeated over and over (e.g., several distinct social behaviors or several hunting behaviors), and this multiplicity of distinct behaviors then allows *comparisons* of neural responses between them. In other words, instead of a human experimenter "controlling" the animal by training it to perform repeatedly two distinct artificial behaviors (e.g., lick left versus lick right), we ought to utilize the natural tendency of animals to repeat *natural* yet distinct behaviors over and over. In this way, the animal will "control" the experiment for us. Such self-initiated natural behaviors can include hunting behaviors, escape behaviors, social behaviors, and a host of other behaviors

that animals repeat time and again—some of which have been discussed throughout this book. Studying such natural reproducible behaviors relinquishes the need for the experimenter to enforce behavioral reproducibility via external intervention. Perhaps such behaviors will not always be as reproducible as in highly controlled laboratory assays, and not as reproducible as the amazing temporal reproducibility of a birdsong, a bat's attack on an insect, or a jamming avoidance response in electric fish (figure 1.2, top right)—but they should be reproducible enough. Importantly, a century of research in ethology and behavioral ecology has demonstrated that one can in fact find many naturally occurring behaviors that are quite reproducible, wherein the control is provided by the animal, of its own volition. It is up to us neuroscientists to find such natural behaviors where the animal is doing the control for us and to study them mechanistically ("how?"), all while keeping an eye on the functional significance of these behaviors ("why?") (table 8.1).

2. Perhaps the variability of behavior is not as problematic as we fear. Maybe some neurons do not care so much about motor variations in behavior—which we can measure and quantify using the advances in automatic behavioral quantification that were outlined in chapter 3; rather, perhaps the neurons mostly care about the *meaning* of the behavioral motifs (i.e., the syntax of behavior). If so, then by segmenting and classifying continuous behavior into repeatable motifs, we may be able to identify repeatable units of behavior to which the neurons respond quite reproducibly, despite the apparent motor variability.
3. Continuous behaviors without any repetitions *are* in fact amenable to neuroscientific research. Some successful examples include place cells and grid cells, which were studied for many years in 2D arenas, where the animal is running *randomly* (or almost randomly), without repeating any movement twice and without any trial structure—and yet wonderful insights have emerged from this line of research. These neurons provide a beautiful example of how brain mechanisms can be studied in continuous behavior, without any repeating trials (chapter 4). Similar examples of neuroscience research using continuous behavior have been discussed for human neuroscience (chapter 7). Many natural behaviors are continuous and nonrepeating—and therefore, an important aspect of natural neuroscience in years to come will be to conduct mechanistic studies of the brain as the animals perform natural continuous nonrepeating behaviors.
4. To account for the high-dimensional, mixed selectivity of neurons, we must develop new computational tools for disentangling neuronal responses to multiple behavioral variables, as discussed previously.

At the end of the day, however, it might be the case that when studying natural behaviors, we will be able to account for only part of the variability in neuronal responses—despite our best efforts. We must be willing to accept this possibility. But at least we shall know that whatever we have accounted for is highly relevant for the animal.

8.3 Summary

Let me finish with a personal note.

The challenges outlined in this book should not deter us from studying the neurobiology of natural behaviors, which is a central goal of brain research. Rather, these challenges should inspire us to work hard to overcome them. In my own research on the neurobiology of navigation, every time that we moved in a more naturalistic direction, we discovered something surprising about the brain (chapter 4). When we studied how three-dimensional (3D) space is represented in the hippocampal formation, we discovered 3D place cells, 3D head-direction cells, which surprisingly used a toroidal coordinate system to represent direction, and 3D grid cells, which did *not* form a hexagonal 3D lattice—a result that strongly argues against current models of grid cells (Yartsev & Ulanovsky 2013; Finkelstein et al. 2015; Ginosar et al. 2021, 2023). When we introduced navigational goals for the animal to navigate to, we discovered neurons that represent vectors to the goal (Sarel et al. 2017). When we studied two bats interacting in one space, we discovered social place cells that represent the position of the other animal (Omer et al. 2018). When we investigated groups of socially interacting bats, we discovered that hippocampal neurons tracked the locations and identities of multiple other individuals and encoded their sex, dominance hierarchy, and social affiliation (Ray et al. 2024). When we allowed bats to dynamically change their behavior, we discovered extremely rapid switching in the neural code (Sarel et al. 2022). When we increased the size of the experimental arena to 200 m, we discovered a multifield, multiscale representation of space—unlike anything seen to date in the mammalian hippocampus (Eliav et al. 2021). I am certain that future naturalistic forays by our lab will keep yielding equally surprising findings.

Likewise, as I outlined in this book, I strongly believe that *every* field of neuroscience should move toward studying evermore natural behaviors. The payoff will be enormous—a true mechanistic understanding of the neuroscience of natural behaviors. Natural behaviors are what brains and neural circuits evolved to do. Natural behaviors are the one thing that we know for certain to be truly relevant for animals. Natural behaviors activate neuronal circuits differently than nonnatural ones. Therefore, we should relentlessly try to understand the neural basis of natural behaviors. This is the goal of natural neuroscience.

Epilogue: Questions and Answers on Natural Neuroscience

While I was discussing this book with colleagues and circulating its drafts, I received many excellent comments, questions, and criticisms—both large and small—which have improved the book tremendously. However, some of these questions and comments were very broad in nature and went beyond the context of any particular chapter of the book. I therefore decided to compile these bird's-eye view questions into this epilogue, since I reasoned that this format would allow me to address these broad critiques in a somewhat informal manner. While I allowed myself some rephrasing of the questions, for enhanced clarity, I made an effort to maintain the spirit of the original question. I hope that this informal epilogue would allow the most direct exposition of the critique—as well as the most unfiltered answers.

The book seems like overkill. You argue very forcefully—and I must say, convincingly—for the need to study the brain under more natural conditions. But isn't that obvious? Is there anyone who will argue that it is* not *the ultimate goal of neuroscience to understand behavior and cognition in the real world? So why write a whole book about it?

No, I don't think it's overkill. Yes, I agree that there are very few neuroscientists who will argue about the importance of understanding the neural basis of behavior and cognition in the real world. However, currently only a small fraction of neuroscientists actually *do* that (i.e., take the natural neuroscience approach and actively study the neural basis of natural behavior in their own laboratories). There is a very big difference between keeping this as a general long-term motivation—namely, that "someday" there will be "someone" who will start studying the brain under natural conditions—and *actually* starting to do natural neuroscience *today*. It's the difference between "somebody else will do it in the far future" versus "I will do it starting today."

The view that someone else will do this is also related to a narrow, confirmatory view of natural neuroscience, which is *not* what I promulgate in this book—namely, that the role of naturalistic studies of the brain is merely to confirm ideas and theories that were developed based on laboratory studies. This notion was eloquently described, for example, by Rust and Movshon (2005) in their paper "In Praise of Artifice." This approach lays out a confirmatory role for naturalistic experiments, whereby the "real work of neuroscience" is done in the lab, in highly controlled experiments, and any research that uses more naturalistic stimuli or naturalistic behaviors is aimed simply to confirm the results and ideas that were generated by "real" laboratory work, with its highly controlled and simplified stimuli and behaviors. My answer to this notion is that beautiful theories indeed have been generated by very smart and creative neuroscientists based on laboratory studies—but if these theories are detached from natural behaviors, they may be fundamentally mistaken, at their very core, for all the reasons that I have tried to convey throughout this book (see in particular chapter 1). Such lab-based experiments and theories may have led us astray, at least in some cases, due to their oversimplified nature. Therefore, the role of naturalistic experiments is *not* merely confirmatory; rather, natural neuroscience should take the lead in how we study the brain experimentally, and it should be at the core of how we generate theories about brain function.

Finally, taking the natural neuroscience approach one step further: if what I claim in this book is correct—namely, that the brain operates very differently under natural and artificial conditions—then neuroscientists should be really worried. This is *not* business as usual for brain research. Instead, this means that we should rethink how we study the brain and start researching it much more extensively under naturalistic conditions, with a focus on natural behaviors.

It feels that the types of questions asked and the types of answers sought after are different in natural neuroscience compared to classical controlled neuroscience. Is this true?

To some degree, yes. Classical neuroscience tends to examine simple isolated behaviors and tries to dig down to lower levels of description in order to produce mechanistic explanations. This is the essence of reductionism: it aims to explain the whole in terms of its parts. Natural neuroscience, on the other hand, is less reductionist, and therefore less mechanistic in nature, so instead it seeks a systems-level understanding. This is because natural neuroscience is part of the complex-systems approach to conducting science, which emphasizes that in complex dynamical systems—which are characterized by highly nonlinear emergent phenomena—the whole is more than the sum of its parts. Therefore, it may be less useful to dig down into describing ever-lower-level components. By "less useful," I mean that digging far down is less likely to produce insight, to produce understanding. This is not to say that natural neuroscience does *not* look at all for underlying mechanisms—it does!—but less so than classical neuroscience. It's a difference in emphasis and focus. So, ultimately, the difference between classical neuroscience and natural neuroscience is not just about methodological differences and how to run experiments; it's a philosophical difference regarding the nature of understanding. What does it mean to understand the brain, or to understand behavior?

On the other hand, not everything between classical and natural neuroscience is different. For example, in both approaches, there is a strong emphasis on characterizing what a single neuron is responding to—on understanding which variables affect the neuron and what its tuning curve is. It's true that in classical neuroscience, one would use simple stimuli and simple behaviors, while in natural neuroscience, one would use natural stimuli and natural behaviors—but in both cases, the goal is to understand what a neuron is responding to. Likewise, many contemporary questions on neuronal population dynamics and coding by ensembles of neurons can be studied under naturalistic conditions and using naturalistic stimuli and behaviors—in a similar manner to the more controlled artificial experiments.

So natural neuroscience is *not* an alien domain of inquiry. On the contrary, actually—some of the questions are the same. In fact, I think that it is very possible and desirable to take concepts and questions from classical neuroscience, such as tuning curves of neurons and ensemble dynamics, and study them using more naturalistic stimuli, natural behaviors, and real-life contexts.

Suppose that I would use a larger, richer, more naturalistic environment for conducting neural recordings or neural manipulation experiments while the animal is behaving in this new and improved environment. My question is: How would I know that the animal's behavior is truly naturalistic?

This is a very good question. Let me accentuate your question using a somewhat contrived example: the falling car analogy (courtesy of Ehud Fonio). Suppose that you decide to test a regular car under free-falling conditions, as it's falling from an airplane toward the ground. Stepping on the gas pedal, for example, will not affect the car's speed, but it will cause the car to pitch (due to Newton's third law); turning the steering wheel, on the other hand, might cause lateral yaw rotation (due to the altered wind resistance to the turned wheels). Although these observed behaviors can teach us something about cars, they have nothing to do with the natural behavioral repertoire and original function of driving cars—hence, it's clearly better to test the car under its normal, naturalistic conditions (namely, on a road). What you are asking basically is: If you didn't know anything about cars, how would you know whether there is something wrong with the behavior of the free-falling car? How would you know that it's not even close to the car's natural behavior? This illustrates what may be called "the measurement fallacy"—the illusion that if you perform the requisite scientific procedure and all the right measurements and controls, then these measurements must be useful and insightful. The measurements I am referring to are the following: (1) you *measure* the system's behavior in great detail (e.g., a car pitching and yawing); and (2) this behavior is *reproducible* (every time you throw a car out of an airplane and press the gas pedal or turn the wheel, the car would behave in the same way); and (3) you characterize in detail the internal *mechanisms* (let's say that you measure the activity of every single electric circuit in the car and measure the torque on every single screw in the car's engine while the car is free-falling). But as I tried to convey in this book, this is a fallacy: The car example demonstrates how testing the behavior outside its natural context can lead to meaningless results because if the behavior is far removed from the normal behavioral repertoire of an animal, then the insights that can be gained are rather limited—they may even mislead the researcher.

How would you know that the behavior that you observe is far removed from the normal behavioral repertoire? By comparing the behavior measured in your experiment to the same behavior in the same species when it's measured outdoors. Now, in our contrived example, if one would study the natural behavior of cars outdoors, it would be pretty easy to see that it is very different from the behavior of the same type of cars in the free-falling experiment. No fancy analyses are needed to see the deep difference. However, if we put aside our car example and come back to actual animal behavior, the problem is that it is not always easy to understand whether the differences between a certain behavior of an animal in the wild and in the lab are large or small, and whether they are meaningful. My answer to this difficulty is: We have to observe this behavior very carefully outdoors, in the wild, and compare to how it's expressed in the lab in order to test the validity of the lab behavior. We have to be field-ethologists in our soul! And if for practical reasons, we are ill equipped to conduct outdoor studies of this behavior ourselves, then we must at least carefully read the relevant ethological literature and consult with ethologists, zoologists, and behavioral ecologists about the behavior that interests us in order to get a truly full picture about how the behavior looks in the wild. If we don't do this, then we run the risk of studying the equivalent of the neural basis of free-falling cars.

You put a lot of emphasis on natural behaviors and stimuli and argue against using artificial stimuli and behaviors, and against heavy training. However, one may counterargue that many human learned behaviors are artificial—for instance, driving a car, using a smartphone, or studying neuroscience. In fact, the ability of the human brain to master these artificial behaviors is what makes us human. Likewise, artificial stimuli can be very important. The reader of your book is now reading these letters, which are highly artificial symbols that are very recent from an evolutionary standpoint (with writing being invented only 5,000 years ago)—and yet they carry a lot of meaning to the reader's brain. The ability to assign meaning to arbitrary stimuli is a hallmark of a system with a high capacity to learn, such as the mammalian brain.

You brought up several fundamentally different examples and issues. Driving a car mostly involves motor skill learning and requires long training. I obviously agree that if one is interested in the neural basis of motor skill learning, which requires heavy training, then it *is* justified to embark on heavy training. I have no argument with that. Your second example—using a smartphone—is a very different case, as it is learned very quickly and does not require heavy training. In fact, many behaviors in humans and animals are learned within a few trials and do *not* require thousands of training repetitions, as is commonly done in neuroscience experiments. So, in these cases, the problems are the *amount* of training involved and the common practice of using extensive training or overtraining of animal subjects in neuroscience. As I discussed in chapter 1, I think that such overtraining should be minimized. Your third example—studying neuroscience—is an abstract thinking capacity, which is probably unique to humans. I would be happy if neuroscience understands one day how the human brain studies itself; but this is beyond the scope of the current book.

Your fourth example—the reading of letters—actually serves to illustrate my point, I believe. Letters are meaningful only insofar as they are strung into meaningful words, sentences, paragraphs, and stories, and are used to construct entire texts with a beginning, middle, and end in order to convey complex messages. I have no problem with brain studies of reading or storytelling—on the contrary, I think that these are great examples of naturalistic studies in humans, and stories are indeed excellent stimuli. My problem, as I outlined in chapter 7, is with the common use of arbitrary random lists of words or syllables, which are often utilized in studies of human memory. These random arbitrary lists do *not* carry any meaning and therefore are much less useful in my mind than using coherent stories as stimuli for memory research. Here lies the exact difference, if you care to go back to figure 1.2: arbitrary random lists of words are controlled but highly artificial stimuli (top-left corner of figure 1.2); by contrast, stories or movies are somewhat less controlled but much more natural stimuli (right side of figure 1.2). I prefer the latter.

Aren't you confusing "control" with "repeatability" in behavioral neuroscience experiments? Some natural behaviors are highly repeatable, but* not *controlled. "Controlled" means that we, as experimenters, can turn a knob to change the behavior of the animal so that we can later* compare *the different conditions (i.e., different behaviors). Without such control, there is no comparison—and without comparison, we cannot understand the neural basis of behavior.

This notion of "control" is close in spirit to the notion of "causality"—a manipulation or perturbation that allows understanding causal relations between neural activity and behavior. I have two issues with this idea. First, as discussed in chapter 1, causality in neuroscience is a tricky business because of the network structure and complexity of both brain and behavior. It is not easy to *truly* prove causality in the brain. Second, it is an illusion that by "turning a knob," you fully control the animal's behavior and get rid of the animal's free will. Even if you take the classical approach to behavioral neuroscience—and you train the animal to do one behavior when you give one cue and a second behavior when you give a second cue—it is still the animal that chooses to perform one behavior or another. Likewise, in natural behaviors: the animal chooses to perform one behavior or another. So what is the fundamental difference? I would argue that there is none—and that the perceived advantage of trained behaviors is illusory. In both cases, the animal chooses to do one of several behaviors, and the researchers are looking at the neural correlates of these different behaviors. I think that if there is no fundamental difference in terms of free will, and all we do is to analyze the neural correlates of freely executed behaviors, then it's better to focus on the animal's *natural* set of behaviors and compare neuronal activities to these natural behaviors. The key requirements for this are that the animal performs several natural behaviors that are different from each other; and each of these behaviors is repeatable, with the animal exhibiting many similar repeats of each. The combination of these two requirements will allow conducting a systematic *comparison* between the neural responses to these behaviors—and as you pointed out, such comparisons are indeed crucial for understanding the neural basis of natural behaviors.

There are increasing numbers of mouse experiments that are done in freely moving mice, albeit in small laboratory setups. Do you consider these to be "natural neuroscience" experiments?

First, in my mind, it is a wonderful progress when *any* experiment in behavioral neuroscience is done in freely moving, freely behaving conditions—that's already much better than head-fixed experiments, and it gets us closer to understanding the neural basis of behavior. Second, however, the answer really depends on which kind of experiment we are talking about. If this is a behavioral task that requires extensive training—and there are *many* of those in today's neuroscience—then it's less wonderful, as we are running into all the problems that were discussed in detail in chapter 1, about overtraining, unwanted plasticity in the brain, and other issues. But if this is a more naturalistic experiment—investigating social behaviors in a seminatural large arena, foraging/hunting behaviors, or escape behaviors—then this is much more in the spirit of natural neuroscience.

Aren't you worried that highly natural, uncontrolled experiments generate data that are impossible to analyze and interpret because of too many confounding factors,

uncontrolled behaviors, or multiple behaviors that are correlated to each other, and these make it nearly impossible to understand what the neuron actually is responding to? In other words, perhaps natural behaviors are too high-dimensional to allow true understanding of their neural mechanisms?

How to analyze neural data during natural behaviors indeed presents a major challenge. There are two answers here. First, not all naturalistic behaviors are high-dimensional. One example is our experiments in which two bats were flying in the 200-m-long flight tunnel and were avoiding collisions with each other (figure 4.5). Here, there were only *two* relevant variables (the recorded bat's position and the distance to the other bat), even though we did allow for expressing a key aspect of natural behavior—namely, switching between different behaviors (in this case, switching between navigation and collision avoidance). Therefore, it is *not* the case that natural neuroscience necessarily means very high-dimensional data. In this example, it meant two-dimensional (2D) data, which are reasonably easy to analyze and interpret. Second, some natural behaviors are indeed high dimensional, but recent years have seen the development of several computational approaches that allow estimating the contribution of each behavioral variable to neuronal firing, despite the high dimensionality of the behavior and despite the behavioral correlations (i.e., correlations between the different behavioral variables). Some of these approaches were mentioned in chapters 4 and 8. Just as one example, in our study of hippocampal activity in groups of 5–10 bats, which was described in chapter 4 (Ray et al. 2024), we used for this purpose a generalized additive model (GAM). This is a powerful method that allows for assessing quantitatively the contribution of each behavioral variable to the neuron's firing. To overcome the problem of correlations between behavioral variables, we found that we can fit two models separately: an allocentric model, with only allocentric variables (self-position, self-head direction, and absolute positions of the other bats); and an egocentric model, with only egocentric variables (relative distance and relative direction to the other bats). Within each model, the variables were *not* strongly correlated to each other, thus overcoming the problem of behavioral correlations. We then used a model-selection approach (based on deviance information criterion) to choose the best-fitting model. This approach overcomes most of the problems that were raised in this question. That said, I fully agree that indeed it is crucial to further develop better computational approaches for deciphering to which behavioral variables the neurons are responding.

What are the biggest challenges facing natural neuroscience? Is it collecting the experimental data? Analyzing them? Or something else?

I think that the experimental methods for data collection are *no longer* the main challenge for studying the neural basis of natural behaviors—although they were some decades ago. That's because, as discussed in chapter 3, we now have robust methods for wireless recording (and manipulation) of the activity of multiple neurons in the brains of freely moving, freely behaving animals; and we also have witnessed an explosion of new methods for measuring the detailed dynamics of the behavior itself. Instead, I believe that the main two challenges are (1) *selecting an appropriate behavior* that is performed in a reproducible manner by the animal, and yet is ethologically interesting; and (2) the *data analysis.*

The key problems for data analysis are threefold. First, as mentioned in the previous question, in many cases natural behavior is high dimensional, comprising many behavioral variables, such as the animal's position, posture, emitted vocalizations, gaze, and interactions with different individuals. Even if we carefully select a reproducible natural behavior, which the animal is repeating many times within a recording session, this still remains a high-dimensional problem. Second, these behavioral variables are correlated: for example, when interacting with a specific individual, the animal may change its posture in a certain way and emit specific social vocalization—with the postural change and vocalization often cooccurring together—and these behavioral correlations make it difficult to disentangle exactly what is the neuron responding to. The third element is sparse behavioral sampling: even if we are interested in a particular subset of N behavioral variables, the natural behavior of the animal during a given experiment is typically sampling only part of this N-dimensional behavioral space (for $N > 3$ or 4). The good news is that, as previously mentioned, recent years have seen a substantial progress in data analysis methods, which now allows for tackling these issues (e.g., by using GAMs). However, although these methods have advanced our ability to decipher the function of individual neurons during natural behaviors, there is still much work to be done. Therefore, I believe that a major challenge for theoretical and computational neuroscience in the coming years would be to develop new data-analysis methods for analyzing neural activity during natural behaviors—specifically, methods for characterizing the multidimensional response properties of neurons in the face of multidimensional, correlated, and sparse behaviors.

You can't seem to make up your mind: Are you for or against controlled laboratory experiments? This leaves the reader with a tension. On the one hand, throughout the book, you provide very systematic arguments **against** ***classical laboratory experiments. On the other hand, in several places, you explicitly write that neuroscience as a field should pursue both the natural neuroscience approach*** **and** ***laboratory-based experiments. So what do you truly believe?***

Great question! I am indeed truly ambivalent. On the one hand, as I detailed throughout the book, I see a long list of fundamental problems with classical laboratory experiments. On the other hand, these controlled experiments do have their value, in three respects. First, they lay the basic groundwork for understanding some of the fundamental phenomena in brain and behavior—albeit with many caveats about the interpretation of such experiments, as detailed in chapter 1. Thus, controlled experiments provide a useful first step for understanding the neural bases of behavior. But of course, this is just a first step; it must be followed by more naturalistic experiments, as detailed in the book. Second, I truly believe that there is great power in combining both types of experiments: e.g., in recording the activity of the *same* neurons under both controlled laboratory conditions *and* during more naturalistic behavior. This compare-and-contrast approach is very useful, in my mind. Third, I think that even the classical laboratory experiments can be improved dramatically by making them somewhat more natural. Therefore, my take-home message is not that people should necessarily abandon entirely their highly controlled setups. They are welcome to abandon them, of course, as I did when I switched from head-fixed experiments to freely moving ones—but even if they opt to stick to controlled experiments, they must strive to improve their naturalness. For example, as detailed in chapters 3 and 7, it is important to improve the realism and naturalness of VR systems that are being used in both animal head-fixed research and human fMRI research. Thus, even if you elect to stick to head-fixed animal experiments or to human fMRI experiments that are conducted inside the magnet, you should make every effort to improve their realism. The same goes for many other neuroscientific experiments. Thus, the message of the book, in my mind, is that we should strive to study the brain in situations closer to the natural behavior of the animal—where "closer" can mean in truly naturalistic experiments, in the spirit of natural neuroscience, or it can mean performing classical laboratory experiments in a somewhat more naturalistic manner. Any improvement in naturalness—large or small—is welcome.

Glossary

Autocorrelation of spike trains: The autocorrelation of a spike train recorded from one neuron shows the probability of having two spikes at a particular time lag (time interval). A periodic structure in the autocorrelation indicates a temporally periodic, oscillatory firing of the neuron (e.g., figure 2.5b, "rat"), while the lack of such a periodic structure indicates lack of such oscillatory firing (e.g., figure 2.5b, "bat").

Collective behaviors: Behaviors of a group of animals, neurons, or physical particles, which display nontrivial synchronized dynamics that cannot be easily explained by the properties of the individual components. Such behaviors often occur in complex systems, such as animal groups (e.g., bird flocks, fish schools, and ant colonies), neural networks, and physical complex systems.

Complex system: A complex system is a system composed of many interacting components, which may create complex collective behaviors that cannot be easily explained by the individual components. Such systems are often described as those for which "the whole is greater than the sum of its parts." Examples of complex systems include neural networks, economic markets, bird flocks, ecosystems, or the Earth's atmosphere.

Cortical column: A group of neurons forming a vertical columnar structure in the cortex, perpendicular to its surface. In monkeys and cats, the neurons within a cortical column often have similar functional properties (e.g., similar tuning curves).

Cortical map: A collection of cortical columns, along which the encoded property changes gradually. A classical example is the map for orientation tuning in the primary visual cortex of monkeys or cats, wherein the preferred orientation of neurons (the orientation to which the neurons respond maximally) changes gradually along the cortical surface.

Declarative memory: An umbrella term encompassing two types of memory: *episodic memory* (memory for personal events, such as whom I met at my fortieth birthday party, and what we talked about); and *semantic memory* (memory for facts about the world, devoid of personal relevance, such as the fact that the Vatican is located in Rome).

Ecologically relevant behaviors: Behaviors that an animal performs in order to survive and reproduce in its ecological niche outdoors. These behaviors are selected by evolution.

Efficient coding theory: A theory suggesting that neurons encode information in an efficient manner. In particular, Horace Barlow proposed that the neural code of sensory neurons efficiently represents sensory information by minimizing the number of spikes required for transmitting the signal; or, equivalently, the neural code maximizes the information per spike. The efficient coding theory is thus tightly linked to notions of optimality in the brain. To achieve this optimality, the neural code must be matched to the sensory statistics of the natural environment of the animal—that is, to its *Umwelt* (Rieke et al. 1996).

Emergent phenomena/emergent properties: Emergent phenomena in a system appear collectively at a macroscopic scale, but they are not exhibited by each of the microscopic parts of the system. Similarly, emergent properties of a system are properties exhibited by the system as a whole, which are not exhibited by any component of that system. Examples include propagating waves of activity in brain networks, circular swimming in fish schools, and the construction of termite mounds by termite colonies.

Labeled-line code: A type of a neural code in which the *identity* of the active neuron (its "label") determines the information that it conveys about a stimulus, which affects how its activity is decoded by downstream circuits.

Mixed selectivity: The notion that single neurons in high brain areas (such as the prefrontal cortex and hippocampus) are tuned to mixtures of multiple behaviorally relevant variables. Each neuron is tuned to a different combination of these variables. Such type of high-dimensional heterogeneous tuning creates a combinatorial, distributed coding of these variables across the neuronal population.

Natural behaviors: *Behaviors* are the ways in which an animal or human performs actions. *Natural behaviors* are behaviors—either innate or learned—that occur in a particular species in its natural environment in the wild. Examples include navigating, exploring, foraging for food, escaping predators, and social behaviors.

Natural neuroscience: The science of studying the neural mechanisms of ecologically relevant natural behaviors in freely moving, unrestrained animals.

Neural coding: The way in which neural activity encodes (or represents) information about a stimulus or a behavioral variable. Some prominent examples of neural codes include rate coding, temporal coding, and labeled-line coding. See **labeled-line code, rate code,** and **temporal code**.

Rate code: A type of neural code in which the *firing rate* of a neuron conveys information about a stimulus (regardless of the precise timing of the spikes). This is considered the most common type of neural code in the brain because many neurons throughout the brain exhibit tuning to particular sensory or behavioral variables, whereby the neuron's firing rate encodes the relevant variable. See **Tuning curve**.

Receptive field: The region of space to which a sensory neuron is responding. When the appropriate stimulus (e.g., a visual, auditory, or somatosensory stimulus) is presented within the receptive field, the neuron responds, but it does not respond to stimuli presented outside the receptive field.

Spike raster: Spikes, or action potentials, are "all-or-none" electrical pulses that are generated by neurons and used to communicate between neurons. A *spike raster* is a type of display that shows, one above the other, the series of spikes that a neuron emitted upon repeated presentations of the same stimulus or the same behavior, aligned on the stimulus or the behavior. The average of the spike raster is called the peri-stimulus time histogram (PSTH), and it captures the average response of the neuron to the particular stimulus or behavior. For examples of spike rasters, see figures 4.2 (dots), 5.13c, and 6.2a (bottom, gray dots). For examples of PSTHs, see figures 4.2 (top, curves) and 6.2a (bottom, black curves).

Systems neuroscience: The branch of neuroscience that deals with understanding the function of brain systems and their underlying neuronal circuits, and how they generate perception, behavior, and cognition. Systems neuroscience tries to elucidate the response properties of individual neurons, as well as the collective emergent activity of neural networks in the brain.

Temporal code: A type of neural code in which the *timing* of spikes (relative to a stimulus, relative to each other, or relative to some ongoing oscillation) conveys information about a stimulus.

Tuning curve: A curve that shows the selectivity of a neuron to a particular continuous variable—either a sensory variable or a behavioral variable. The tuning curve is depicted as a plot of the firing rate of the neuron as a function of the relevant variable. Examples include (1) the orientation tuning curve: tuning to line orientation in visual neurons in the primary visual cortex; (2) the frequency tuning curve: tuning to the frequency of an auditory tone in auditory neurons; and (3) the head-direction tuning curve of head-direction cells (see figure 4.1, right column; and figure 4.3b). The parameter value at which the firing rate is maximal is often called the "preferred parameter" (e.g., preferred orientation, preferred frequency, or preferred direction).

Umwelt: The full sensory world and behavioral repertoire of a species. The *Umwelt* may vary dramatically among different species, which see different wavelengths, hear different frequencies, smell the world differently, and behave in fundamentally different ways. The *Umwelt* often also changes during development (ontogeny).

Notes

Chapter 1

1. Reductionism can be divided into two distinct concepts: *ontological reductionism,* the notion that the explanation of all natural phenomena can be reduced to a small number of simple entities and a few basic laws of nature; and *methodological reductionism,* which pertains to how experiments should be done: specifically, the need to simplify and control the experimental conditions. In practice, these two types of reductionism are often tightly interwoven because if one believes that ontological reductionism should be applied to a certain scientific problem, then it would usually lead to the application of methodological reductionism to that problem. Therefore, the term "reductionism" will be applied throughout the book to both concepts interchangeably, although it will mostly imply methodological reductionism.
2. There are other fundamental difficulties with optogenetic-activation experiments (see also chapter 3). First, the neuronal class X_1 may itself contain millions of neurons, whose natural, complex, spatiotemporal dynamics are highly asynchronous and complex—very different from the synchronized unison that is imposed by classical population-wide optogenetic activation. Such synchronized activation thus necessarily distorts the normal activity of neuronal ensembles. Second, because neuronal networks exhibit articulated dynamic trajectories in neural activity-space (Laurent et al. 2001; Mante et al. 2013; Vyas et al. 2020)—dynamics that form the very essence of what we ought to understand about how neural networks control behavior—then the classical synchronized optogenetic population activation provides an "off-manifold" perturbation for the neural ensemble, which distorts the neural dynamics. This may then lead to abnormal behavior by the animal (Jazayeri & Afraz 2017). Synchronized optogenetic inactivation (shutdown) suffers from the same problem of distorting the neural and behavioral dynamics. These distortions in turn may lead to erroneous conclusions about causality between neural activity and behavior (Jazayeri & Afraz 2017). That said, it is noteworthy that off-manifold perturbations can be useful if one wishes to erase a working-memory pattern or to examine the relaxation of the network back to its manifold. Moreover, recent years have seen major developments in patterned optogenetic photoactivation, which allow desynchronized, more natural activations of neuronal ensembles—and potentially even on-manifold perturbations.
3. The reductionist approach also has deep-rooted philosophical underpinnings in *physical determinism*: the philosophical notion that everything in the world—from atoms and single cells to human behavior and complex societies—can be fully and deterministically understood (at least in principle) from simpler, lower-level components and processes, if only we could measure all the details of these components and processes. This notion, which reigned during the Age of Enlightenment, has necessarily led to the idea that one should aim to study the detailed mechanisms of the lower-level processes—namely, that one should always be drilling down to ever lower levels: a mechanistic-reductionist approach to science. But, as pointed out by a number of philosophers in the twentieth century—most notably by Karl Popper (1966) in his essay "Of Clouds and Clocks"—most systems in the world are likely *not* deterministic in this sense (in Popper's language, they are not perfect, predictable "clocks"). This indeterminism is due in part to noise, and in part it stems from the feedback loops and complex networks of nonlinear components that lead to emergent natural phenomena that are neither deterministic nor completely noisy (Popper did not use the term "emergence," but he has put substantial emphasis on multiple feedback loops, which today we know lead to emergent phenomena that cannot be explained easily by classical reductionism). Brain and behavior are two paramount examples of Popper's point. Although a full discussion of these philosophical points is beyond the scope of this book, it is noteworthy that Popper's attack on determinism—one based on philosophical grounds—is also an attack on excessive reductionism. Thus, Popper's stance provides support for natural neuroscience.

4. *Top-down causation* (or downward causation) is the concept that the higher levels of a system causally affect the lower levels—a causal directional arrow that is the opposite of classical mechanistic reductionism, in which simple building blocks at lower levels (e.g., molecules, cells) causally determine what happens at higher levels (e.g., behavior). Examples of top-down causation include the following: In physics and engineering, a thermostat—whereby rotation of the thermostat knob (high level of complexity) causally affects the movements of individual molecules at lower levels (e.g., faster molecular motion is caused by turning the thermostat knob to increase the temperature). In ecology, the behavior of apex predators, located at the top of the "food web" hierarchy (or "food network")—such as sharks or big cats—can dramatically affect the behavior of many other species and individual organisms lower down in the ecosystem, and may cascade all the way down to affect even the smallest organisms at the bottom of the hierarchy (Terborgh & Estes 2010). And in neuroscience, cognitive processes at high-level brain areas were shown to exert a major effect on the activity of individual neurons at lower-level brain areas, as will be discussed in chapters 5 and 7.
5. Interestingly, Gao and Ganguli (2017) also showed that the number of neurons N that one needs to record in order to reconstruct the neuronal dynamics, increases only logarithmically with the neural task complexity (NTC): $N \alpha \log NTC$. This result is very important because it suggests that as we make our behavioral tasks more complex, we also need to concomitantly increase the number of recorded neurons, but this increase does *not* need to be astronomical, and in fact it can be very moderate (*logarithmic*). Actually, a steady increase in the number of recorded neurons is being driven by continuous technological advances anyway (see chapter 3). In other words, in addition to recording more neurons, what is even more important for advancing neuroscience is to substantially increase the behavioral task complexity—which linearly affects the recoverable neuronal dimensionality—and as promulgated here, the way to do this is to study naturalistic behaviors. That said, it is also important to record from multiple brain regions when studying complex tasks because different brain regions may have distinct contributions to behavior (according to the "localization of function" idea).
6. A related point is that *naturalistic experiments can overcome the curse of dimensionality.* The real world contains many input variables and feedbacks. It is impossible to systematically investigate their effects on neural circuits one by one, in all possible combinations, because this would require an exponentially large number of combination experiments—which is not feasible because the number of combinations is huge. This is the so-called curse of dimensionality. By contrast, natural experiments sample those combinations of input variables that are relevant to the real world—a much smaller number of combinations, which makes this approach much more tractable. By analogy, this is the reason why software developers are using beta versions of software in the real world: they cannot rely on quality-assurance (QA) engineers to test all possible combinations in controlled testing. The ultimate test is to study the system in the real world.
7. It is noteworthy that recent years have seen the development of voluntary head-restraint devices that allow rodents to put their heads into a locking port of their own volition, and then the mechanism automatically locks into place a head plate that is mounted on the subject (Scott et al. 2015). Such voluntary head-restraint devices are clearly less stressful than classical head-restraint and provide a step toward performing head-fixed experiments with less stress (but notably, the other critiques outlined in this chapter still apply).
8. While research on both birdsong and place cells capitalizes on the animals' natural behavior—the tendency of male birds to sing and of rats to explore the space around them—there is also a noteworthy difference between these two examples. Birdsong is a "controlled" behavior, in the sense that it is a highly reproducible, stereotyped behavior, at least in some bird species such as zebra finches, and this allows for averaging over many repetitions of almost the same behavior. This ability to average over many repetitions constitutes a key advantage of studying stereotyped behaviors (top-right corner of figure 1.2). By contrast, rats' exploration of an open environment is much less stereotyped—in fact, in typical place-cell experiments, the rat explores space rather randomly, passing through each location from many directions—and this nonstereotypy required John O'Keefe to conduct multiple control experiments to show that place cells indeed encode place and not a simple sensory stimulus, such as a local odor or a visual cue (O'Keefe & Nadel 1978).
9. There are additional possible meanings of the term "natural behavior." For a geneticist, "natural behavior" may mean the diverse behavioral phenotypes of a particular behavior, as expressed across an entire population of a wild species. Inbred lab colonies carrying one set of genetic alleles would represent only part of the distribution of natural behavior. For an evolutionary biologist, "natural behavior" may mean behavior shaped by natural selection. Domesticated animals such as lab mice are shaped by artificial selection, and therefore they may have behavioral repertoires that only partially overlap with those of wild animals. Both of these problems mean that we should consider our model organisms very carefully, and if we study mice, for example, then it is very useful to study wild-derived mice. I will discuss both of these problems in section 2.2 of chapter 2.

Chapter 2

1. Two caveats are noteworthy here. First, Suga discovered these remarkable specialized cortical fields in the mustached bat, a species that uses particularly sophisticated echolocation calls; however, such developed cortical topography was not found in some other bat species that use simpler echolocation calls (Ulanovsky & Moss

2008). Second, although Suga discovered neurons with amazing specificity to echolocation sounds, later work showed that some of these neurons were not as specialized as originally thought, but rather responded to a wider array of sounds, including the bat's communication calls (Ohlemiller et al. 1996; Esser et al. 1997).

2. Another key motivation to study the brains of invertebrates such as insects is that they perform extraordinarily complex behaviors—such as long-distance navigation, courtship, kin recognition, and tool use—with a tiny fraction of the neurons possessed by a vertebrate that performs an almost identical behavior. It may be easier to unravel the underlying neuronal principles in a bee than in a mouse or a rat. Further, it's a fascinating question: How do insects perform all these behaviors with such tiny brains?
3. An additional important avenue of future research is to elucidate the *genetic* and *molecular* bases of natural behaviors. A number of studies have made important discoveries in these directions—for example, in social behaviors (Lim et al. 2004), burrowing behaviors (Weber et al. 2013), and others. Further studies are needed, and it would be particularly valuable to link the various levels of description of the mechanisms of behavior (i.e., to link the molecular and genetic levels to the circuit and systems levels, and ultimately to the levels of neural coding, computation, and dynamics of neuronal populations).
4. Another characteristic of today's neuroscientific studies in mice is that they are performed almost exclusively on males. As part of the diversification of neuroscience, it would also be important to increase the study of the female mouse brain (and the female brain in general), with two aims: first, to create more unbiased knowledge base about brain function; and second, to elucidate the neural bases of differences in behavior between females and males (neural bases of sexual dimorphism).
5. Another drawback is that colonies of inbred mice undergo *genetic drift,* which changes the genetic makeup of the colony over the course of years. As a result of this genetic drift, if one compares today the exact same inbred strain from two different colonies—for instance, two different providers or two different labs—it turns out that their genomes are not identical anymore (Stevens et al. 2007).
6. Dogs also provide an example of the dramatic changes in behavior that are brought about by domestication. While dogs are not laboratory animals, they do provide a case in point: various inbred domesticated strains (breeds) of dogs are famous for exhibiting behaviors that are drastically different between strains—and drastically different from their wild wolf ancestors.
7. The difference is also not due to differences in brain size. It is true that the brains of rats are smaller than those of cats or macaque monkeys. However, the brain of the gray squirrel (a rodent) is *larger* than the brain of the tree shrew (a species closely related to primates), and yet squirrels do not have orientation columns in V1, while tree shrews do (Kaschube et al. 2010). Even more extreme examples come from a tiny primate, the mouse lemur, which has orientation columns in V1—while a huge rodent, the agouti, has salt-and-pepper organization of V1 (Schmidt and Wolf, 2021). Together, these examples suggest that salt-and-pepper organization cannot be explained by brain size, visual acuity, or foraging ecology. Instead, the explanation is phylogenetic: Rodents are truly diffrent.
8. A comment is due here regarding studying a diversity of animal species. Different animal species perceive the world differently from each other, and differently from us. Mice, for example, see the world in a very blurry manner; they hear ultrasonic frequencies that we cannot hear, but they cannot hear low frequencies that we hear very well; and they can smell certain odors at concentrations that we cannot detect. Many insects see ultraviolet wavelengths of light and can even see polarized light. Water shrews can smell underwater by emitting bubbles from their nose and then inhaling those bubbles back in after they touch objects of interest. The list of fascinating animals and their captivating worldviews goes on, and on, and on. Each species has its own *Umwelt*—a German word that denotes the unique way in which each species perceives and interacts with the world. It is therefore fascinating to study each creature in its own right as part of the scientific quest to understand the natural world. This diversity across species also serves as a warning: It is absolutely crucial that we, as researchers, understand very well the *Umwelt* of the organism that we are studying—and we should *not* assume that our animal model is simply a little human being that perceives the world similar to us. It isn't.

Chapter 3

1. Notably, the geometry of imaging with miniscopes versus recordings with Neuropixels probes is very different—hence, each of these two methods is better suited for particular brain areas. Imaging with miniscopes is done in a 2D focal plane, and therefore particularly large numbers of cells can be obtained in brain areas that are organized on a 2D sheet, like hippocampal area CA1, where the microscope can be lowered *perpendicular* to the 2D sheet, such that many cells are in the focal plane of the microscope. By contrast, Neuropixels probes record neurons along a 1D line, which is useful primarily in cortical areas like the entorhinal cortex and prefrontal cortex, in which the probe can be easily inserted *parallel* to the structure.
2. It is noteworthy that table-top two-photon microscopes allow imaging much larger neuronal populations in head-fixed animals, such as imaging hundreds of thousands of neurons simultaneously in mice (Demas et al. 2021). Such high-throughput, head-fixed imaging methods are compatible with VR behavioral paradigms,

which are less natural than the behaviors of freely moving animals, but they offer distinct advantages for studying neuronal population dynamics.

3. High-capacity rechargeable batteries are a key technological front that has matured only recently. Specifically, lithium-polymer batteries are particularly important: these are today's smallest high-capacity rechargeable batteries, with the highest specific energy—and are thus crucial for wireless systems used in neuroscience research, such as wireless electrophysiology, wireless imaging, and wireless optogenetics. These batteries were first introduced in the 1990s but reached maturity only in the 2000s—and this enabled the gamut of miniaturized systems that can be carried by small animals, as surveyed in this chapter. For example, today's tiny lithium-polymer batteries enable wireless neural loggers to record 64 channels of high-bandwidth spiking data for several hours continuously without replacing the battery, with systems that weigh just a few grams, including the logger, battery, and casing (Eliav et al. 2021)—something that would have been unimaginable a decade ago. This battery miniaturization has thus been essential for launching the natural neuroscience research program.

Chapter 4

1. Nearly all the research conducted on spatial representations in the hippocampus has used small boxes or short linear tracks, 1–2 m long. Only two studies have used substantially larger environments. One study (Kjelstrup et al. 2008) used an 18-m linear track, but this track passed through several rooms, the largest of which was < 10 m in length. The second study (Rich et al. 2014) used a 48-m zig-zagging track composed of about 1–2-m segments (compartments). Thus, until recently, the largest single-compartment environment in which hippocampal place cells were recorded was < 10 m in length.

Chapter 5

1. While almost all sensory neuroscientists in the 1950s and 1960s (and indeed well into the 1990s) used the head-fixed, anesthetized approach, it is noteworthy that early in his career, David Hubel pushed for neural recordings in behaving animals and pioneered methods for studying the visual cortex in awake, unrestrained cats. To this end, he developed in 1957 two methodological innovations: the tungsten microelectrode (Hubel 1957b) and the hydraulic microdrive (Hubel 1957a), which together allowed him to record from the V1 of cats that were loosely body-restrained but otherwise were free to move their head and behave. A couple of years later, he published a little-known paper that summarized the results of these experiments (Hubel 1959). Hubel wrote in this paper: "These methods eliminate anaesthetics, paralyzing drugs, brain-stem lesions, and other acute experimental procedures. They make it possible to record electrical events in the higher central nervous system with the animal in a normal state, and to correlate these electrical events with such variables as waking state, attention, learning, and motor activity." And then Hubel went on to describe the activity of V1 neurons as a function of the sleep-wake cycle, as well as response properties of V1 neurons under this "normal state." Hubel found "on" and "off" types of receptive fields in V1, similar to the retina. Surprisingly, he found that many neurons preferred moving stimuli—such as waving the hand through the receptive field—and some of these neurons preferred one movement direction over the other. These early results from awake cats thus formed clear precursors for the subsequent discovery that V1 neurons are tuned to the orientation of bars that are moving in specific directions. Finally, Hubel and colleagues used their new method to record from the *auditory* cortex of awake, behaving cats and discovered there a population of neurons that were active only when the cat attended to the sound source (Hubel et al. 1959). He made a point in that study that such high-level cognitive effects can be studied *only* in awake, behaving animals.

2. This is not to say that sophisticated and detailed visual perception is impossible in a passive state. For example, when we watch a movie, we stay passive, with our head almost fixed, and yet we get rich, complex visual perception. However, as will be discussed in this chapter, even in this passive case, behavior plays in fact a crucial role: in particular, to perceive the movie, we must move our eyes to scan the movie screen behaviorally with our gaze. Further, it is also possible to have visual perception even without any eye movements whatsoever—for example, during dreaming or hallucinations—but this type of visual perception lacks the spatial detail of real-world, natural vision. It is thus consistent with the theme that will come up later in this chapter—namely, that eye movements and other active behaviors are specifically required for high-acuity vision, while low-acuity vision can be done more passively (see section 5.2.2).

3. Two words of caution are needed here. First, although Leinweber et al. (2017) clearly showed that the *numbers* of neurons projecting from the retrosplenial, motor, and auditory cortices into V1 are several times larger than the number of neurons projecting from dLGN into V1 (figure 5.2b), this does not necessarily mean that the *strength* of these projections is as large. The strength of a projection also depends on the size and efficacy of the synapses, as well as on their location on the dendritic trees in V1, which is generally unknown for all these regions. Perhaps dLGN synapses are stronger or are closer to the cell body of V1 neurons than the synapses of the top-down feedback inputs into V1. Therefore, the distribution of synapses and the synaptic strength need to be investigated for each of the input-regions into V1. Second, another factor that may increase the efficacy of the dLGN input is *timing*. Perhaps the fast synchronized wave of activity that arrives from

dLGN upon sensory stimulation—whereby many dLGN neurons fire synchronously within a few tens of milliseconds—is more efficient in driving V1 responses than the (presumably) less-synchronized feedback inputs into V1. But here again, nothing is really known about the level of synchronicity for the inputs coming into V1 from each of these brain regions, and hence quantitative comparisons cannot be made at present. Therefore, the level of synchronicity also needs to be investigated for each of the input regions into V1.

4. A word of caution: it is possible that the V1 of mice corresponds to a higher visual cortical area in monkeys. In this case, perhaps strong attentional effects *do* exist in V1 of mice, and hence *can* explain the results by Niell and Stryker (2010).

5. A word of caution: The study by Saleem et al. (2018) was done in a VR setup, where visual information was the only sensory information available to the animal about its position, so in this situation, the visual cortex may perhaps encode the animal's position more strongly than in more naturalistic situations, which are multisensory.

6. A third major category of eye movements is smooth pursuit (Lisberger 2010), a specialized behavior, supported by specialized brain circuits, that allows the eyes to perform smooth visual tracking of a moving target. Further discussion of smooth pursuit is beyond the scope of this chapter.

7. While this section focuses on vision, the issue of brief stimuli that suddenly appear out of nowhere is prevalent throughout sensory neuroscience—and equally problematic for many other senses, such as somatosensation or olfaction. Natural perception is dynamic (i.e., requires time and relies on active interaction of the subject with the environment), while the classic passive methodology tends to eliminate dynamics by "flashing" the stimulus onto the sensory organ. The aim is to eliminate behavioral noise, but as we argue here, behavior (movement) is *not* noise; rather, it is crucial for active perception. The classic approach also stems from the implicit assumption that sensation is the input to the system, while motion is the output—and the classical "flashing paradigm" is therefore designed to separate the input (sensation), the neural processing, and the output stage (movement); but in natural real-life situations, these stages in fact are intertwined in a loop.

8. Researchers *did* consider the related problem of how is it possible that we perceive the visual world as stable, despite the fact that our eyes are constantly moving—the conundrum of perceptual stability (Melcher & Colby 2008; Melcher 2011). This line of research focused on questions such as how we perceive the world smoothly despite the discontinuous retinal input (trans-saccadic perception); how perception and action are coordinated to achieve this smoothness; and how perception and memory interact. At the neuronal level, much work has been done on neurons in the parietal cortex (LIP), frontal cortex (frontal eye fields) and superior colliculus, which shift their receptive field position during the saccade in the direction of the eye movement. This receptive-field shift (remapping) could help stabilize the image relative to these receptive fields.

9. Two critiques can be raised against this challenge. First, perhaps hyperacuity is a result of overtraining in experts and is not a general phenomenon across naive humans. However, good hyperacuity is found even in completely naive subjects, although they improve somewhat over time (G. Westheimer, personal communication), which suggests that hyperacuity is not achieved purely from strong overtraining. Second, it was shown that hyperacuity can be achieved even using 1.3-ms flashed stimuli (Westheimer & McKee 1977), so perhaps only the first few milliseconds of the stimulus are important for the hyperacuity phenomenon—a brief time during which no eye movement and no smearing occur. However, eye drifts can be as fast as 10 degrees per second (Ahissar et al. 2015), creating substantial smear even within 1 ms, which should degrade hyperacuity; but it doesn't. So the mystery remains. It would be interesting to measure eye movements during hyperacuity experiments in naive subjects versus experts to see if the improvement in the experts is accompanied by a reduction in fixational eye movements, and hence a reduction in retinal smear—something that has not been tested to date. This highlights again the importance of measuring active-sensing behavior (in this case, eye movements) to understand sensory perception.

10. The main controversy concerns the question whether drift and tremor are essentially noise (e.g., produced by spontaneous spiking in the motor neurons controlling the eye muscles) or centrally controlled by brain circuits. One experiment showed that the statistics of drift are influenced by the visual context, suggesting that drift is not noise but is neuronally controlled in a closed-loop manner (Gruber & Ahissar 2020). Another experiment showed that drift is not random but is compensating rather precisely for microscopic head movements—presumably to better stabilize the image on the retina (Poletti et al. 2015). A third experiment showed that ocular drift precisely positions the image on the retina to enhance visual acuity (Intoy & Rucci 2020). However, to date, there is no direct causal evidence that the *detailed trajectories* of eye drift are controlled centrally; it is quite possible that, at least to some degree, drift trajectories reflect neuronal noise in the oculomotor system. Nevertheless, regardless of the question of what generates the fixational eye movements, it seems clear that these movements are causally related to perception, as demonstrated by the muscle-paralysis experiments and image-stabilization experiments discussed here.

11. This notion—that eye movements and behavior are necessary for high-acuity foveal vision, but not for low-acuity peripheral vision—answers one of the main critiques of the active-vision approach. This critique points out that one can have visual perception even without any eye movements whatsoever (e.g., during dreaming or hallucinations). The answer to this is that dreaming and hallucinations lack the spatial detail of real natural vision, and therefore are likely mediated by low-acuity peripheral visual mechanisms. By

contrast, to see full spatial detail, it is necessary to use high-acuity visual mechanisms, which in turn require active behavior and eye movements.

12. It is noteworthy that V1 peripheral neurons *were* shown to represent eye gaze direction during large saccades, as well as during large-amplitude smooth pursuit (Trotter & Celebrini 1999; Morris & Krekelberg 2019), but this gaze-direction representation had a coarse accuracy of a few degrees. This accuracy is three to four orders of magnitude coarser than the ultraprecise 0.05-arcminute representation of gaze direction that would be required to explain visual hyperacuity in the face of fixational eye movements.

13. Because of their highly developed whisking behaviors, rodents are often regarded as one of nature's "vibrissa specialists" (Brecht et al. 1997; Prescott et al. 2011). Another notable vibrissa-specialist is the seal, which uses its whiskers to track the underwater hydrodynamic trails that are left behind swimming fish, which persist for several minutes in their wake (Dehnhardt et al. 2001).

Chapter 6

1. An STRF represents a linear spectrotemporal estimate of the temporal responses of the neuron to pure tones at all possible frequencies. In other words, it is a linear kernel showing the neuron's firing rate as a function of frequency × time. The STRF can be used to predict the responses of the same neuron to new sounds by convolving the STRF kernel with the spectrograms of new incoming sound stimuli. Importantly, the computation of the STRF and the convolution operation assume that the neuron is *linear*—and the prediction will fail if the neuron is strongly nonlinear. Historically, the STRF was originally computed using rich sets of complex artificial sounds as stimuli (deCharms et al. 1998; Depireux et al. 2001), but subsequently researchers also started computing STRFs based on natural sounds (Theunissen & Elie 2014). There are some subtle technical issues with how to compute STRFs correctly from natural sounds, such as the fact that natural sounds cover only a subspace of all the possible stimulus space, and moreover, do not cover this subspace uniformly. Another problem is that the input space is high-dimensional (large number of pixels in the frequency × time kernel), which may cause overfitting unless appropriate regularization techniques are used. These issues, and their common solutions, were reviewed by Theunissen and Elie (2014). Importantly, Machens et al. (2004) directly addressed most of these issues. Later studies addressed more fully these and other technical challenges of estimating STRFs and found strong nonlinearities in cortical processing of natural sounds, as described later in this section.

2. By "natural sounds," I mean any sound that occurs naturally in the world: both environmental sounds, such as the sounds recorded in forests or meadows, but also human speech and music. Interestingly, all of those exhibit the same $1/f$ power law in their temporal envelope spectrum.

3. It is noteworthy that the approach of using natural or naturalistic stimuli to probe visual neurons has not gone uncriticized. In their paper "In Praise of Artifice," Rust and Movshon (2005) highlighted the difficulties in using natural visual stimuli for building mechanistic models of V1 neuronal processing; instead, they suggested that we should use simple artificial stimuli to construct models of neuronal responses, and then use natural stimuli to test those models. In principle, I agree with them: if simple, well-controlled stimuli can be used to build satisfactory models, they should be used. We should indeed utilize simple, well-controlled stimuli to measure neuronal responses and gradually build better models that capture evermore neuronal nonlinearities, dynamics, and context sensitivity until, hopefully, we obtain mechanistic models that provide a satisfactory explanation for responses to a variety of stimuli, including natural stimuli. However, three important comments are due here. First, when these models fail—as they often do—we should try to use natural or naturalistic stimuli for both model-building and testing V1 neurons. In fact, I think we should use a diversity of stimuli—both simple/controlled and complex/natural, at both the model-building stage and the testing stage, in order to better elucidate V1 response properties. Second, even the greatest critics of the natural-stimulus approach, such as Movshon, agree that natural stimuli are the ultimate gold standard of sensory neuroscience: they are the stimuli that the brain evolved to process; and they are the stimuli against which models of visual processing should always be tested (Rust & Movshon 2005). Third and finally, we should always keep in mind that a central goal of brain research is to understand how the brain processes natural stimuli and is generating natural behaviors and natural cognitive processes—in short, how the brain deals with the real world. Using simple stimuli and simple experiments is a means to this end, but it is only a means. The ultimate goal is to understand the neural basis of natural sensory processing and natural behaviors.

4. A study by Vinje and Gallant (2000) came close to the proposed "perfect experiment." They found sparse coding and decorrelation between V1 neurons when monkeys were presented with moving natural images. However, this study fell short of the proposed "perfect experiment" in two important ways. First, in that study, the monkeys were not allowed to move their eyes; rather, they were passively presented with images that moved according to a prerecorded scan path of a monkey's eyes. Second, the experimenters did not track the small fixational eye movements of the animals they recorded from (which requires a precise eye tracker).

5. The need to cancel the effects of self-motion (i.e., subtract a negative image) is of general importance across all sensory systems, well beyond the auditory system. Examples include (1) the visual system, where we perceive a stable visual world despite self-motion of the body and eyes (discussed in chapter 5); (2) electrolocation in

weakly electric fish, where the unwanted effects of self-motion are canceled precisely by subtracting a negative image (as discussed in chapter 2); (3) the vestibular system, where beautiful work by Kathleen Cullen and colleagues has shown how efference copies of self-movement are used to cancel out the active self-motion signals in the vestibular system, in order to retain only the passive vestibular signals (such as the signals produced when you unintentionally trip and fall over, reviewed in Cullen 2019).

6. The matching of animals' visual systems to their environments can be much more intricate than the presence of a visual streak or a fovea in the retina. Examples of some incredible adaptations—especially in the visual systems of insects—were reviewed by Rüdiger Wehner (1987). In this famous paper, he also introduced the notion of a "matched filter"—namely, the idea that the structure of the animal's sensory organs, as well as of the peripheral sensory system, are matched to the structure of an animal's world and to the behavioral needs of the animal. Wehner argued that such matched filters, although inflexible, provide a good enough solution for performing fast computations, and thus allow executing behavior very rapidly and automatically. I refer the reader to Wehner's paper for some beautiful examples and discussion.
7. Mirror camouflage is the phenomenon wherein silvery fish, such as sardines, are camouflaged by their silver-colored scales, which act as nearly perfect mirrors (Dusenbery 1992). The principle of this camouflage is as follows. At depths of > 30 m, there are multiple reflections of photons from water molecules (as larger depths entail more reflections). These reflections create a relatively omnidirectional illumination in the horizontal plane. Consequently, the light arriving at a predator's eyes from behind a silvery fish is almost equal to the light reflected by its mirror scales—thus equalizing the appearance of the fish and its background, and creating highly effective camouflage.
8. An important distinction to make is between actual body movement and a point-of-view movement. The latter can occur even in the absence of body movement—for example, in movies, where the camera is being moved, which allows the viewer to sit statically in a chair but nevertheless to obtain a rather realistic perception of the 3D structure of the cinematic world. Thus, the crucial aspect seems to be the point-of-view movement. Its effects on brain activity can be studied either by allowing the animal (or human) to actually move through space or, in more classical laboratory setups, by using highly realistic movies that include camera movement. However, a word of caution is needed here: presenting movies to stationary subjects may still miss key mechanisms because we *know* that the visual cortex contains very strong motor signals indicating self-movement (discussed in chapter 5)—and these motor signals would be eliminated in such movie experiments. Therefore, the best approach probably would be to compare and contrast neuronal responses recorded in self-movement experiments against the responses of the same neurons recorded during the presentation of movies with similar point-of-view dynamics (camera dynamics), as exhibited by the animal during actual locomotion.

Chapter 7

1. Note that figure 7.1 shows brain voxels that were consistently activated across the thirty-two studies during memory recall, according to a standard statistical criterion used for meta-analysis of many studies (McDermott et al. 2009). As any statistical criterion, it creates thresholding, so potentially there could be somewhat larger areas of the brain that are activated but did not cross the statistical threshold. This would mean that there could be somewhat larger subthreshold overlap between areas activated by cued recall versus those activated by autobiographical recall. However, this would still mean that in those brain voxels in which there is strong activation for autobiographical recall, there is very weak activation for cued recall (even if not complete absence of activation), and vice versa. Therefore, this does *not* change the key conclusion—namely: that the brain areas activated in autobiographical recall versus lab-based cued recall are very different.
2. Alternative conclusions are also possible. First, the difference may be due to the specific task used in the laboratory tests. Figure 7.1 suggests that cued recall of words from a random word list (a classical task where subjects are asked a question such as, "Did the word 'car' appear in the list that you were given 10 minutes ago?") bears little relevance to cued recall of autobiographical memories. One could test other options for cued recall in the real world in a different manner than the studies summarized by McDermott et al. (2009): for example, test cued recall of semantic memory ("What is the capital of Papua New Guinea? It starts with "Port . . ."). It is also conceivable that laboratory tests of free recall of random word lists (where subjects are asked to recall freely as many words as they remember from the list) might be closer to autobiographical free recall. This remains to be tested. Second, the difference may be in the timescales: very short timescale in classical laboratory tests versus much longer timescales in real-life memories. Importantly, these alternative interpretations still raise doubts regarding the relevance of classical cued-recall tasks for understanding human real-life memory: Wouldn't we expect at least some overlap between brain activations in lab-based cued memory recall versus real-world cued memory recall? Aren't there *any* shared processes and brain areas? The almost complete lack of overlap in figure 7.1 is too striking to ignore, regardless of the underlying explanation.
3. Complex versus simple memoranda differ in fundamental ways. For example, Koriat and Goldsmith (1996) noted that laboratory tests at that time tended to be very *discrete*, focusing on whether the subject

remembered a certain item (1 if remembered, 0 if not) and typically measured the *number* of remembered items. By contrast, in real-life scenarios, what determines the quality of memory is the *graded* level of detail with which you remember a certain event: for instance, how well you remember a certain interesting conversation that you had at dinner seven months ago, or how well an eyewitness remembers the details of a crime scene—both of which can be evaluated by testing the *accuracy* of the memory. Koriat and Goldsmith emphasized that this graded nature of real-life complex memories is fundamentally different from the discrete nature of laboratory-based memory tests, and they argued that this major difference reflects two very different metaphors for thinking about memory. One is of *memory as a storehouse of discrete items* ("the storehouse metaphor"), which then leads to experimental tests for the *number* of remembered items; and another is of *memory as a correspondence to the original event* ("the correspondence metaphor"), which then leads to tests of the *accuracy* of complex, real-life memories. They argued that these two very different ways of thinking about memory also lead to different standpoints in the laboratory versus real-life memory debate. Proponents of the storehouse metaphor tend to embrace the laboratory style of experiments using highly controlled random lists of discrete simple items, while proponents of the correspondence metaphor embrace a real-life, naturalistic approach to studying complex memories. Interestingly, today, three decades after Koriat and Goldsmith's paper, the naturalistic approach has penetrated even into the most classical laboratory experiments, as many current studies do measure a full graded scale of memory.

4. The example of working memory serves to illustrate another problem with current neuroscientific thought as part of the reductionist approach. We researchers tend to separate phenomena and categorize them into "boxes"—for example, a box for "sensory processing," a box for "working memory," a box for "long-term memory," and so on. Indeed, there is clear evidence that these processes are separate, to some degree. But this amount of separation is much smaller than we tend to think (Hasson et al. 2015). First, both sensory processing and working memory are influenced by past history over *multiple timescales*, from milliseconds to many minutes (Ulanovsky et al. 2004; Hasson et al. 2008b; Lerner et al. 2011). The fact that sensory processing exhibits very long timescales implies that in real life, accumulation of information over time is key for the sensory processing of natural stimuli (as discussed in chapter 6). Further, since both sensory processing and working memory have overlapping timescales—that is, there is no separation of timescales between the two—it is unclear whether there *is* a real separation of sensory processing and working memory. Indeed, the two are functionally highly entangled: sensory processing is the input to working memory, and in turn working memory strongly affects sensory processing. Thus, the two form a dynamic loop and are completely intertwined (Fuster 1997; Hasson et al. 2015). We can try to separate them into two boxes by designing carefully controlled laboratory experiments, but this separation is quite artificial and is moot in real life. Thus, it might be incorrect to think about ongoing sensory processing and working memory as two completely separate processes. Second, *each neuron in the brain is a processing unit, a working-memory unit, and a long-term memory unit at the same time*. It is a processing unit because it integrates input, performs complex dendritic computations, and issues output (London & Häusser 2005). It is a working-memory unit because (as part of a neural network) it can store information over multiple timescales (Ulanovsky et al. 2004; Mongillo et al. 2008). And it is also a long-term memory unit because each neuron (again, as part of a network of neurons) stores long-term memories via modifications in the strength of its input synapses. Therefore, since every single neuron in our brain mixes sensory processing, working memory, and long-term memory, this calls for rethinking our concept of dividing the three into separate and distinct boxes. Important to the theme of this book, the classical neuroscientific thinking about boxes emerged, to a large degree, from the reductionist approach, which seeks to atomize the world into separate boxes, each of which is an individual entity and can be separately controlled (chapter 1). However, once you start thinking about natural real-life behavior, it becomes clear that the reductionist way of thinking, despite its obvious advantages, also has substantial limitations—in particular, it can lead us to focus on the wrong concepts (such as separate boxes) and consequently to formulate ideas that have little relevance to the real world. Such box-and-arrow type of thinking may be relevant for modeling isolated behaviors in the lab—but not for modeling natural behaviors in the real world.

5. The multilayered networks in the visual system—both in the dorsal and in the ventral stream—are reminiscent of deep neural networks. Many studies have tried to compare deep neural networks to the brain and to draw conclusions about the brain (Yamins & Dicarlo 2016; Kriegeskorte & Douglas 2018; Kell et al. 2018; Bashivan et al. 2019; Richards et al. 2019; Schyns et al. 2022; Goldstein et al. 2022). I think that the jury is still out on whether such comparisons are useful—and if so, what can we learn from them about the brain. While this discussion is beyond the scope of this chapter, I do wish to make one point, returning to an issue that I already touched upon in chapter 1 (see also Hasson et al. 2020). Classical theoretical neuroscience has focused on rather simple models, with only a few parameters. These simple models allow us the researchers to *understand* the model, even though the behavioral performance of the model often does not recapitulate fully the neuronal and behavioral phenomena of interest. By contrast, deep neural networks offer a radically different type of model, with a very different objective. Their objective is to allow good behavioral performance, even if it comes at the expense of understanding—a totally opposite objective from the classical types of models. With their millions of parameters, these new, ultracomplex models indeed defy understanding, thus frustrating many neuroscientists, who seek understanding (as all scientists do). On the other hand, these models offer increasingly good behavioral performance: modern deep neural networks can drive cars, play games, translate between languages, generate literary texts, draw art, identify objects and faces—and in general, they

rapidly approach (and in some cases even surpass) human performance, at least in certain aspects. These models do not "understand" the world themselves, and they do not allow researchers to understand the model in detail. They are simply too complicated. But what these models increasingly allow is for mobile agents to perform and behave and act in the real world—much as Gibson envisioned with his action-oriented approach to perception. These models are thus quite ecological. However, contrary to Gibson's more controversial assertion on direct perception, these models do perform computations and exhibit internal representations—but these computations and representations seem to be very different from the classical computations and representations that were considered for many decades in both theoretical neuroscience and computer science.

6. The two main criticisms of Norman's idea are as follows. First, there is direct evidence against Norman's suggestion that the dorsal stream does not require conscious awareness. For example, Rafael Malach and colleagues have conducted a human fMRI experiment showing that BOLD activity is suppressed for images that the subject is unaware of, compared to images that reach conscious awareness—and, importantly, this suppression had similar magnitude in the dorsal and ventral visual streams (Hesselmann & Malach 2011). That said, it may still be true that the dorsal stream requires *less* awareness than the ventral stream—even if some awareness is still required—and more generally, the dorsal stream may require less complex cognitive processing. Second, and more fundamentally, some studies have suggested that the dorsal and ventral visual streams are not two separate processing streams, but rather their functions are more intermixed than initially thought. For example, one study reported that damage to the ventral stream in humans causes deficits in motion perception, even for nonform motion such as random moving dots (Gilaie-Dotan et al. 2013). Other studies used visual illusions to show that there are some interactions between the two streams, indicating that they are not entirely segregated (Goodale & Westwood 2004). Finally, there have been suggestions for more than two visual streams (Rolls et al. 2023).

7. Beyond the *interaction* ("ping-pong") of perception and cognition, there is also growing evidence that cognition exerts direct top-down influences on what we see. There is evidence that beliefs, emotions, motivations, intentions, and semantic knowledge about the world all directly affect how and what we see in the world, in the most literal sense (Ullman 1996; Balcetis & Dunning 2010; Riccio et al. 2013; Collins & Olson 2014; Vetter & Newen 2014; Goldstone et al. 2015; Lupyan 2015a,b). This line of work proposes that perception and cognition are *not* two separate faculties that merely interact—rather, cognition penetrates perception. In other words, cognition is an inseparable part of perception. By contrast, other studies have pointed out various pitfalls that may undermine the claimed top-down influences of cognition on sensory perception and argued against the penetration of cognition into perception (Firestone & Scholl 2016). This debate rages on. It is clear, however, that in order to adjudicate between these two camps, it is important to conduct experiments that allow the normal involvement of cognition in perceptual experiments—that is, more naturalistic experiments.

8. Two notes are apt to make here. First, to clarify Gibson's statement: of course we *can* move in the dark, but this still requires some nonvisual sensory information, such as somatosensory, vestibular, or proprioceptive. It would be very difficult to move without *any* sensory perception; hence, we must perceive in order to move. And, as discussed in chapter 5 on active sensing, we must move in order to perceive. Second, the idea that perception and action are tightly linked in functional loops was strongly promoted by Gibson, but in fact it is a much older idea, originating at least from John Dewey at the end of the nineteenth century (Dewey 1896; Wallach et al. 2016), and likely even earlier than that.

9. Similar experiments can be done in other sensory systems, such as the somatosensory whisker system of rodents, electrolocation system of fish, or echolocation system of bats. In all these systems, one should compare neuronal responses in freely moving animals that are actively inspecting objects in various directions to the responses of the same neurons when the animal is passively presented with the same objects from a variety of angles.

10. While I do not discuss it in detail here, the *neural generation and processing of language* constitute another domain of neuroscientific research that aligns with the natural neuroscience approach, because language production and understanding are highly natural for humans yet allow controlled investigation (which places it at the top-right corner of figure 1.2). Language is obviously also universally important for real-life human behavior. There has been a long history of research in *psycholinguistics,* the branch of research that investigates the psychological processes underlying the use of language (Traxler 2011; Sedivy 2019). Recent years have also seen some beautiful examples of investigations in *neurolinguistics*—studies that aim to understand the neuronal basis of the use of language (Poeppel 2014; Bhaya-Grossman & Chang 2022). Language research is a field where natural behaviors and naturalistic experiments can help us elucidate human cognition and the brain mechanisms that underlie it. More work needs to be done on natural conversations, which have rarely been used in either psycholinguistics or neurolinguistics despite the fact that free conversation is the most common form of language use in real life. It seems though that the field is going in that direction, as well as in the direction of detailed mechanistic neuronal investigation of language production and comprehension (Hamilton et al. 2021; Bhaya-Grossman & Chang 2022; Chung et al. 2022; Leonard et al. 2023). In the future, we can expect to see rapid growth of neuroscientific investigations into this important human natural behavior. This growth will certainly be accelerated further by the recent explosion of interest in large language models, such as GPT-4 and GPT-5 (Bubeck et al. 2023).

11. Another disadvantage of classical, single-person fMRI experiments is in the field of *social neuroscience,* where the core demand is *actual* interactions among people. To address this problem, a number of groups have employed a "hyperscanning" approach, wherein two subjects are being scanned simultaneously in two fMRI magnets as they interact in real time (Schilbach et al. 2013; Czeszumski et al. 2020). Such experiments have revealed enhanced synchronization between specific brain regions in the two subjects, and the degree of synchronization was correlated with enhanced verbal communication, enhanced story comprehension, or joint attention (Saito et al. 2010; Stephens et al. 2010b; Czeszumski et al. 2020). Another study (Xie et al. 2020) scanned triplets of subjects simultaneously as they engaged in a collaborative game of Pictionary, in which the trio had to collaborate in order to draw word definitions that were interpreted by a fourth person (the observer). They found that the collaborative team performance was correlated with intersubject correlations in the right temporo-parietal junction and the adjacent angular gyrus—well-known core areas within the "theory-of-mind brain network." However, while the rationale for this study was that people behave fundamentally differently in groups of three or more people as compared to pairs (dyads), the authors did not actually compare pairwise-collaboration and triple-collaboration paradigms—something that would be very interesting to do. Overall, it is clear that future experiments utilizing multiperson fMRI hyperscanning can lead to important insights into the social neuroscience of collaboration, mediation, social rejection, and many other types of group interactions.
12. Additional important topics for naturalistic neuroimaging experiments outdoors are real-life experiments in *neuroeconomics* and in the neurobiology of *social preferences* and *pro-social behaviors.* It is important to test these under naturalistic conditions because behavioral studies on these topics by economist John List and others have demonstrated that very different results are often found in laboratory experiments versus more naturalistic experiments outdoors (Levitt & List 2007). In particular, in studies of experimental-economics games (such as the Dictator and Ultimatum games), very different social-preference results were found in the lab versus outdoors, due to several factors: (1) *selection bias* (bias in the subjects that are willing to participate in psychology experiments in the lab); (2) *scrutiny* (unintended effects that the presence of an experimenter exerts on the social-preference choices of scrutinized subjects); and (3) *laboratory context* (being a participant in a laboratory experiment often causes people to behave differently than they do in the real world). The distorting effects of selection bias, scrutiny, and laboratory context have been demonstrated behaviorally (Levitt & List 2007); it would thus be very interesting (albeit challenging) to investigate the neural correlates of some of these fascinating cognitive distortions using mobile neuroimaging.

Chapter 8

1. Some aspects of long-distance navigation can in fact be studied in confined laboratory spaces. Many bird species that are long-distance migrants exhibit *migratory restlessness* when kept in a cage, whereby they try to walk in the direction in which migration would normally occur outdoors. This allows studying aspects of migratory behaviors in restricted laboratory conditions—most notably, compass mechanisms. For example, when using magnetic coils to rotate the Earth's magnetic field, the direction of the bird's migratory restlessness changes accordingly, suggesting a biological magnetic compass; likewise, rotating the pattern of stars in a planetarium rotates the direction of migratory restlessness in night-migratory birds, suggesting a star compass (Emlen 1975; Wiltschko & Wiltschko 2005; Chernetsov 2017; Mouritsen 2018). However, this approach has some major limitations. First, it is unclear how well the direction of migratory restlessness in a small (30-cm diameter) cage truly represents the bird's full migration trajectory outdoors, over thousands of kilometers. Second, most mammalian migratory species are *not* known to exhibit clear migratory restlessness, and therefore small-scale laboratory assays cannot be used to study long-distance migration in mammals. One notable exception is provided by some migratory bat species (Lindecke et al. 2019). Third, many animals exhibit remarkable navigation abilities over distances of many kilometers or even hundreds of kilometers, which are unrelated to seasonal migration—but instead are related to foraging abilities or to homing (Davis 1966; Wallraff 2005; Tsoar et al. 2011). Thus, migration is far from the only example of long-distance navigation behavior. These nonmigratory navigation behaviors cannot be studied in the lab.
2. As previously mentioned, Marr's three levels of analysis—the functional (computational), algorithmic, and implementation levels—map nicely onto Tinbergen's four questions. In particular, the algorithmic level sits right between the proximate question of mechanisms and the ultimate question of utility (function). Because of this, the algorithmic level is arguably the most important of Marr's levels. Notably, *evolution operates by natural selection on all three levels—the functional, algorithmic, and implementation levels.* However, natural selection likely acts particularly strongly on the performance of the algorithm. For example, natural selection acts to improve the spatial accuracy of the particular navigational algorithm that the animal uses to reach its foraging site and then to return home. It acts on the decision-making algorithms for choosing a mate or escaping from a predator. And it acts on the algorithm of a beaver for building an efficient dam. The precise performance of these algorithms during evolution was likely the key factor for natural selection. Of course, we can seek to study the *development of the algorithm during ontogeny,* as well as the *brain mechanisms underlying the algorithm.* These kinds of questions on algorithms are rarely asked in

neuroscience—indeed, they are difficult to study—but they should be, because as pointed out by Carandini (2012), the algorithmic level forms a critical midlevel link between neural circuits and behavior.

3. Many of the factors in equation (8.1) are dynamic and may change over time. For example, during a foraging trip, the rate of energy loss from travel might increase over time as the animal eats and gains weight; in parallel, hunger levels go down over time as the animal eats. Likewise, competition and predation risk also may change dynamically along the foraging trip because usually both competitors and predators concentrate at specific locations—often at the best foraging sites. Such dynamism may be modeled by assuming that the various β_i's in equation (8.1) are not static constants but are functions of time: $\beta_i(t)$. Here, this was ignored for the sake of simplicity.
4. These findings are reminiscent of Itzhak Fried's classical study about the hippocampal formation of humans, where he and his colleagues discovered neurons that selectively encode the identity of specific individuals, such as Jennifer Aniston or Halle Berry. These neurons responded both to the presentation of these people's photograph and to their written names (Quiroga et al. 2005). Identity-coding neurons were recently found also in the hippocampus of marmosets by the group of Cory Miller (Tyree et al. 2023). However, an important caveat of these studies was that these neurons were found for *passively* presented images, but they were not looked for during *actual* social behaviors—unlike the newer studies from Ziv Williams and colleagues (e.g., Li et al. 2022), which were conducted during actual social interactions. See also section 4.3, regarding social responses and identity coding in the hippocampus of bats during free social interactions (Ray et al. 2024).
5. The reductionism in behavioral ecology—just like the reductionism in neuroscience—has two separate sources. One is the philosophical belief that examining individual factors one by one, in a controlled manner, is the correct approach; and likewise, that examining the behaviors of individual animals or dyadic interactions will explain the behavior of entire animal groups. In short, understanding the parts will explain the whole. The critique that I outlined in chapters 1 and 2 against the controlled-reductionist approach in neuroscience applies fundamentally to the reductionism of behavioral ecology as well. Bradbury and Vehrencamp (2014) summarized nicely the limitations of the reductionist approach in behavioral ecology, emphasizing the importance of complexity, network structure of animal groups, nonlinear interactions, rich dynamics, and emergent properties—none of which are captured by the traditional approach of behavioral ecology. In particular, they highlighted the need to study the natural behavior of large naturalistic social groups rather than studying only pairwise social interactions. In other words, behavioral ecology, just like neuroscience, needs to look at more complex natural behaviors. The second source of reductionism in behavioral ecology is the ubiquity of technical limitations in studying natural behaviors of animals outdoors. However, experimental methods for field research have undergone a revolution in recent years, with the advent of new methods that allow for measuring and quantifying the details of animal behavior outdoors, ranging from high-resolution GPS tracking to on-board audio recordings, motion sensors, on-animal cameras, and much more (Tsoar et al. 2011; Kays et al. 2015; Strandburg-Peshkin et al. 2015; Egert-Berg et al. 2018; Kano et al. 2018; Brandes et al. 2021; Couzin & Heins 2023). These novel technologies have enabled exciting new discoveries about a variety of behaviors outdoors, including animal navigation, hunting behaviors, collective motion, and social behaviors (such as group foraging and group decision-making). Thus, just as new technologies were cardinal for the advent of natural neuroscience (chapter 3), the new technologies for outdoors field studies will likely revolutionize behavioral ecology.
6. The holding conditions of animals should also be made much more naturalistic and enriched. Experimental animals (of all species) should be held in large cages, with complex structures, diverse objects, and ample opportunity for social interactions. This is crucial because the use of standard, impoverished cages in animal facilities around the world leads to poor dendritic morphology, deficits in learning and memory, and abnormalities in many social behaviors (Lewis 2004). We do not really know how much of our knowledge of brain function and structure has been biased by the use of animals living in impoverished cages. Therefore, it is crucial to emphasize the importance of switching to enriched holding conditions, regardless of the neuroscientific question being investigated (except, of course, in studies whose explicit goal is to compare the effects of impoverished versus enriched conditions on brain and behavior).
7. An example that was already mentioned in chapter 3 comes from a study by Alon Chen and colleagues (Anpilov et al. 2020), who used a wireless-optogenetics device to stimulate oxytocin neurons in groups of mice interacting in a complex, seminatural arena. This study found that oxytocin sometimes had pro-social effects—in line with the classical view on oxytocin—but sometimes it had *anti*-social effects. These results depict a more nuanced view of the social role of oxytocin and highlight the importance of using seminatural arenas, which yielded quite different results from the traditional, highly simplified social assays that typically find prosocial-only effects of oxytocin.
8. It is noteworthy that funding agencies also have started to acknowledge that measuring and quantifying naturalistic animal behaviors are of crucial importance for neuroscience. For example, the National Institutes of Health (NIH) BRAIN 2.0 Initiative explicitly stated that one of its main aims is "Develop tools to analyze naturalistic (untrained) and trained behaviors." Further, the NIH has issued in 2022 the Brain Behavioral Quantification and Synchronization (BBQS) funding program, which focuses on quantifying

natural behaviors concomitantly with neuronal recordings. This indicates the community's recognition of the importance of studying the neural basis of natural behaviors.

9. In addition to measuring brain and behavior, it would be useful to measure heart rate and gut content. While these factors change on much slower timescales than changes in brain activity or behavior, there is mounting evidence for cardiovascular-brain interactions during behavior (Fujimoto et al. 2021; Legrand et al. 2022), as well as gut-brain interactions (Cryan & O'Mahony 2011; Champagne-Jorgensen & McVey Neufeld 2021). Therefore, one would obtain a more integrative understanding of behavior by measuring also body physiology, whereby the concerted contributions of both brain and body are taken into account.

10. Beyond the specific case of decision-making, it is true more broadly that behavior and cognition are continuous and nondiscrete processes—and yet they are usually tested using trial-based, discretized paradigms. Alex Huk and colleagues (2018) argued strongly against the sole use of trial-based experimental paradigms in sensory neuroscience, behavioral neuroscience, and cognitive neuroscience. They pointed out that there are *multiple timescales* in natural sensory stimuli, natural behaviors, and ongoing brain activity, which pose major problems for trial-based experimental designs that typically have a single timescale (the trial duration). Further, there are strong temporal correlations in brain activity, which induce temporal correlations in perception and behavior—a result that was documented by many studies. These correlations are extremely long-lasting—up to *minutes*—which again poses a problem for trial-based designs, which fundamentally rely on the assumption of trial-to-trial independence; this assumption is simply not supported by the data. Therefore, Huk and colleagues proposed to go beyond trial-based designs and forced-choice tasks, and instead to use continuous stimuli and continuous behaviors; and they offered analytical approaches for analyzing such experimental data: for example, using Kalman filter–based analysis or nonlinear state-space models.

11. One may argue that it is also necessary to develop better experimental methods for conducting longer neuronal recordings in order to allow the animal to fill more densely the multidimensional behavioral space while the experimenter is stably recording from an ensemble of neurons. However, although improving the recording duration may be useful, I would argue that this is not critical because current electrophysiological methods (e.g., tetrodes) already allow rather stable recordings of single neurons in freely moving animals over several hours—which is in fact usually longer than the duration of time that animals are willing to behave in a particular way. In other words, the stability of neuronal recordings today is *already* better than the stability of animal behavior, so increasing the duration of neuronal recordings will be of little help. Therefore, the challenge remains primarily theoretical: given the limited behavioral and neuronal data that can be collected during the time period over which animal behavior is roughly consistent, can we develop better theoretical methods to disentangle the relations of multidimensional neurons to multidimensional behaviors?

12. An animal may navigate from point A to point B for a number of very different reasons. (1) It may be hungry and navigating toward a familiar food source at location B (*memory-based navigation for foraging*). (2) It may be hungry but does not know where food might be found, so it is searching for it randomly (*random foraging*). (3) It is escaping from a predator and trying to reach its roost site, B (e.g., its burrow) as fast as possible (*memory-based predator escape*). (4) It is navigating toward its roost site, location B, after having eaten, so it is not hungry and is not in any immediate danger (*homing*). (5) It is navigating to find a mate (*navigation for reproduction*). (6) It is navigating as part of its seasonal migration (*migration*). (7) It is navigating following another conspecific, who is the leader and knows the way (*following*).

    I would propose that the *reason* for which an animal is navigating may be affecting spatial representations in the brain, in particular in the hippocampal formation. For example, at a given location, hippocampal place cells may fire differently when the animal has a specific navigational goal (e.g., reason 1 or 4) versus when the animal has no specific goal (reason 2); in fact, we know this to be the case, as demonstrated by findings of prospective coding in the hippocampus, whereby hippocampal neurons activate differently in the *same* location depending on the animal's navigational goal (Wood et al. 2000; Ferbinteanu & Shapiro 2003). Place cells also may be affected by the fearful experience of a predator (reason 3), as supported indeed by the demonstration that hippocampal place fields in mice undergo remapping following exposure to the odor of a predator, the coyote (Wang et al. 2012). By contrast, very little is known about how hippocampal neurons encode self-positioning in social situations in the presence of other individuals—for instance, when navigating to mate or following a leader (reasons 5 and 7). The discovery of hippocampal neurons that represent the positions of other individuals—the so-called *social place cells* found in the hippocampuses of bats and rats (Danjo et al. 2018; Omer et al. 2018) (chapter 4)—highlights the importance of studying this question, especially given the significance of social-spatial interactions for almost all animal species.

    Key open questions for future research include the following: How does the hippocampus represent simultaneously the animal's self-position and the positions of other individuals? And how does the hippocampus represent the positions of *multiple* individuals? Our recent experiments in bats have started to provide answers (Ray et al. 2024). Next, we may ask: Does the hunger state affect hippocampal spatial codes (reason 1)? This may sound unlikely, but we truly do not know that, because typical experiments on hippocampal place cells do not systematically manipulate the hunger state. In fact, several studies have shown that the hunger state of a mouse can have dramatic effects on neuronal responses in multiple brain regions, including the postrhinal cortex (Burgess et al. 2016, 2018), which in turn directly projects to the medial entorhinal cortex, and then to the hippocampus. This suggests that the hunger state may affect hippocampal representations. Indeed, a recent study showed a metabolic role for hippocampal sharp-wave ripples in regulating peripheral glucose homeostasis (Tingley et al. 2021), which indicates an interaction between hunger state and hippocampal function. In truth, we can expect that most of the factors listed in this discussion may shape the activity of hippocampal neurons in the context of navigation behavior.

References

Abdelfattah, A. S., Zheng, J., Singh, A., et al. 2023. Sensitivity optimization of a rhodopsin-based fluorescent voltage indicator. *Neuron* 111: 1547–1563.

Abeles, M. 1982. *Local cortical circuits: an electrophysiological study.* Springer.

Accanto, N., Blot, F. G. C., Lorca-Cámara, A., et al. 2023. A flexible two-photon fiberscope for fast activity imaging and precise optogenetic photostimulation of neurons in freely moving mice. *Neuron* 111: 176–189.

Acharya, L., Aghajan, Z. M., Vuong, C., Moore, J. J., & Mehta, M. R. 2016. Causal influence of visual cues on hippocampal directional selectivity. *Cell* 164: 197–207.

Adam, Y., Kim, J. J., Lou, S., et al. 2019. Voltage imaging and optogenetics reveal behaviour-dependent changes in hippocampal dynamics. *Nature* 569: 413–417.

Adamantidis, A., Arber, S., Bains, J. S., et al. 2015. Optogenetics: 10 years after ChR2 in neurons—views from the community. *Nature Neuroscience* 18: 1202–1212.

Adolphs, R. 2009. The social brain: Neural basis of social knowledge. *Annual Review of Psychology* 60: 693–716.

Adrian, E. D. 1928. *The basis of sensation: The action of the sense organs.* W. W. Norton & Co.

Agarwal, A., Sarel, A., Derdikman, D., Ulanovsky, N., & Gutfreund, Y. 2023. Spatial coding in the hippocampus and hyperpallium of flying owls. *Proceedings of the National Academy of Sciences of the USA* 120: e2212418120.

Aghajan, Z. M., Schuette, P., Fields, T. A., et al. 2017. Theta oscillations in the human medial temporal lobe during real-world ambulatory movement. *Current Biology* 27: 3743–3751.

Aghajan, Z. M., Acharya, L., Moore, J. J., Cushman, J. D., Vuong, C., & Mehta, M. R. 2015. Impaired spatial selectivity and intact phase precession in two-dimensional virtual reality. *Nature Neuroscience* 18: 121–128.

Aharoni, D., & Hoogland, T. M. 2019. Circuit investigations with open-source miniaturized microscopes: Past, present and future. *Frontiers in Cellular Neuroscience* 13: 141.

Aharoni, D., Khakh, B. S., Silva, A. J., & Golshani, P. 2019. All the light that we can see: A new era in miniaturized microscopy. *Nature Methods* 16: 11–13.

Ahissar, E., & Arieli, A. 2001. Figuring space by time. *Neuron* 32: 185–201.

Ahissar, E., & Arieli, A. 2012. Seeing via miniature eye movements: A dynamic hypothesis for vision. *Frontiers in Computational Neuroscience* 6: 89.

Ahissar, E., Arieli, A., Fried, M., & Bonneh, Y. 2016. On the possible roles of microsaccades and drifts in visual perception. *Vision Research* 118: 25–30.

Ahissar, E., & Assa, E. 2016. Perception as a closed-loop convergence process. *eLife* 5: e12830.

Ahissar, E., & Knutsen, P. M. 2008. Object localization with whiskers. *Biological Cybernetics* 98: 449–458.

Ahissar, E., Ozana, S., & Arieli, A. 2015. 1-D vision: Encoding of eye movements by simple receptive fields. *Perception* 44: 986–994.

Ahissar, M., & Hochstein, S. 1997. Task difficulty and the specificity of perceptual learning. *Nature* 387: 401–406.

Ahrens, M. B., Linden, J. F., & Sahani, M. 2008. Nonlinearities and contextual influences in auditory cortical responses modeled with multilinear spectrotemporal methods. *Journal of Neuroscience* 28: 1929–1942.

Albright, T. D. 1984. Direction and orientation selectivity of neurons in visual area MT of the macaque. *Journal of Neurophysiology* 52: 1106–1130.

Andersen, P., Morris, R. G. M., Amaral, D. G., Bliss, T. V., & O'Keefe, J., eds. 2007. *The hippocampus book.* Oxford University Press.

Anderson, D. J., & Perona, P. 2014. Toward a science of computational ethology. *Neuron* 84: 18–31.

Anderson, P. W. 1972. More is different. *Science* 177: 393–396.

Angelaki, D. E. 2014. How optic flow and inertial cues improve motion perception. *Cold Spring Harbor Symposia on Quantitative Biology* 79: 141–148.

Angelaki, D. E., Ng, J., Abrego, A. M., et al. 2020. A gravity-based three-dimensional compass in the mouse brain. *Nature Communications* 11: 1855.

Anpilov, S., Shemesh, Y., Eren, N., et al. 2020. Wireless optogenetic stimulation of oxytocin neurons in a semi-natural setup dynamically elevates both pro-social and agonistic behaviors. *Neuron* 107: 644–655.

Aravanis, A. M., Wang, L. P., Zhang, F., et al. 2007. An optical neural interface: In vivo control of rodent motor cortex with integrated fiberoptic and optogenetic technology. *Journal of Neural Engineering* 4: S143–S156.

Arieli, A., Sterkin, A., Grinvald, A., & Aertsen, A. 1996. Dynamics of ongoing activity: Explanation of the large variability in evoked cortical responses. *Science* 273: 1868–1871.

Aronov, D., & Tank, D. W. 2014. Engagement of neural circuits underlying 2D spatial navigation in a rodent virtual reality system. *Neuron* 84: 442–456.

Asari, H., & Zador, A. M. 2009. Long-lasting context dependence constrains neural encoding models in rodent auditory cortex. *Journal of Neurophysiology* 102: 2638–2656.

Austad, S. N. 1993. The comparative perspective and choice of animal models in aging research. *Aging* 5: 259–267.

Aytekin, M., Victor, J. D., & Rucci, M. 2014. The visual input to the retina during natural head-free fixation. *Journal of Neuroscience* 34: 12701–12715.

Bacon, F. 1620. *Novum organum.* 2015 annotated ed. Jazzybee Verlag.

Baddeley, A., & Hitch, G. J. 1974. Working memory. In *The psychology of learning and motivation: Advances in research and theory,* ed. G. A. Bower, pp. 47–89. Academic Press.

Baddeley, A., & Hitch, G. J. 2010. Working memory. *Scholarpedia* 5: 3015.

Báez-Mendoza, R., Mastrobattista, E. P., Wang, A. J., & Williams, Z. M. 2021. Social agent identity cells in the prefrontal cortex of interacting groups of primates. *Science* 374: eabb4149.

Bagdasarian, K., Szwed, M., Knutsen, P. M., et al. 2013. Pre-neuronal morphological processing of object location by individual whiskers. *Nature Neuroscience* 16: 622–631.

Bagi, B., Brecht, M., & Sanguinetti-Scheck, J. I. 2022. Unsupervised discovery of behaviorally relevant brain states in rats playing hide-and-seek. *Current Biology* 32: 2640–2653.

Balcetis, E., & Dunning, D. 2010. Wishful seeing: More desired objects are seen as closer. *Psychological Science* 21: 147–152.

Baldassano, C., Hasson, U., & Norman, K. A. 2018. Representation of real-world event schemas during narrative perception. *Journal of Neuroscience* 38: 9689–9699.

Ballentine, B., Hyman, J., & Nowicki, S. 2004. Vocal performance influences female response to male bird song: An experimental test. *Behavioral Ecology* 15: 163–168.

Balzani, E., Lakshminarasimhan, K. J., Angelaki, D. E., & Savin, C. 2022. Efficient estimation of neural tuning during naturalistic behavior. *Advances in Neural Information Processing Systems* 33: 12604–12614.

Ban, H. Y., Barrett, G. M., Borisevich, A., et al. 2022. Kernel flow: A high channel count scalable time-domain functional near-infrared spectroscopy system. *Journal of Biomedical Optics* 27: 074710.

Banaji, M. R., & Crowder, R. G. 1989. The bankruptcy of everyday memory. *American Psychologist* 44: 1185–1193.

Bar-Yosef, O., & Nelken, I. 2007. The effects of background noise on the neural responses to natural sounds in cat primary auditory cortex. *Frontiers in Computational Neuroscience* 1: 3.

Bar-Yosef, O., Rotman, Y., & Nelken, I. 2002. Responses of neurons in cat primary auditory cortex to bird chirps: Effects of temporal and spectral context. *Journal of Neuroscience* 22: 8619–8632.

Barrett, R. D. H., Laurent, S., Mallarino, R., et al. 2019. Linking a mutation to survival in wild mice. *Science* 363: 499–504.

Barry, C., Bush, D., O'Keefe, J., & Burgess, N. 2012. Models of grid cells and theta oscillations. *Nature* 488: E1–E2; discussion E2–E3.

Barry, D. N., Tierney, T. M., Holmes, N., et al. 2019. Imaging the human hippocampus with optically-pumped magnetoencephalography. *Neuroimage* 203: 116192.

Bartlett, F. C. 1932. *Remembering.* Cambridge University Press.

Bashivan, P., Kar, K., & DiCarlo, J. J. 2019. Neural population control via deep image synthesis. *Science* 364: eaav9436.

Bastos, A. M., Usrey, W. M., Adams, R. A., Mangun, G. R., Fries, P., & Friston, K. J. 2012. Canonical microcircuits for predictive coding. *Neuron* 76: 695–711.

Bateson, P., & Laland, K. N. 2013. Tinbergen's four questions: An appreciation and an update. *Trends in Ecology & Evolution* 28: 712–718.

Beetz, M. J., Hechavarria, J. C., & Kössl, M. 2016. Temporal tuning in the bat auditory cortex is sharper when studied with natural echolocation sequences. *Scientific Reports* 6: 29102.

Bell, C., Bodznick, D., Montgomery, J., & Bastian, J. 1997. The generation and subtraction of sensory expectations within cerebellum-like structures. *Brain, Behavior and Evolution* 50 (Suppl 1): 17–31.

Bell, C. C., Han, V., & Sawtell, N. B. 2008. Cerebellum-like structures and their implications for cerebellar function. *Annual Review of Neuroscience* 31: 1–24.

Benjamini, Y., Fonio, E., Galili, T., Havkin, G. Z., & Golani, I. 2011. Quantifying the buildup in extent and complexity of free exploration in mice. *Proceedings of the National Academy of Sciences of the USA* 108 (Suppl 3): 15580–15587.

Ben-Yishay, E., Krivoruchko, K., Ron, S., Ulanovsky, N., Derdikman, D., & Gutfreund, Y. 2021. Directional tuning in the hippocampal formation of birds. *Current Biology* 31: 2592–2602.

Berdahl, A. M., Kao, A. B., Flack, A., et al. 2018. Collective animal navigation and migratory culture: From theoretical models to empirical evidence. *Philosophical Transactions of the Royal Society B: Biological Sciences* 373: 20170009.

Berg, R. W., & Kleinfeld, D. 2003. Rhythmic whisking by rat: Retraction as well as protraction of the vibrissae is under active muscular control. *Journal of Neurophysiology* 89: 104–117.

Berkes, P., Orbán, G., Lengyel, M., & Fiser, J. 2011. Spontaneous cortical activity reveals hallmarks of an optimal internal model of the environment. *Science* 331: 83–87.

Berman, G. J., Bialek, W., & Shaevitz, J. W. 2016. Predictability and hierarchy in Drosophila behavior. *Proceedings of the National Academy of Sciences of the USA* 113: 11943–11948.

Berman, G. J., Choi, D. M., Bialek, W., & Shaevitz, J. W. 2014. Mapping the stereotyped behaviour of freely moving fruit flies. *Journal of the Royal Society Interface* 11: 20140672.

Bernstein, J. G., & Boyden, E. S. 2011. Optogenetic tools for analyzing the neural circuits of behavior. *Trends in Cognitive Sciences* 15: 592–600.

Berthold, P., Kaatz, M., & Querner, U. 2004. Long-term satellite tracking of white stork (Ciconia ciconia) migration: Constancy versus variability. *Journal of Ornithology* 145: 356–359.

Bhagavatula, P. S., Claudianos, C., Ibbotson, M. R., & Srinivasan, M. V. 2011. Optic flow cues guide flight in birds. *Current Biology* 21: 1794–1799.

Bhaya-Grossman, I., & Chang, E. F. 2022. Speech computations of the human superior temporal gyrus. *Annual Review of Psychology* 73: 79–102.

Bialek, W. 2012. *Biophysics: Searching for principles.* Princeton University Press.

Bisley, J. W., & Goldberg, M. E. 2010. Attention, intention, and priority in the parietal lobe. *Annual Review of Neuroscience* 33: 1–21.

Bittner, K. C., Milstein, A. D., Grienberger, C., Romani, S., & Magee, J. C. 2017. Behavioral time scale synaptic plasticity underlies CA1 place fields. *Science* 357: 1033–1036.

Bizley, J. K., & Cohen, Y. E. 2013. The what, where and how of auditory-object perception. *Nature Reviews Neuroscience* 14: 693–707.

Blanchard, R. J., Hebert, M. A., Ferrari, P. F., et al. 1998. Defensive behaviors in wild and laboratory (Swiss) mice: The mouse defense test battery. *Physiology & Behavior* 65: 201–209.

Bobrov, E., Wolfe, J., Rao, R. P., & Brecht, M. 2014. The representation of social facial touch in rat barrel cortex. *Current Biology* 24: 109–115.

Bocca, E., Antonelli, A. R., & Mosciaro, O. 1965. Mechanical co-factors in olfactory stimulation. *Acta Oto-Laryngologica* 59: 243–247.

Boccara, C. N., Sargolini, F., Thoresen, V. H., et al. 2010. Grid cells in pre- and parasubiculum. *Nature Neuroscience* 13: 987–994.

Bolhuis, J. J., & Gahr, M. 2006. Neural mechanisms of birdsong memory. *Nature Reviews Neuroscience* 7: 347–357.

Bonnen, K., Czuba, T. B., Whritner, J. A., Kohn, A., Huk, A. C., & Cormack, L. K. 2020. Binocular viewing geometry shapes the neural representation of the dynamic three-dimensional environment. *Nature Neuroscience* 23: 113–121.

Borst, A., & Helmstaedter, M. 2015. Common circuit design in fly and mammalian motion vision. *Nature Neuroscience* 18: 1067–1076.

Boto, E., Hill, R. M., Rea, M., et al. 2021. Measuring functional connectivity with wearable MEG. *NeuroImage* 230: 117815.

Boto, E., Holmes, N., Leggett, J., et al. 2018. Moving magnetoencephalography towards real-world applications with a wearable system. *Nature* 555: 657–661.

Boto, E., Seedat, Z. A., Holmes, N., et al. 2019. Wearable neuroimaging: Combining and contrasting magnetoencephalography and electroencephalography. *NeuroImage* 201: 116099.

Bradbury, J. W., & Vehrencamp, S. L. 2014. Complexity in behavioral ecology. *Behavioral Ecology* 25: 435–442.

Brand, A., Behrend, O., Marquardt, T., McAlpine, D., & Grothe, B. 2002. Precise inhibition is essential for microsecond interaural time difference coding. *Nature* 417: 543–547.

Brandes, S., Sicks, F., & Berger, A. 2021. Behaviour classification on giraffes (Giraffa camelopardalis) using machine learning algorithms on triaxial acceleration data of two commonly used GPS devices and its possible application for their management and conservation. *Sensors (Basel)* 21: 2229.

Brecht, M., Preilowski, B., & Merzenich, M. M. 1997. Functional architecture of the mystacial vibrissae. *Behavioural Brain Research* 84: 81–97.

Bregman, A. S. 1994. *Auditory scene analysis: The perceptual organization of sound.* MIT Press

Brenner, S. 2002. Life sentences: Detective Rummage investigates. *Genome Biology* 3: 1013.1.

Brenowitz, E. A., & Zakon, H. H. 2015. Emerging from the bottleneck: Benefits of the comparative approach to modern neuroscience. *Trends in Neurosciences* 38: 273–278.

Briggs, F. 2020. Role of feedback connections in central visual processing. *Annual Review of Vision Science* 6: 313–334.

Briggs, F., & Usrey, W. M. 2011. Corticogeniculate feedback and visual processing in the primate. *Journal of Physiology* 589: 33–40.

Broekmans, O. D., Rodgers, J. B., Ryu, W. S., & Stephens, G. J. 2016. Resolving coiled shapes reveals new reorientation behaviors in *C. elegans*. *eLife* 5: e17227.

Brosch, M., & Schreiner, C. E. 1997. Time course of forward masking tuning curves in cat primary auditory cortex. *Journal of Neurophysiology* 77: 923–943.

Brown, A. E. X., & de Bivort, B. L. 2018. Ethology as a physical science. *Nature Physics* 14: 653–657.

Bruce, C., Desimone, R., & Gross, C. G. 1981. Visual properties of neurons in a polysensory area in superior temporal sulcus of the macaque. *Journal of Neurophysiology* 46: 369–384.

Bruce, D. 1985. The how and why of ecological memory. *Journal of Experimental Psychology: General* 114: 78–90.

Brunner, C., Grillet, M., Urban, A., Roska, B., Montaldo, G., & Macé, E. 2021. Whole-brain functional ultrasound imaging in awake head-fixed mice. *Nature Protocols* 16: 3547–3571.

Bubeck, S., Chandrasekaran, V., Eldan, R., et al. 2023. Sparks of artificial general intelligence: Early experiments with GPT-4. *arXiv*: 2303.12712.

Buccino, A. P., Hurwitz, C. L., Garcia, S., et al. 2020. SpikeInterface, a unified framework for spike sorting. *eLife* 9: e61834.

Burak, Y., Rokni, U., Meister, M., & Sompolinsky, H. 2010. Bayesian model of dynamic image stabilization in the visual system. *Proceedings of the National Academy of Sciences of the USA* 107: 19525–19530.

Burgess, C. R., Livneh, Y., Ramesh, R. N., & Andermann, M. L. 2018. Gating of visual processing by physiological need. *Current Opinion in Neurobiology* 49: 16–23.

Burgess, C. R., Ramesh, R. N., Sugden, A. U., et al. 2016. Hunger-dependent enhancement of food cue responses in mouse postrhinal cortex and lateral amygdala. *Neuron* 91: 1154–1169.

Burgess, N., Barry, C., & O'Keefe, J. 2007. An oscillatory interference model of grid cell firing. *Hippocampus* 17: 801–812.

Burkett, J. P., Andari, E., Johnson, Z. V., et al. 2016. Oxytocin-dependent consolation behavior in rodents. *Science* 351: 375–378.

Buschke, H. 1984. Cued recall in amnesia. *Journal of Clinical Neuropsychology* 6: 433–440.

Buzsáki, G. 2006. *Rhythms of the brain*. Oxford University Press.

Buzsáki, G., & Moser, E. I. 2013. Memory, navigation and theta rhythm in the hippocampal-entorhinal system. *Nature Neuroscience* 16: 130–138.

Cai, D. J., Aharoni, D., Shuman, T., et al. 2016. A shared neural ensemble links distinct contextual memories encoded close in time. *Nature* 534: 115–118.

Camhi, J. 1984. *Neuroethology: Nerve cells and the natural behavior of animals.* Sinauer.

Campagner, D., Evans, M. H., Bale, M. R., Erskine, A., & Petersen, R. S. 2016. Prediction of primary somatosensory neuron activity during active tactile exploration. *eLife* 5: e10696.

Campagner, D., Vale, R., Tan, Y. L., et al. 2023. A cortico-collicular circuit for orienting to shelter during escape. *Nature* 613: 111–119.

Carandini, M. 2012. From circuits to behavior: A bridge too far? *Nature Neuroscience* 15: 507–509.

Carandini, M., Demb, J. B., Mante, V., et al. 2005. Do we know what the early visual system does? *Journal of Neuroscience* 25: 10577–10597.

Carew, T. J. 2004. *Behavioral neurobiology: The cellular organization of natural behavior.* Sinauer.

Carr, C. E., & Konishi, M. 1988. Axonal delay lines for time measurement in the owl's brainstem. *Proceedings of the National Academy of Sciences of the USA* 85: 8311–8315.

Carr, C. E., & Konishi, M. 1990. A circuit for detection of interaural time differences in the brain stem of the barn owl. *Journal of Neuroscience* 10: 3227–3246.

Carter, G. G., Farine, D. R., Crisp, R. J., et al. 2020. Development of new food-sharing relationships in vampire bats. *Current Biology* 30: 1275–1279.

Carter, G. G., & Wilkinson, G. S. 2013. Food sharing in vampire bats: Reciprocal help predicts donations more than relatedness or harassment. *Proceedings of the Royal Society B: Biological Sciences* 280: 20122573.

Carter, G. G., & Wilkinson, G. S. 2015. Social benefits of non-kin food sharing by female vampire bats. *Proceedings of the Royal Society B: Biological Sciences* 282: 20152524.

Catania, K. C. 2006. Underwater 'sniffing' by semi-aquatic mammals. *Nature* 444: 1024–1025.

Chalfin, L., Dayan, M., Levy, D. R., et al. 2014. Mapping ecologically relevant social behaviours by gene knockout in wild mice. *Nature Commununications* 5: 4569.

Champagne-Jorgensen, K., & McVey Neufeld, K. A. 2021. The effect of microbiota on behaviour. *Modern Trends in Psychiatry* 32: 58–67.

Chang, L., & Tsao, D. Y. 2017. The code for facial identity in the primate brain. *Cell* 169: 1013–1028.

Chang, S. W., Brent, L. J., Adams, G. K., et al. 2013. Neuroethology of primate social behavior. *Proceedings of the National Academy of Sciences of the USA* 110 (Suppl 2): 10387–10394.

Chaudhuri, R., Gerçek, B., Pandey, B., Peyrache, A., & Fiete, I. 2019. The intrinsic attractor manifold and population dynamics of a canonical cognitive circuit across waking and sleep. *Nature Neuroscience* 22: 1512–1520.

Chernetsov, N. 2017. Compass systems. *Journal of Comparative Physiology A* 203: 447–453.

Chettih, S. N., Mackevicius, E. L., Hale, S., & Aronov, D. 2024. Barcoding of episodic memories in the hippocampus of a food-caching bird. *Cell* 187: 1922–1935.

Cheung, A., & Vickerstaff, R. 2010. Finding the way with a noisy brain. *PLoS Computational Biology* 6: e1000992.

Chung, J. E., Joo, H. R., Fan, J. L., et al. 2019. High-density, long-lasting, and multi-region electrophysiological recordings using polymer electrode arrays. *Neuron* 101: 21–31.

Chung, J. E., Sellers, K. K., Leonard, M. K., et al. 2022. High-density single-unit human cortical recordings using the Neuropixels probe. *Neuron* 110: 2409–2421.

Cisek, P., & Green, A. M. 2024. Toward a neuroscience of natural behaviors. *Current Opinion in Neurobiology* 86: 102859.

Clavagnier, S., Falchier, A., & Kennedy, H. 2004. Long-distance feedback projections to area V1: Implications for multisensory integration, spatial awareness, and visual consciousness. *Cognitive, Affective, & Behavioral Neuroscience* 4: 117–126.

Clayton, N. S., & Dickinson, A. 1998. Episodic-like memory during cache recovery by scrub jays. *Nature* 395: 272–274.

Clayton, N. S., & Krebs, J. R. 1995. Memory in food-storing birds: From behaviour to brain. *Current Opinion in Neurobiology* 5: 149–154.

Clutton-Brock, T. 2016. *Mammal societies.* Wiley-Blackwell.

Coen, P., Clemens, J., Weinstein, A. J., Pacheco, D. A., Deng, Y., & Murthy, M. 2014. Dynamic sensory cues shape song structure in Drosophila. *Nature* 507: 233–237.

Cohen, G., & Conway, M. A. 2007. *Memory in the real world.* 3rd ed. Psychology Press.

Cohen, J. Y., Haesler, S., Vong, L., Lowell, B. B., & Uchida, N. 2012. Neuron-type-specific signals for reward and punishment in the ventral tegmental area. *Nature* 482: 85–88.

Cohen, Y., Nicholson, D. A., Sanchioni, A., Mallaber, E. K., Skidanova, V., & Gardner, T. J. 2022. Automated annotation of birdsong with a neural network that segments spectrograms. *eLife* 11: e63853.

Cohen, Y., Shen, J., Semu, D., et al. 2020. Hidden neural states underlie canary song syntax. *Nature* 582: 539–544.

Colgin, L. L., Moser, E. I., & Moser, M.-B. 2008. Understanding memory through hippocampal remapping. *Trends in Neurosciences* 31: 469–477.

Collins, J. A., & Olson, I. R. 2014. Knowledge is power: How conceptual knowledge transforms visual cognition. *Psychonomic Bulletin & Review* 21: 843–860.

Constantinidis, C., & Klingberg, T. 2016. The neuroscience of working memory capacity and training. *Nature Reviews Neuroscience* 17: 438–449.

Coppola, D., & Purves, D. 1996. The extraordinarily rapid disappearance of entoptic images. *Proceedings of the National Academy of Sciences of the USA* 93: 8001–8004.

Corcoran, A. J., Barber, J. R., & Conner, W. E. 2009. Tiger moth jams bat sonar. *Science* 325: 325–327.

Corcoran, A. J., & Conner, W. E. 2014. Bats jamming bats: Food competition through sonar interference. *Science* 346: 745–747.

Couzin, I. D., & Heins, C. 2023. Emerging technologies for behavioral research in changing environments. *Trends in Ecology & Evolution* 38: 346–354.

Couzin, I. D., Krause, J., Franks, N. R., & Levin, S. A. 2005. Effective leadership and decision-making in animal groups on the move. *Nature* 433: 513–516.

Crall, J. D., Switzer, C. M., Oppenheimer, R. L., et al. 2018. Neonicotinoid exposure disrupts bumblebee nest behavior, social networks, and thermoregulation. *Science* 362: 683–686.

Crawley, J. N., Belknap, J. K., Collins, A., et al. 1997. Behavioral phenotypes of inbred mouse strains: Implications and recommendations for molecular studies. *Psychopharmacology (Berlin)* 132: 107–124.

Cronin, T. W., Johnsen, S., Marshall, N. J., & Warrant, E. J. 2014. *Visual ecology.* Princeton University Press.

Cryan, J. F., & O'Mahony, S. M. 2011. The microbiome-gut-brain axis: From bowel to behavior. *Neurogastroenterology & Motility* 23: 187–192.

Cudeiro, J., & Sillito, A. M. 2006. Looking back: Corticothalamic feedback and early visual processing. *Trends in Neuroscience* 29: 298–306.

Cui, X., Bray, S., Bryant, D. M., Glover, G. H., & Reiss, A. L. 2011. A quantitative comparison of NIRS and fMRI across multiple cognitive tasks. *NeuroImage* 54: 2808–2821.

Cullen, K. E. 2019. Vestibular processing during natural self-motion: Implications for perception and action. *Nature Reviews Neuroscience* 20: 346–363.

Curtis, J. C., & Kleinfeld, D. 2006. Seeing what the mouse sees with its vibrissae: A matter of behavioral state. *Neuron* 50: 524–526.

Czeszumski, A., Eustergerling, S., Lang, A., et al. 2020. Hyperscanning: A valid method to study neural inter-brain underpinnings of social interaction. *Frontiers in Human Neuroscience* 14: 39.

Dagnew, R., Lin, Y. Y., Agatep, J., et al. 2017. CerebraLux: A low-cost, open-source, wireless probe for optogenetic stimulation. *Neurophotonics* 4: 045001.

Danjo, T., Toyoizumi, T., & Fujisawa, S. 2018. Spatial representations of self and other in the hippocampus. *Science* 359: 213–218.

Datta, S. R., Anderson, D. J., Branson, K., Perona, P., & Leifer, A. 2019. Computational neuroethology: A call to action. *Neuron* 104: 11–24.

David, S. V., & Shamma, S. A. 2013. Integration over multiple timescales in primary auditory cortex. *Journal of Neuroscience* 33: 19154–19166.

Davidson, J. D., & El Hady, A. 2019. Foraging as an evidence accumulation process. *PLoS Computational Biology* 15: e1007060.

Davies, N. B., Krebs, J. R., & West, S. A. 2012. *An introduction to behavioral ecology.* 4th ed. Wiley-Blackwell.

Davis, R. 1966. Homing performance and homing ability in bats. *Ecological Monographs* 36: 201–237.

Dawkins, R. 1976. *The selfish gene.* Oxford University Press.

deCharms, R. C., Blake, D. T., & Merzenich, M. M. 1998. Optimizing sound features for cortical neurons. *Science* 280: 1439–1443.

Deffieux, T., Demene, C., Pernot, M., & Tanter, M. 2018. Functional ultrasound neuroimaging: A review of the preclinical and clinical state of the art. *Current Opinion in Neurobiology* 50: 128–135.

Dehnhardt, G., Mauck, B., Hanke, W., & Bleckmann, H. 2001. Hydrodynamic trail-following in harbor seals (Phoca vitulina). *Science* 293: 102–104.

Deisseroth, K. 2015. Optogenetics: 10 years of microbial opsins in neuroscience. *Nature Neuroscience* 18: 1213–1225.

Demas, J., Manley, J., Tejera, F., et al. 2021. High-speed, cortex-wide volumetric recording of neuroactivity at cellular resolution using light beads microscopy. *Nature Methods* 18: 1103–1111.

Dennis, E. J., El Hady, A., Michaiel, A., et al. 2021. Systems neuroscience of natural behaviors in rodents. *Journal of Neuroscience* 41: 911–919.

Depireux, D. A., Simon, J. Z., Klein, D. J., & Shamma, S. A. 2001. Spectro-temporal response field characterization with dynamic ripples in ferret primary auditory cortex. *Journal of Neurophysiology* 85: 1220–1234.

Descartes, R. 1644. *Principles of philosophy* (2009 reprint). SMK Books.

Deschênes, M., Moore, J., & Kleinfeld, D. 2012. Sniffing and whisking in rodents. *Current Opinion in Neurobiology* 22: 243–250.

de Waal, F. B., & Suchak, M. 2010. Prosocial primates: Selfish and unselfish motivations. *Philosophical Transactions of the Royal Society B: Biological Sciences* 365: 2711–2722.

de Waal, F. B. M., & Preston, S. D. 2017. Mammalian empathy: Behavioural manifestations and neural basis. *Nature Reviews Neuroscience* 18: 498–509.

Dewey, J. 1896. The reflex arc concept in psychology. *Psychological Review* 3: 357–370.

Diba, K., & Buzsáki, G. 2007. Forward and reverse hippocampal place-cell sequences during ripples. *Nature Neuroscience* 10: 1241–1242.

di Pellegrino, G., Fadiga, L., Fogassi, L., Gallese, V., & Rizzolatti, G. 1992. Understanding motor events: A neurophysiological study. *Experimental Brain Research* 91: 176–180.

Ditchburn, R. W. 1973. *Eye movements and visual perception.* Clarendon Press.

Dolensek, N., Gehrlach, D. A., Klein, A. S., & Gogolla, N. 2020. Facial expressions of emotion states and their neuronal correlates in mice. *Science* 368: 89–94.

Doupe, A. J., & Konishi, M. 1991. Song-selective auditory circuits in the vocal control system of the zebra finch. *Proceedings of the National Academy of Sciences of the USA* 88: 11339–11343.

Doupe, A. J., Solis, M. M., Kimpo, R., & Boettiger, C. A. 2004. Cellular, circuit, and synaptic mechanisms in song learning. *Annals of the New York Academy of Sciences* 1016: 495–523.

Dudai, Y. 2002. *Memory from A to Z.* Oxford University Press.

Duffy, C. J., & Wurtz, R. H. 1991. Sensitivity of MST neurons to optic flow stimuli. I. A continuum of response selectivity to large-field stimuli. *Journal of Neurophysiology* 65: 1329–1345.

Duffy, C. J., & Wurtz, R. H. 1995. Response of monkey MST neurons to optic flow stimuli with shifted centers of motion. *Journal of Neuroscience* 15: 5192–5208.

Dugatkin, L. A. 2020. *Principles of animal behavior.* 4th ed. University of Chicago Press.

Dunn, T. W., Marshall, J. D., Severson, K. S., et al. 2021. Geometric deep learning enables 3D kinematic profiling across species and environments. *Nature Methods* 18: 564–573.

Dusenbery, D. B. 1992. *Sensory ecology.* Freeman.

Ebbesen, C. L., & Froemke, R. C. 2022. Automatic mapping of multiplexed social receptive fields by deep learning and GPU-accelerated 3D videography. *Nature Communications* 13: 593.

Ebner, F. F., & Kaas, J. H. 2015. Somatosensory system. In: *The rat nervous system* (4th ed.), ed. G. Paxinos, pp. 675–701. Elsevier.

Eccles, J. C. 1964. *The physiology of synapses.* Springer.

Edelson, M., Sharot, T., Dolan, R. J., & Dudai, Y. 2011. Following the crowd: Brain substrates of long-term memory conformity. *Science* 333: 108–111.

Egert-Berg, K., Hurme, E. R., Greif, S., et al. 2018. Resource ephemerality drives social foraging in bats. *Current Biology* 28: 3667–3673.

Egnor, S. E. R., & Branson, K. 2016. Computational analysis of behavior. *Annual Review of Neuroscience* 39: 217–236.

Eichenbaum, H. 2015. Perspectives on 2014 Nobel Prize. *Hippocampus* 25: 679–681.

Eichenbaum, H., & Cohen, N. J. 2014. Can we reconcile the declarative memory and spatial navigation views on hippocampal function? *Neuron* 83: 764–770.

Ekstrom, A. D., Caplan, J. B., Ho, E., Shattuck, K., Fried, I., & Kahana, M. J. 2005. Human hippocampal theta activity during virtual navigation. *Hippocampus* 15: 881–889.

Ekstrom, A. D., Huffman, D. J., & Starrett, M. 2017. Interacting networks of brain regions underlie human spatial navigation: A review and novel synthesis of the literature. *Journal of Neurophysiology* 118: 3328–3344.

Eliades, S. J., & Wang, X. 2008. Neural substrates of vocalization feedback monitoring in primate auditory cortex. *Nature* 453: 1102–1106.

Elias, D. O., Mason, A. C., Maddison, W. P., & Hoy, R. R. 2003. Seismic signals in a courting male jumping spider (Araneae: Salticidae). *Journal of Experimental Biology* 206: 4029–4039.

Eliav, T., Geva-Sagiv, M., Yartsev, M. M., et al. 2018. Nonoscillatory phase coding and synchronization in the bat hippocampal formation. *Cell* 175: 1119–1130.

Eliav, T., Maimon, S. R., Aljadeff, J., et al. 2021. Multiscale representation of very large environments in the hippocampus of flying bats. *Science* 372: eabg4020.

Ellis, G. 2016. *How can physics underlie the mind? Top-down causation in the human context.* Springer.

Emara, M. S., Pisanello, M., Sileo, L., De Vittorio, M., & Pisanello, F. 2019. A wireless head-mountable device with tapered optical fiber-coupled laser diode for light delivery in deep brain regions. *IEEE Transactions on Biomedical Engineering* 66, 1996–2009.

Emlen, S. T. 1975. The stellar-orientation system of a migratory bird. *Scientific American* 233: 102–111.

Ennaceur, A., Michalikova, S., van Rensburg, R., & Chazot, P. L. 2008. Are benzodiazepines really anxiolytic? Evidence from a 3D maze spatial navigation task. *Behavioural Brain Research* 188: 136–153.

Errico, C., Osmanski, B. F., Pezet, S., Couture, O., Lenkei, Z., & Tanter, M. 2016. Transcranial functional ultrasound imaging of the brain using microbubble-enhanced ultrasensitive Doppler. *NeuroImage* 124: 752–761.

Esch, H. E., Zhang, S., Srinivasan, M. V., & Tautz, J. 2001. Honeybee dances communicate distances measured by optic flow. *Nature* 411: 581–583.

Esser, K. H., Condon, C. J., Suga, N., & Kanwal, J. S. 1997. Syntax processing by auditory cortical neurons in the FM-FM area of the mustached bat Pteronotus parnellii. *Proceedings of the National Academy of Sciences of the USA* 94: 14019–14024.

Etienne, A. S., Maurer, R., Berlie, J., et al. 1998. Navigation through vector addition. *Nature* 396: 161–164.

Evans, D. A., Stempel, A. V., Vale, R., Ruehle, S., Lefler, Y., & Branco, T. 2018. A synaptic threshold mechanism for computing escape decisions. *Nature* 558: 590–594.

Fanselow, E. E., & Nicolelis, M. A. 1999. Behavioral modulation of tactile responses in the rat somatosensory system. *Journal of Neuroscience* 19: 7603–7616.

Fei-Fei, L., Iyer, A., Koch, C., & Perona, P. 2007. What do we perceive in a glance of a real-world scene? *Journal of Vision* 7: 10.

Feldman, D. E. 2009. Synaptic mechanisms for plasticity in neocortex. *Annual Review of Neuroscience* 32: 33–55.

Feng, A. S., Hall, J. C., & Gooler, D. M. 1990. Neural basis of sound pattern recognition in anurans. *Progress in Neurobiology* 34: 313–329.

Ferbinteanu, J., & Shapiro, M. L. 2003. Prospective and retrospective memory coding in the hippocampus. *Neuron* 40: 1227–1239.

Ferezou, I., Bolea, S., & Petersen, C. C. 2006. Visualizing the cortical representation of whisker touch: Voltage-sensitive dye imaging in freely moving mice. *Neuron* 50: 617–629.

Fiete, I. R., Burak, Y., & Brookings, T. 2008. What grid cells convey about rat location. *Journal of Neuroscience* 28: 6858–6871.

Finkelstein, A., Derdikman, D., Rubin, A., Foerster, J. N., Las, L., & Ulanovsky, N. 2015. Three-dimensional head-direction coding in the bat brain. *Nature* 517: 159–164.

Finkelstein, A., Las, L., & Ulanovsky, N. 2016. 3-D maps and compasses in the brain. *Annual Review of Neuroscience* 39: 171–196.

Firestone, C., & Scholl, B. J. 2016. Cognition does not affect perception: Evaluating the evidence for "top-down" effects. *Behavioral and Brain Sciences* 39: e229.

Fiser, J., Chiu, C., & Weliky, M. 2004. Small modulation of ongoing cortical dynamics by sensory input during natural vision. *Nature* 431: 573–578.

Fishman, Y. I., Arezzo, J. C., & Steinschneider, M. 2004. Auditory stream segregation in monkey auditory cortex: Effects of frequency separation, presentation rate, and tone duration. *Journal of the Acoustical Society of America* 116: 1656–1670.

Fishman, Y. I., Reser, D. H., Arezzo, J. C., & Steinschneider, M. 2001. Neural correlates of auditory stream segregation in primary auditory cortex of the awake monkey. *Hearing Research* 151: 167–187.

Fitzpatrick, D., & Ulanovsky, N. 2014. Neural maps. *Current Opinion in Neurobiology* 24: iv–vi.

Flash, T., & Hochner, B. 2005. Motor primitives in vertebrates and invertebrates. *Current Opinion in Neurobiology* 15: 660–666.

Flavell, S. W., Raizen, D. M., & You, Y. J. 2020. Behavioral states. *Genetics* 216: 315–332.

Flusberg, B. A., Nimmerjahn, A., Cocker, E. D., et al. 2008. High-speed, miniaturized fluorescence microscopy in freely moving mice. *Nature Methods* 5: 935–938.

Fonio, E., Benjamini, Y., & Golani, I. 2009. Freedom of movement and the stability of its unfolding in free exploration of mice. *Proceedings of the National Academy of Sciences of the USA* 106: 21335–21340.

Fonio, E., Benjamini, Y., & Golani, I. 2012a. Short and long term measures of anxiety exhibit opposite results. *PLoS One* 7: e48414.

Fonio, E., Golani, I., & Benjamini, Y. 2012b. Measuring behavior of animal models: Faults and remedies. *Nature Methods* 9: 1167–1170.

Fonio, E., Gordon, G., Barak, N., et al. 2016. Coordination of sniffing and whisking depends on the mode of interaction with the environment. *Israel Journal of Ecology and Evolution* 61: 95–105.

Forkosh, O., Karamihalev, S., Roeh, S., et al. 2019. Identity domains capture individual differences from across the behavioral repertoire. *Nature Neuroscience* 22: 2023–2028.

Forli, A., & Yartsev, M. M. 2023. Hippocampal representation during collective spatial behaviour in bats. *Nature* 621: 796–803.

Foster, D. J., & Wilson, M. A. 2006. Reverse replay of behavioural sequences in hippocampal place cells during the awake state. *Nature* 440: 680–683.

Fritz, J., Mishkin, M., & Saunders, R. C. 2005. In search of an auditory engram. *Proceedings of the National Academy of Sciences of the USA* 102: 9359–9364.

Fritz, J., Shamma, S., Elhilali, M., & Klein, D. 2003. Rapid task-related plasticity of spectrotemporal receptive fields in primary auditory cortex. *Nature Neuroscience* 6: 1216–1223.

Fritz, J. B., Elhilali, M., & Shamma, S. A. 2007. Adaptive changes in cortical receptive fields induced by attention to complex sounds. *Journal of Neurophysiology* 98: 2337–2346.

Froemke, R. C., & Young, L. J. 2021. Oxytocin, neural plasticity, and social behavior. *Annual Review of Neuroscience* 44: 359–381.

Frostig, R. D. 2006. Functional organization and plasticity in the adult rat barrel cortex: Moving out-of-the-box. *Current Opinion in Neurobiology* 16: 445–450.

Fujimoto, A., Murray, E. A., & Rudebeck, P. H. 2021. Interaction between decision-making and interoceptive representations of bodily arousal in frontal cortex. *Proceedings of the National Academy of Sciences of the USA* 118: e2014781118.

Fukushima, M., Chao, Z. C., & Fujii, N. 2015. Studying brain functions with mesoscopic measurements: Advances in electrocorticography for non-human primates. *Current Opinion in Neurobiology* 32: 124–131.

Furtak, S. C., Wei, S. M., Agster, K. L., & Burwell, R. D. 2007. Functional neuroanatomy of the parahippocampal region in the rat: The perirhinal and postrhinal cortices. *Hippocampus* 17: 709–722.

Fusi, S., Miller, E. K., & Rigotti, M. 2016. Why neurons mix: High dimensionality for higher cognition. *Current Opinion in Neurobiology* 37: 66–74.

Fuster, J. M. 1997. Network memory. *Trends in Neurosciences* 20: 451–459.

Gallant, J. L., Connor, C. E., & Van Essen, D. C. 1998. Neural activity in areas V1, V2 and V4 during free viewing of natural scenes compared to controlled viewing. *Neuroreport* 9: 2153–2158.

Gao, P., & Ganguli, S. 2015. On simplicity and complexity in the brave new world of large-scale neuroscience. *Current Opinion in Neurobiology* 32: 148–155.

Gao, P., Trautmann, E., Yu, B., et al. 2017. A theory of multineuronal dimensionality, dynamics and measurement. *bioRxiv*. https://doi.org/10.1101/214262.

Garcia-Lazaro, J. A., Ahmed, B., & Schnupp, J. W. 2006. Tuning to natural stimulus dynamics in primary auditory cortex. *Current Biology* 16: 264–271.

Garcia-Lazaro, J. A., Ahmed, B., & Schnupp, J. W. 2011. Emergence of tuning to natural stimulus statistics along the central auditory pathway. *PLoS One* 6: e22584.

Gardner, R. J., Hermansen, E., Pachitariu, M., et al. 2022. Toroidal topology of population activity in grid cells. *Nature* 602: 123–128.

Geffen, M. N., Broome, B. M., Laurent, G., & Meister, M. 2009. Neural encoding of rapidly fluctuating odors. *Neuron* 61: 570–586.

Gentner, T. Q., & Hulse, S. H. 2000. Female European starling preference and choice for variation in conspecific male song. *Animal Behaviour* 59: 443–458.

Gerardin, P., Kourtzi, Z., & Mamassian, P. 2010. Prior knowledge of illumination for 3D perception in the human brain. *Proceedings of the National Academy of Sciences of the USA* 107: 16309–16314.

Geva-Sagiv, M., Las, L., Yovel, Y., & Ulanovsky, N. 2015. Spatial cognition in bats and rats: From sensory acquisition to multiscale maps and navigation. *Nature Reviews Neuroscience* 16: 94–108.

Ghose, K., & Moss, C. F. 2003. The sonar beam pattern of a flying bat as it tracks tethered insects. *Journal of the Acoustical Society of America* 114: 1120–1131.

Ghosh, K. K., Burns, L. D., Cocker, E. D., et al. 2011. Miniaturized integration of a fluorescence microscope. *Nature Methods* 8: 871–878.

Gibson, J. J. 1962. Observations on active touch. *Psychological Review* 69: 477–491.

Gibson, J. J. 1979. *The ecological approach to visual perception.* Houghton Mifflin.

Gilad, A., Oz, R., & Slovin, H. 2017. Uncovering the spatial profile of contour integration from fixational saccades: Evidence for widespread processing in V1. *Cerebral Cortex* 27: 5261–5273.

Gilaie-Dotan, S., Saygin, A. P., Lorenzi, L. J., Egan, R., Rees, G., & Behrmann, M. 2013. The role of human ventral visual cortex in motion perception. *Brain* 136: 2784–2798.

Gilbert, C. D., & Li, W. 2013. Top-down influences on visual processing. *Nature Reviews Neuroscience* 14: 350–363.

Gilchrist, I. D., Brown, V., & Findlay, J. M. 1997. Saccades without eye movements. *Nature* 390: 130–131.

Ginosar, G., Aljadeff, J., Burak, Y., Sompolinsky, H., Las, L., & Ulanovsky, N. 2021. Locally ordered representation of 3D space in the entorhinal cortex. *Nature* 596: 404–409.

Ginosar, G., Aljadeff, J., Las, L., Derdikman, D., & Ulanovsky, N. 2023. Are grid cells used for navigation? On local metrics, subjective spaces, and black holes. *Neuron* 111: 1858–1875.

Goffinet, J., Brudner, S., Mooney, R., & Pearson, J. 2021. Low-dimensional learned feature spaces quantify individual and group differences in vocal repertoires. *eLife* 10: e67855.

Goldstein, A., Zada, Z., Buchnik, E., et al. 2022. Shared computational principles for language processing in humans and deep language models. *Nature Neuroscience* 25: 369–380.

Goldstone, R. L., de Leeuw, J. R., & Landy, D. H. 2015. Fitting perception in and to cognition. *Cognition* 135: 24–29.

Gollisch, T., & Meister, M. 2008. Rapid neural coding in the retina with relative spike latencies. *Science* 319: 1108–1111.

Gollisch, T., & Meister, M. 2010. Eye smarter than scientists believed: Neural computations in circuits of the retina. *Neuron* 65: 150–164.

Gomez-Marin, A., & Ghazanfar, A. A. 2019. The life of behavior. *Neuron* 104: 25–36.

Gomez-Marin, A., Paton, J. J., Kampff, A. R., Costa, R. M., & Mainen, Z. F. 2014. Big behavioral data: Psychology, ethology and the foundations of neuroscience. *Nature Neuroscience* 17: 1455–1462.

Goodale, M. A., & Westwood, D. A. 2004. An evolving view of duplex vision: Separate but interacting cortical pathways for perception and action. *Current Opinion in Neurobiology* 14: 203–211.

Goodwin, N. L., Nilsson, S. R. O., Choong, J. J., & Golden, S. A. 2022. Toward the explainability, transparency, and universality of machine learning for behavioral classification in neuroscience. *Current Opinion in Neurobiology* 73: 102544.

Gould, S. J., & Lewontin, R. C. 1979. The spandrels of San Marco and the Panglossian paradigm: A critique of the adaptationist programme. *Proceedings of the Royal Society B: Biological Sciences* 205: 581–598.

Grant, R. A., Sperber, A. L., & Prescott, T. J. 2012. The role of orienting in vibrissal touch sensing. *Frontiers in Behavioral Neuroscience* 6: 39.

Gray, S., & Hurst, J. 1995. The effects of cage cleaning on aggression within groups of male laboratory mice. *Animal Behaviour* 49: 821–826.

Graziano, M. S. A. 2016. Ethological action maps: A paradigm shift for the motor cortex. *Trends in Cognitive Sciences* 20: 121–132.

Green, J., Adachi, A., Shah, K. K., Hirokawa, J. D., Magani, P. S., & Maimon, G. 2017. A neural circuit architecture for angular integration in Drosophila. *Nature* 546: 101–106.

Green, L., Tingley, D., Rinzel, J., & Buzsáki, G. 2022. Action-driven remapping of hippocampal neuronal populations in jumping rats. *Proceedings of the National Academy of Sciences of the USA* 119: e2122141119.

Greenberg, D. S., Wallace, D. J., & Kerr, J. N. 2014. Imaging neuronal population activity in awake and anesthetized rodents. *Cold Spring Harbor Protocols* 2014: 912–922.

Greene, M. R., & Oliva, A. 2009. The briefest of glances: The time course of natural scene understanding. *Psychological Science* 20: 464–472.

Greif, S., & Yovel, Y. 2019. Using on-board sound recordings to infer behaviour of free-moving wild animals. *Journal of Experimental Biology* 222: jeb184689.

Grienberger, C., & Magee, J. C. 2022. Entorhinal cortex directs learning-related changes in CA1 representations. *Nature* 611: 554–562.

Grieves, R. M., Jedidi-Ayoub, S., Mishchanchuk, K., Liu, A., Renaudineau, S., & Jeffery, K. J. 2020. The place-cell representation of volumetric space in rats. *Nature Communications* 11: 789.

Grieves, R. M., Jedidi-Ayoub, S., Mishchanchuk, K., et al. 2021. Irregular distribution of grid cell firing fields in rats exploring a 3D volumetric space. *Nature Neuroscience* 24: 1567–1573.

Griffin, D. R. 1958. *Listening in the dark.* Yale University Press.

Grimsley, J. M., Shanbhag, S. J., Palmer, A. R., & Wallace, M. N. 2012. Processing of communication calls in guinea pig auditory cortex. *PLoS One* 7: e51646.

Groschner, L. N., Chan Wah Hak, L., Bogacz, R., DasGupta, S., & Miesenböck, G. 2018. Dendritic integration of sensory evidence in perceptual decision-making. *Cell* 173: 894–905.

Gross, C. G. 2009. Three before their time: Neuroscientists whose ideas were ignored by their contemporaries. *Experimental Brain Research* 192: 321–334.

Gruber, L. Z., & Ahissar, E. 2020. Closed loop motor-sensory dynamics in human vision. *PLoS One* 15: e0240660.

Gruters, K. G., Murphy, D. L. K., Jenson, C. D., Smith, D. W., Shera, C. A., & Groh, J. M. 2018. The eardrums move when the eyes move: A multisensory effect on the mechanics of hearing. *Proceedings of the National Academy of Sciences of the USA* 115: E1309–1318.

Gu, Y., Angelaki, D. E., & Deangelis, G. C. 2008. Neural correlates of multisensory cue integration in macaque MSTd. *Nature Neuroscience* 11: 1201–1210.

Guénet, J. L., & Bonhomme, F. 2003. Wild mice: An ever-increasing contribution to a popular mammalian model. *Trends in Genetics* 19: 24–31.

Guić-Robles, E., Valdivieso, C., & Guajardo, G. 1989. Rats can learn a roughness discrimination using only their vibrissal system. *Behavioural Brain Research* 31: 285–289.

Guo, C., Blair, G. J., Sehgal, M., et al. 2023. Miniscope-LFOV: A large-field-of-view, single-cell-resolution, miniature microscope for wired and wire-free imaging of neural dynamics in freely behaving animals. *Science Advances* 9: eadg3918.

Hafting, T., Fyhn, M., Molden, S., Moser, M.-B., & Moser, E. I. 2005. Microstructure of a spatial map in the entorhinal cortex. *Nature* 436: 801–806.

Hahnloser, R. H., Kozhevnikov, A. A., & Fee, M. S. 2002. An ultra-sparse code underlies the generation of neural sequences in a songbird. *Nature* 419: 65–70.

Hamilton, L. S., Oganian, Y., Hall, J., & Chang, E. F. 2021. Parallel and distributed encoding of speech across human auditory cortex. *Cell* 184: 4626–4639.

Hamm, J. P., & Yuste, R. 2016. Somatostatin interneurons control a key component of mismatch negativity in mouse visual cortex. *Cell Reports* 16: 597–604.

Hammond, T. T., Springthorpe, D., Walsh, R. E., & Berg-Kirkpatrick, T. 2016. Using accelerometers to remotely and automatically characterize behavior in small animals. *Journal of Experimental Biology* 219: 1618–1624.

Hanazawa, A., & Komatsu, H. 2001. Influence of the direction of elemental luminance gradients on the responses of V4 cells to textured surfaces. *Journal of Neuroscience* 21: 4490–4497.

Hardcastle, K., Maheswaranathan, N., Ganguli, S., & Giocomo, L. M. 2017. A multiplexed, heterogeneous, and adaptive code for navigation in medial entorhinal cortex. *Neuron* 94: 375–387.

Harland, B., Contreras, M., Souder, M., & Fellous, J. M. 2021. Dorsal CA1 hippocampal place cells form a multi-scale representation of megaspace. *Current Biology* 31: 2178–2190.

Haroush, K., & Williams, Z. M. 2015. Neuronal prediction of opponent's behavior during cooperative social interchange in primates. *Cell* 160: 1233–1245.

Harper, J. M. 2008. Wild-derived mouse stocks: An underappreciated tool for aging research. *Age* 30: 135–145.

Harten, L., Katz, A., Goldshtein, A., Handel, M., & Yovel, Y. 2020. The ontogeny of a mammalian cognitive map in the real world. *Science* 369: 194–197.

Hartline, H. K., Wagner, H. G., & Ratliff, F. 1956. Inhibition in the eye of Limulus. *Journal of General Physiology* 39: 651–673.

Harvey, C. D., Coen, P., & Tank, D. W. 2012. Choice-specific sequences in parietal cortex during a virtual-navigation decision task. *Nature* 484: 62–68.

Hasson, U., Chen, J., & Honey, C. J. 2015. Hierarchical process memory: Memory as an integral component of information processing. *Trends in Cognitive Sciences* 19: 304–313.

Hasson, U., Furman, O., Clark, D., Dudai, Y., & Davachi, L. 2008a. Enhanced intersubject correlations during movie viewing correlate with successful episodic encoding. *Neuron* 57: 452–462.

Hasson, U., Nastase, S. A., & Goldstein, A. 2020. Direct fit to nature: An evolutionary perspective on biological and artificial neural networks. *Neuron* 105: 416–434.

Hasson, U., Nir, Y., Levy, I., Fuhrmann, G., & Malach, R. 2004. Intersubject synchronization of cortical activity during natural vision. *Science* 303: 1634–1640.

Hasson, U., Yang, E., Vallines, I., Heeger, D. J., & Rubin, N. 2008b. A hierarchy of temporal receptive windows in human cortex. *Journal of Neuroscience* 28: 2539–2550.

Havermann, K., Cherici, C., Rucci, M., & Lappe, M. 2014. Fine-scale plasticity of microscopic saccades. *Journal of Neuroscience* 34: 11665–11672.

Hawking, S. 2009. Galileo and the birth of modern science. *American Heritage of Invention and Technology* 24: 36.

Headley, D. B., Kanta, V., Kyriazi, P., & Paré, D. 2019. Embracing complexity in defensive networks. *Neuron* 103: 189–201.

Healy, S. D., Gwinner, E., & Krebs, J. R. 1996. Hippocampal volume in migratory and non-migratory warblers: Effects of age and experience. *Behavioural Brain Research* 81: 61–68.

Heidenreich, M., & Zhang, F. 2016. Applications of CRISPR-Cas systems in neuroscience. *Nature Reviews Neuroscience* 17: 36–44.

Heiligenberg, W. 1977. *Principles of electrolocation and jamming avoidance in electric fish*. Springer.

Heiligenberg, W. 1991a. The neural basis of behavior: A neuroethological view. *Annual Review of Neuroscience* 14: 247–267.

Heiligenberg, W. 1991b. *Neural nets in electric fish*. MIT Press.

Hein, A. M., Altshuler, D. L., Cade, D. E., Liao, J. C., Martin, B. T., & Taylor, G. K. 2020. An algorithmic approach to natural behavior. *Current Biology* 30: R663–675.

Helmchen, F., Fee, M. S., Tank, D. W., & Denk, W. 2001. A miniature head-mounted two-photon microscope: High-resolution brain imaging in freely moving animals. *Neuron* 31: 903–912.

Hemberger, M., Pammer, L., & Laurent, G. 2016. Comparative approaches to cortical microcircuits. *Current Opinion in Neurobiology* 41: 24–30.

Hennig, M. H., Hurwitz, C., & Sorbaro, M. 2019. Scaling spike detection and sorting for next-generation electrophysiology. *Advances in Neurobiology* 22: 171–184.

Hentschke, H., Haiss, F., & Schwarz, C. 2006. Central signals rapidly switch tactile processing in rat barrel cortex during whisker movements. *Cerebral Cortex* 16: 1142–1156.

Hesse, J. K., & Tsao, D. Y. 2020. The macaque face patch system: A turtle's underbelly for the brain. *Nature Reviews Neuroscience* 21: 695–716.

Hesselmann, G., & Malach, R. 2011. The link between fMRI-BOLD activation and perceptual awareness is "stream-invariant" in the human visual system. *Cerebral Cortex* 21: 2829–2837.

Heukamp, A. S., Warwick, R. A., & Rivlin-Etzion, M. 2020. Topographic variations in retinal encoding of visual space. *Annual Review of Vision Science* 6: 237–259.

Heys, J. G., MacLeod, K. M., Moss, C. F., & Hasselmo, M. E. 2013. Bat and rat neurons differ in theta-frequency resonance despite similar coding of space. *Science* 340: 363–367.

Hill, D. E. 1979. Orientation by jumping spiders of the genus Phidippus (Araneae, Salticidae) during the pursuit of prey. *Behavioral Ecology and Sociobiology* 5: 301–322.

Hitti, F. L., & Siegelbaum, S. A. 2014. The hippocampal CA2 region is essential for social memory. *Nature* 508: 88–92.

Hochstein, S., & Ahissar, M. 2002. View from the top: Hierarchies and reverse hierarchies in the visual system. *Neuron* 36: 791–804.

Hodgkin, A. L., & Huxley, A. F. 1952. A quantitative description of membrane current and its application to conduction and excitation in nerve. *Journal of Physiology* 117: 500–544.

Hoffmann, S., Trost, L., Voigt, C., et al. 2019. Duets recorded in the wild reveal that interindividually coordinated motor control enables cooperative behavior. *Nature Communications* 10: 2577.

Holderied, M. W., & von Helversen, O. 2003. Echolocation range and wingbeat period match in aerial-hawking bats. *Proceedings of the Royal Society B: Biological Sciences* 270: 2293–2299.

Holland, R. A., & Waters, D. A. 2005. Echolocation signals and pinnae movements in the fruit bat Rousettus aegyptiacus. *Acta Chiropterologica* 7: 83–90.

Hong, G., & Lieber, C. M. 2019. Novel electrode technologies for neural recordings. *Nature Reviews Neuroscience* 20: 330–345.

Howard, L. R., Javadi, A. H., Yu, Y., et al. 2014. The hippocampus and entorhinal cortex encode the path and Euclidean distances to goals during navigation. *Current Biology* 24: 1331–1340.

Howe, C. Q., & Purves, D. 2002. Range image statistics can explain the anomalous perception of length. *Proceedings of the National Academy of Sciences of the USA* 99: 13184–13188.

Høydal, O. A., Skytøen, E. R., Andersson, S. O., Moser, M.-B., & Moser, E. I. 2019. Object-vector coding in the medial entorhinal cortex. *Nature* 568: 400–404.

Hoyle, G. 1984. The scope of neuroethology. *Behavioral and Brain Sciences* 7: 367–412.

Huang, L., Ledochowitsch, P., Knoblich, U., et al. 2021. Relationship between simultaneously recorded spiking activity and fluorescence signal in GCaMP6 transgenic mice. *eLife* 10: e51675.

Hubel, D. H. 1957a. Single unit activity in visual cortex of the unanesthetized cat. *Federation Proceedings* 16: 63.

Hubel, D. H. 1957b. Tungsten microelectrode for recording from single units. *Science* 125: 549–550.

Hubel, D. H. 1959. Single unit activity in striate cortex of unrestrained cats. *Journal of Physiology* 147: 226–238.

Hubel, D. H., Henson, C. O., Rupert, A., & Galambos, R. 1959. Attention units in the auditory cortex. *Science* 129: 1279–1280.

Hubel, D. H., & Wiesel, T. N. 1959. Receptive fields of single neurones in the cat's striate cortex. *Journal of Physiology* 148: 574–591.

Hubel, D. H., & Wiesel, T. N. 1962. Receptive fields, binocular interaction and functional architecture in the cat's visual cortex. *Journal of Physiology* 160: 106–154.

Hubel, D. H., & Wiesel, T. N. 1963a. Receptive fields of cells in striate cortex of very young, visually inexperienced kittens. *Journal of Neurophysiology* 26: 994–1002.

Hubel, D. H., & Wiesel, T. N. 1963b. Shape and arrangement of columns in cat's striate cortex. *Journal of Physiology* 165: 559–568.

Hubel, D. H., & Wiesel, T. N. 1968. Receptive fields and functional architecture of monkey striate cortex. *Journal of Physiology* 195: 215–243.

Huck, U. W., & Price, E. O. 1976. Effect of the post-weaning environment on the climbing behaviour of wild and domestic Norway rats. *Animal Behaviour* 24: 364–371.

Huerta, P. T., & Lisman, J. E. 1993. Heightened synaptic plasticity of hippocampal CA1 neurons during a cholinergically induced rhythmic state. *Nature* 364: 723–725.

Hughes, A. 1971. Topographical relationships between the anatomy and physiology of the rabbit visual system. *Documenta Ophthalmologica* 30: 33–159.

Hughes, A. 1977. The topography of vision in mammals of contrasting lifestyle: Comparative optics and retinal organization. In *Handbook of sensory physiology,* Vol. VII/5, ed. F. Crescitelli, pp. 613–756. Springer.

Huk, A., Bonnen, K., & He, B. J. 2018. Beyond trial-based paradigms: Continuous behavior, ongoing neural activity, and natural stimuli. *Journal of Neuroscience* 38: 7551–7558.

Huppert, T. J., Hoge, R. D., Diamond, S. G., Franceschini, M. A., & Boas, D. A. 2006. A temporal comparison of BOLD, ASL, and NIRS hemodynamic responses to motor stimuli in adult humans. *NeuroImage* 29: 368–382.

Hurlstone, M. J., Hitch, G. J., & Baddeley, A. D. 2014. Memory for serial order across domains: An overview of the literature and directions for future research. *Psychological Bulletin* 140: 339–373.

Huston, S. J., Stopfer, M., Cassenaer, S., Aldworth, Z. N., & Laurent, G. 2015. Neural encoding of odors during active sampling and in turbulent plumes. *Neuron* 88: 403–418.

Huxter, J., Burgess, N., & O'Keefe, J. 2003. Independent rate and temporal coding in hippocampal pyramidal cells. *Nature* 425: 828–832.

Ibbotson, M., & Jung, Y. J. 2020. Origins of functional organization in the visual cortex. *Frontiers in Systems Neuroscience* 14: 10.

Ilg, E., Mayer, N., Saikia, T., Keuper, M., Dosovitskiy, A., & Brox, T. 2017. FlowNet 2.0: evolution of optical flow estimation with deep networks. *IEEE Conference on Computer Vision and Pattern Recognition (CVPR)*: 1647–1655.

Intoy, J., & Rucci, M. 2020. Finely tuned eye movements enhance visual acuity. *Nature Communications* 11: 795.

Inzelberg, L., & Hanein, Y. 2018. Electrophysiology meets printed electronics: The beginning of a beautiful friendship. *Frontiers in Neuroscience* 12: 992.

Irvine, D. R. F. 2017. Plasticity in the auditory system. *Hearing Research* 362: 61–73.

Ito, M., & Gilbert, C. D. 1999. Attention modulates contextual influences in the primary visual cortex of alert monkeys. *Neuron* 22: 593–604.

Izawa, S., Chowdhury, S., Miyazaki, T., et al. 2019. REM sleep-active MCH neurons are involved in forgetting hippocampus-dependent memories. *Science* 365: 1308–1313.

Jadhav, S. P., Wolfe, J., & Feldman, D. E. 2009. Sparse temporal coding of elementary tactile features during active whisker sensation. *Nature Neuroscience* 12: 792–800.

Jakobsen, L., Ratcliffe, J. M., & Surlykke, A. 2013. Convergent acoustic field of view in echolocating bats. *Nature* 493: 93–96.

Jakobsen, L., & Surlykke, A. 2010. Vespertilionid bats control the width of their biosonar sound beam dynamically during prey pursuit. *Proceedings of the National Academy of Sciences of the USA* 107: 13930–13935.

Jakubowski, M., & Terkel, J. 1982. Infanticide and caretaking in non-lactating mus musculus. *Animal Behaviour* 30: 1029–1035.

Javadi, A. H., Emo, B., Howard, L. R., et al. 2017. Hippocampal and prefrontal processing of network topology to simulate the future. *Nature Communications* 8: 14652.

Jazayeri, M., & Afraz, A. 2017. Navigating the neural space in search of the neural code. *Neuron* 93: 1003–1014.

Jeffery, K. J. 2023. Symmetries and asymmetries in the neural encoding of 3D space. *Philosophical Transactions of the Royal Society B: Biological Sciences* 378: 20210452.

Jeffery, K. J., Wilson, J. J., Casali, G., & Hayman, R. M. 2015. Neural encoding of large-scale three-dimensional space—properties and constraints. *Frontiers in Psychology* 6: 927.

Jeffress, L. A. 1948. A place theory of sound localization. *Journal of Comparative Psychology* 41: 35–39.

Ji, N., Maden, G. K., Fabre, G. I., et al. 2021. A neural circuit for flexible control of persistent behavioral states. *eLife* 10: e62889.

Joëls, M., & Baram, T. Z. 2009. The neuro-symphony of stress. *Nature Reviews Neuroscience* 10: 459–466.

Jones, G. 2005. Echolocation. *Current Biology* 15: R484–488.

Jones, G., & Holderied, M. W. 2007. Bat echolocation calls: Adaptation and convergent evolution. *Proceedings of the Royal Society B: Biological Sciences* 274: 905–912.

Jones, R. M., Webster, J. G., & Keesey, U. T. 1972. An active feedback system for stabilizing visual images. *IEEE Transactions on Biomedical Engineering* 19: 29–33.

Joshi, A., Denovellis, E. L., Mankili, A., et al. 2023. Dynamic synchronization between hippocampal representations and stepping. *Nature* 617: 125–131.

Jun, J. J., Steinmetz, N. A., Siegle, J. H., et al. 2017. Fully integrated silicon probes for high-density recording of neural activity. *Nature* 551: 232–236.

Jutras, M. J., Fries, P., & Buffalo, E. A. 2013. Oscillatory activity in the monkey hippocampus during visual exploration and memory formation. *Proceedings of the National Academy of Sciences of the USA* 110: 13144–13149.

Kabra, M., Robie, A. A., Rivera-Alba, M., Branson, S., & Branson, K. 2013. JAABA: Interactive machine learning for automatic annotation of animal behavior. *Nature Methods* 10: 64–67.

Kandel, E. R., & Schwartz, J. H. 1982. Molecular biology of learning: modulation of transmitter release. *Science* 218: 433–443.

Kano, F., Walker, J., Sasaki, T., & Biro, D. 2018. Head-mounted sensors reveal visual attention of free-flying homing pigeons. *Journal of Experimental Biology* 221: jeb183475.

Kaschube, M., Schnabel, M., Löwel, S., Coppola, D. M., White, L. E., & Wolf, F. 2010. Universality in the evolution of orientation columns in the visual cortex. *Science* 330: 1113–1116.

Kawasaki, M., Margoliash, D., & Suga, N. 1988. Delay-tuned combination-sensitive neurons in the auditory cortex of the vocalizing mustached bat. *Journal of Neurophysiology* 59: 623–635.

Kay, K., & Frank, L. M. 2019. Three brain states in the hippocampus and cortex. *Hippocampus* 29: 184–238.

Kays, R., Crofoot, M. C., Jetz, W., & Wikelski, M. 2015. Terrestrial animal tracking as an eye on life and planet. *Science* 348: aaa2478.

Kayser, C., Körding, K. P., & König, P. 2004. Processing of complex stimuli and natural scenes in the visual cortex. *Current Opinion in Neurobiology* 14: 468–473.

Kell, A. J. E., Yamins, D. L. K., Shook, E. N., Norman-Haignere, S. V., & McDermott, J. H. 2018. A task-optimized neural network replicates human auditory behavior, predicts brain responses, and reveals a cortical processing hierarchy. *Neuron* 98: 630–644.

Keller, G. B., Bonhoeffer, T., & Hübener, M. 2012. Sensorimotor mismatch signals in primary visual cortex of the behaving mouse. *Neuron* 74: 809–815.

Keller, G. B., & Mrsic-Flogel, T. D. 2018. Predictive processing: A canonical cortical computation. *Neuron* 100: 424–435.

Kenet, T., Bibitchkov, D., Tsodyks, M., Grinvald, A., & Arieli, A. 2003. Spontaneously emerging cortical representations of visual attributes. *Nature* 425: 954–956.

Kepecs, A., Uchida, N., & Mainen, Z. F. 2006. The sniff as a unit of olfactory processing. *Chemical Senses* 31: 167–179.

Kepecs, A., Uchida, N., & Mainen, Z. F. 2007. Rapid and precise control of sniffing during olfactory discrimination in rats. *Journal of Neurophysiology* 98: 205–213.

Khan, A. G., & Hofer, S. B. 2018. Contextual signals in visual cortex. *Current Opinion in Neurobiology* 52: 131–138.

Kilpatrick, Z. P., Davidson, J. D., & El Hady, A. 2021. Uncertainty drives deviations in normative foraging decision strategies. *Journal of the Royal Society Interface* 18: 20210337.

Kim, J. J., & Diamond, D. M. 2002. The stressed hippocampus, synaptic plasticity and lost memories. *Nature Reviews Neuroscience* 3: 453–462.

Kim, S. S., Rouault, H., Druckmann, S., & Jayaraman, V. 2017. Ring attractor dynamics in the Drosophila central brain. *Science* 356: 849–853.

Kingsbury, L., Huang, S., Wang, J., et al. 2019. Correlated neural activity and encoding of behavior across brains of socially interacting animals. *Cell* 178: 429–446.

Kjelstrup, K. B., Solstad, T., Brun, V. H., et al. 2008. Finite scale of spatial representation in the hippocampus. *Science* 321: 140–143.

Klatzky, R. L. 1991. Let's be friends. *American Psychologist* 46: 43–45.

Kleinfeld, D., Ahissar, E., & Diamond, M. E. 2006. Active sensation: Insights from the rodent vibrissa sensorimotor system. *Current Opinion in Neurobiology* 16: 435–444.

Klioutchnikov, A., Wallace, D. J., Frosz, M. H., et al. 2020. Three-photon head-mounted microscope for imaging deep cortical layers in freely moving rats. *Nature Methods* 17: 509–513.

Knudsen, E. I., & Brainard, M. S. 1991. Visual instruction of the neural map of auditory space in the developing optic tectum. *Science* 253: 85–87.

Knudsen, E. I., & Brainard, M. S. 1995. Creating a unified representation of visual and auditory space in the brain. *Annual Review of Neuroscience* 18: 19–43.

Knudsen, E. I., du Lac, S., & Esterly, S. D. 1987. Computational maps in the brain. *Annual Review of Neuroscience* 10: 41–65.

Knudsen, E. I., & Konishi, M. 1978. A neural map of auditory space in the owl. *Science* 200: 795–797.

Knutsen, P. M., & Ahissar, E. 2009. Orthogonal coding of object location. *Trends in Neurosciences* 32: 101–109.

Ko, H., Hofer, S. B., Pichler, B., Buchanan, K. A., Sjöström, P. J., & Mrsic-Flogel, T. D. 2011. Functional specificity of local synaptic connections in neocortical networks. *Nature* 473: 87–91.

Koger, B., Deshpande, A., Kerby, J. T., Graving, J. M., Costelloe, B. R., & Couzin, I. D. 2023. Quantifying the movement, behaviour and environmental context of group-living animals using drones and computer vision. *Journal of Animal Ecology* 92: 1357–1371.

Kohn, A., Coen-Cagli, R., Kanitscheider, I., & Pouget, A. 2016. Correlations and neuronal population information. *Annual Review of Neuroscience* 39: 237–256.

Koriat, A., & Goldsmith, M. 1996. Memory metaphors and the real-life/laboratory controversy: Correspondence versus storehouse conceptions of memory. *Behavioral and Brain Sciences* 19: 167–228.

Kössl, M., Hechavarria, J. C., Voss, C., Macias, S., Mora, E. C., & Vater, M. 2014. Neural maps for target range in the auditory cortex of echolocating bats. *Current Opinion in Neurobiology* 24: 68–75.

Kothari, N. B., Wohlgemuth, M. J., & Moss, C. F. 2018. Dynamic representation of 3D auditory space in the midbrain of the free-flying echolocating bat. *eLife* 7: e29053.

Krahe, R., & Maler, L. 2014. Neural maps in the electrosensory system of weakly electric fish. *Current Opinion in Neurobiology* 24: 13–21.

Krakauer, J. W., Ghazanfar, A. A., Gomez-Marin, A., MacIver, M. A., & Poeppel, D. 2017. Neuroscience needs behavior: Correcting a reductionist bias. *Neuron* 93: 480–490.

Krapp, H. G., & Hengstenberg, R. 1996. Estimation of self-motion by optic flow processing in single visual interneurons. *Nature* 384: 463–466.

Krauzlis, R. J., Goffart, L., & Hafed, Z. M. 2017. Neuronal control of fixation and fixational eye movements. *Philosophical Transactions of the Royal Society B: Biological Sciences* 372: 20160205.

Kravitz, E. A., Potter, D. D., & Van Gelder, N. M. 1962. Gamma-aminobutyric acid and other blocking substances extracted from crab muscle. *Nature* 194: 382–383.

Krebs, J. R., Sherry, D. F., Healy, S. D., Perry, V. H., & Vaccarino, A. L. 1989. Hippocampal specialization of food-storing birds. *Proceedings of the National Academy of Sciences of the USA* 86: 1388–1392.

Kriegeskorte, N., & Douglas, P. K. 2018. Cognitive computational neuroscience. *Nature Neuroscience* 21: 1148–1160.

Krishnaswamy, P., Obregon-Henao, G., Ahveninen, J., et al. 2017. Sparsity enables estimation of both subcortical and cortical activity from MEG and EEG. *Proceedings of the National Academy of Sciences of the USA* 114: E10465–10474.

Krogh, A. 1929. The progress of physiology. *Science* 70: 200–204.

Kuang, X., Poletti, M., Victor, J. D., & Rucci, M. 2012. Temporal encoding of spatial information during active visual fixation. *Current Biology* 22: 510–514.

Kubie, J. L., Levy, E. R. J., & Fenton, A. A. 2020. Is hippocampal remapping the physiological basis for context? *Hippocampus* 30: 851–864.

Kuffler, S. W. 1953. Discharge patterns and functional organization of mammalian retina. *Journal of Neurophysiology* 16: 37–68.

La Chioma, A., Bonhoeffer, T., & Hübener, M. 2019. Area-specific mapping of binocular disparity across mouse visual cortex. *Current Biology* 29: 2954–2960.

Ladouce, S., Donaldson, D. I., Dudchenko, P. A., & Ietswaart, M. 2019. Mobile EEG identifies the re-allocation of attention during real-world activity. *Scientific Reports* 9: 15851.

Lander, E. S. 2016. The heroes of CRISPR. *Cell* 164: 18–28.

Langston, R. F., Ainge, J. A., Couey, J. J., et al. 2010. Development of the spatial representation system in the rat. *Science* 328: 1576–1580.

Las, L., Stern, E. A., & Nelken, I. 2005. Representation of tone in fluctuating maskers in the ascending auditory system. *Journal of Neuroscience* 25: 1503–1513.

Las, L., & Ulanovsky, N. 2014. Hippocampal neurophysiology across species. In: *Space, time and memory in the hippocampal formation*, ed. D. Derdikman & J. J. Knierim, pp. 431–461. Springer.

Las, L., & Ulanovsky, N. 2024. Bats. *Nature Methods* 31: 1135–1137.

Lauer, J., Zhou, M., Ye, S., et al. 2022. Multi-animal pose estimation, identification and tracking with DeepLabCut. *Nature Methods* 19: 496–504.

Laurens, J., & Angelaki, D. E. 2019. A model-based reassessment of the three-dimensional tuning of head direction cells in rats. *Journal of Neurophysiology* 122: 1274–1287.

Laurens, J., Kim, B., Dickman, J. D., & Angelaki, D. E. 2016. Gravity orientation tuning in macaque anterior thalamus. *Nature Neuroscience* 19: 1566–1568.

Laurent, G. 2002. Olfactory network dynamics and the coding of multidimensional signals. *Nature Reviews Neuroscience* 3: 884–895.

Laurent, G. 2020. On the value of model diversity in neuroscience. *Nature Reviews Neuroscience* 21: 395–396.

Laurent, G., Stopfer, M., Friedrich, R. W., Rabinovich, M. I., Volkovskii, A., & Abarbanel, H. D. 2001. Odor encoding as an active, dynamical process: Experiments, computation, and theory. *Annual Review of Neuroscience* 24: 263–297.

Lecoq, J., Orlova, N., & Grewe, B. F. 2019. Wide. Fast. Deep: Recent advances in multiphoton microscopy of in vivo neuronal activity. *Journal of Neuroscience* 39: 9042–9052.

Lee, D. N., & Reddish, P. E. 1981. Plummeting gannets: A paradigm of ecological optics. *Nature* 293: 293–294.

Legrand, N., Nikolova, N., Correa, C., et al. 2022. The heart rate discrimination task: A psychophysical method to estimate the accuracy and precision of interoceptive beliefs. *Biological Psychology* 168: 108239.

Lehky, S. R., & Sereno, A. B. 2007. Comparison of shape encoding in primate dorsal and ventral visual pathways. *Journal of Neurophysiology* 97: 307–319.

Leinweber, M., Ward, D. R., Sobczak, J. M., Attinger, A., & Keller, G. B. 2017. A sensorimotor circuit in mouse cortex for visual flow predictions. *Neuron* 95: 1420–1432.

Lenschow, C., & Brecht, M. 2015. Barrel cortex membrane potential dynamics in social touch. *Neuron* 85: 718–725.

Leonard, M. K., Gwilliams, L., Sellers, K. K., et al. 2023. Large-scale single-neuron speech sound encoding across the depth of human cortex. *Nature* 626: 593–602.

Lepage, K. Q., Macdonald, C. J., Eichenbaum, H., & Eden, U. T. 2012. The statistical analysis of partially confounded covariates important to neural spiking. *Journal of Neuroscience Methods* 205: 295–304.

Lerner, Y., Honey, C. J., Silbert, L. J., & Hasson, U. 2011. Topographic mapping of a hierarchy of temporal receptive windows using a narrated story. *Journal of Neuroscience* 31: 2906–2915.

Lever, C., Burton, S., Jeewajee, A., O'Keefe, J., & Burgess, N. 2009. Boundary vector cells in the subiculum of the hippocampal formation. *Journal of Neuroscience* 29: 9771–9777.

Levi-Montalcini, R. 1998. The saga of the nerve growth factor. *Neuroreport* 9: R71–R83.

Levitt, S. D., & List, J. A. 2007. What do laboratory experiments measuring social preferences reveal about the real world? *Journal of Economic Perspectives* 21: 153–174.

Lewicki, M. S., Olshausen, B. A., Surlykke, A., & Moss, C. F. 2014. Scene analysis in the natural environment. *Frontiers in Psychology* 5: 199.

Lewis, M. H. 2004. Environmental complexity and central nervous system development and function. *Mental Retardation and Developmental Disabilities Research Reviews* 10: 91–95.

Li, S. W., Zeliger, O., Strahs, L., et al. 2022. Frontal neurons driving competitive behaviour and ecology of social groups. *Nature* 603: 661–666.

Li, W., Piech, V., & Gilbert, C. D. 2004. Perceptual learning and top-down influences in primary visual cortex. *Nature Neuroscience* 7: 651–657.

Li, Y., Mathis, A., Grewe, B. F., et al. 2017. Neuronal representation of social information in the medial amygdala of awake behaving mice. *Cell* 171: 1176–1190.

Liberti, W. A., Perkins, L. N., Leman, D. P., & Gardner, T. J. 2017. An open source, wireless capable miniature microscope system. *Journal of Neural Engineering* 14: 045001.

Lim, M. M., Wang, Z., Olazábal, D. E., Ren, X., Terwilliger, E. F., & Young, L. J. 2004. Enhanced partner preference in a promiscuous species by manipulating the expression of a single gene. *Nature* 429: 754–757.

Lindecke, O., Elksne, A., Holland, R. A., Pētersons, G., & Voigt, C. C. 2019. Experienced migratory bats integrate the sun's position at dusk for navigation at night. *Current Biology* 29: 1369–1373.

Lisberger, S. G. 2010. Visual guidance of smooth-pursuit eye movements: Sensation, action, and what happens in between. *Neuron* 66: 477–491.

Liu, Y., Piazza, E. A., Simony, E., et al. 2017. Measuring speaker-listener neural coupling with functional near infrared spectroscopy. *Scientific Reports* 7: 43293.

Livingstone, M. S., Freeman, D. C., & Hubel, D. H. 1996. Visual responses in V1 of freely viewing monkeys. *Cold Spring Harbor Symposia on Quantitative Biology* 61: 27–37.

Livingstone, M. S., & Hubel, D. H. 1981. Effects of sleep and arousal on the processing of visual information in the cat. *Nature* 291: 554–561.

Loftus, E. F. 1979. The malleability of human memory. *American Scientist* 67: 312–320.

London, M., & Häusser, M. 2005. Dendritic computation. *Annual Review of Neuroscience* 28: 503–532.

Long, M. A., & Fee, M. S. 2008. Using temperature to analyse temporal dynamics in the songbird motor pathway. *Nature* 456: 189–194.

Long, M. A., Jin, D. Z., & Fee, M. S. 2010. Support for a synaptic chain model of neuronal sequence generation. *Nature* 468: 394–399.

Lorenz, K. Z. 1982. *The foundations of ethology: The principal ideas and discoveries in animal behavior.* Simon and Schuster.

Luebbert, L., & Pachter, L. 2024. The miscalibration of the honeybee odometer. *arXiv*: 2405.12998.

Lu, K., & Vicario, D. S. 2017. Familiar but unexpected: Effects of sound context statistics on auditory responses in the songbird forebrain. *Journal of Neuroscience* 37: 12006–12017.

Lupyan, G. 2015a. Cognitive penetrability of perception in the age of prediction: predictive systems are penetrable systems. *Review of Philosophy and Psychology* 6: 547–569.

Lupyan, G. 2015b. Object knowledge changes visual appearance: Semantic effects on color afterimages. *Acta Psychologica (Amsterdam)* 161: 117–130.

Luxem, K., Mocellin, P., Fuhrmann, F., et al. 2022. Identifying behavioral structure from deep variational embeddings of animal motion. *Communications Biology* 5: 1267.

Luxem, K., Sun, J. J., Bradley, S. P., et al. 2023. Open-source tools for behavioral video analysis: Setup, methods, and best practices. *eLife* 12: e79305.

Machens, C. K., Wehr, M. S., & Zador, A. M. 2004. Linearity of cortical receptive fields measured with natural sounds. *Journal of Neuroscience* 24: 1089–1100.

Maguire, E. A., Frackowiak, R. S., & Frith, C. D. 1997. Recalling routes around London: Activation of the right hippocampus in taxi drivers. *Journal of Neuroscience* 17: 7103–7110.

Mainland, J., & Sobel, N. 2006. The sniff is part of the olfactory percept. *Chemical Senses* 31: 181–196.

Malach, R., Amir, Y., Harel, M., & Grinvald, A. 1993. Relationship between intrinsic connections and functional architecture revealed by optical imaging and in vivo targeted biocytin injections in primate striate cortex. *Proceedings of the National Academy of Sciences of the USA* 90: 10469–10473.

Mallory, C. S., Hardcastle, K., Campbell, M. G., et al. 2021. Mouse entorhinal cortex encodes a diverse repertoire of self-motion signals. *Nature Communications* 12: 671.

Manning, C. J., Dewsbury, D. A., Wakeland, E. K., & Potts, W. K. 1995. Communal nesting and communal nursing in house mice, Mus musculus domesticus. *Animal Behaviour* 50: 741–751.

Mante, V., Sussillo, D., Shenoy, K. V., & Newsome, W. T. 2013. Context-dependent computation by recurrent dynamics in prefrontal cortex. *Nature* 503: 78–84.

Mao, D., Avila, E., Caziot, B., Laurens, J., Dickman, J. D., & Angelaki, D. E. 2021. Spatial modulation of hippocampal activity in freely moving macaques. *Neuron* 109: 3521–3534.

Marder, E. 2002. Non-mammalian models for studying neural development and function. *Nature* 417: 318–321.

Margoliash, D. 1997. Functional organization of forebrain pathways for song production and perception. *Journal of Neurobiology* 33: 671–693.

Markowitz, J. E., Gillis, W. F., Beron, C. C., et al. 2018. The striatum organizes 3D behavior via moment-to-moment action selection. *Cell* 174: 44–58.

Markowitz, J. E., Gillis, W. F., Jay, M., et al. 2023. Spontaneous behaviour is structured by reinforcement without explicit reward. *Nature* 614: 108–117.

Marr, D. 1982. *Vision: A computational investigation into the human representation and processing of visual information*. Freeman.

Marshall, J. D., Aldarondo, D. E., Dunn, T. W., Wang, W. L., Berman, G. J., & Ölveczky, B. P. 2021. Continuous whole-body 3D kinematic recordings across the rodent behavioral repertoire. *Neuron* 109: 420–437.

Martinez-Conde, S., & Macknik, S. L. 2008. Fixational eye movements across vertebrates: Comparative dynamics, physiology, and perception. *Journal of Visualization* 8: 28.

Martinez-Conde, S., Macknik, S. L., & Hubel, D. H. 2000. Microsaccadic eye movements and firing of single cells in the striate cortex of macaque monkeys. *Nature Neuroscience* 3: 251–258.

Martinez-Conde, S., Macknik, S. L., & Hubel, D. H. 2004. The role of fixational eye movements in visual perception. *Nature Reviews Neuroscience* 5: 229–240.

Martinez-Conde, S., Macknik, S. L., Troncoso, X. G., & Dyar, T. A. 2006. Microsaccades counteract visual fading during fixation. *Neuron* 49: 297–305.

Mathis, A., Herz, A. V., & Stemmler, M. B. 2012. Resolution of nested neuronal representations can be exponential in the number of neurons. *Physical Review Letters* 109: 018103.

Mathis, A., Mamidanna, P., Cury, K. M., et al. 2018. DeepLabCut: Markerless pose estimation of user-defined body parts with deep learning. *Nature Neuroscience* 21: 1281–1289.

Mathis, A., Schneider, S., Lauer, J., & Mathis, M. W. 2020. A primer on motion capture with deep learning: Principles, pitfalls, and perspectives. *Neuron* 108: 44–65.

Mathis, M. W., & Mathis, A. 2020. Deep learning tools for the measurement of animal behavior in neuroscience. *Current Opinion in Neurobiology* 60: 1–11.

Maunsell, J. H., & Cook, E. P. 2002. The role of attention in visual processing. *Philosophical Transactions of the Royal Society B: Biological Sciences* 357: 1063–1072.

Mavros, P., Austwick, M. Z., & Smith, A. H. 2016. Geo-EEG: Towards the use of EEG in the study of urban behaviour. *Applied Spatial Analysis and Policy* 9: 191–212.

Mayr, E. 1961. Cause and effect in biology. *Science* 134: 1501–1506.

Mazzocchi, F. 2008. Complexity in biology: Exceeding the limits of reductionism and determinism using complexity theory. *EMBO Reports* 9: 10–14.

McCamy, M. B., Otero-Millan, J., Di Stasi, L. L., Macknik, S. L., & Martinez-Conde, S. 2014. Highly informative natural scene regions increase microsaccade production during visual scanning. *Journal of Neuroscience* 34: 2956–2966.

McCamy, M. B., Otero-Millan, J., Macknik, S. L., et al. 2012. Microsaccadic efficacy and contribution to foveal and peripheral vision. *Journal of Neuroscience* 32: 9194–9204.

McDermott, K. B., Szpunar, K. K., & Christ, S. E. 2009. Laboratory-based and autobiographical retrieval tasks differ substantially in their neural substrates. *Neuropsychologia* 47: 2290–2298.

McNaughton, B. L., Battaglia, F. P., Jensen, O., Moser, E. I., & Moser, M.-B. 2006. Path integration and the neural basis of the "cognitive map." *Nature Reviews Neuroscience* 7: 663–678.

Mehta, M. R., Lee, A. K., & Wilson, M. A. 2002. Role of experience and oscillations in transforming a rate code into a temporal code. *Nature* 417: 741–746.

Melcher, D. 2011. Visual stability. *Philosophical Transactions of the Royal Society B: Biological Sciences* 366: 468–475.

Melcher, D., & Colby, C. L. 2008. Trans-saccadic perception. *Trends in Cognitive Sciences* 12: 466–473.

Mendelsohn, A., Furman, O., Navon, I., & Dudai, Y. 2009. Subjective vs. documented reality: A case study of long-term real-life autobiographical memory. *Learning & Memory* 16: 142–146.

Meyer, A. F., O'Keefe, J., & Poort, J. 2020. Two distinct types of eye-head coupling in freely moving mice. *Current Biology* 30: 2116–2130.

Meyer, A. F., Poort, J., O'Keefe, J., Sahani, M., & Linden, J. F. 2018. A head-mounted camera system integrates detailed behavioral monitoring with multichannel electrophysiology in freely moving mice. *Neuron* 100: 46–60.

Michaiel, A. M., Abe, E. T., & Niell, C. M. 2020. Dynamics of gaze control during prey capture in freely moving mice. *eLife* 9: e57458.

Milinski, M., & Heller, R. 1978. Influence of a predator on the optimal foraging behaviour of sticklebacks (Gasterosteus aculeatus L.). *Nature* 275: 642–644.

Miller, C. T., Gire, D., Hoke, K., et al. 2022. Natural behavior is the language of the brain. *Current Biology* 32: R482–493.

Miller, G. A. 1956. The magical number seven plus or minus two: Some limits on our capacity for processing information. *Psychological Review* 63: 81–97.

Miller, M. W., & Vogt, B. A. 1984. Direct connections of rat visual cortex with sensory, motor, and association cortices. *Journal of Comparative Neurology* 226: 184–202.

Milner, A. D., & Goodale, M. A. 1995. *The visual brain in action*. Oxford University Press.

Mimica, B., Dunn, B. A., Tombaz, T., Bojja, V., & Whitlock, J. R. 2018. Efficient cortical coding of 3D posture in freely behaving rats. *Science* 362: 584–589.

Minderer, M., Harvey, C. D., Donato, F., & Moser, E. I. 2016. Neuroscience: Virtual reality explored. *Nature* 533: 324–325.

Mitchinson, B., Martin, C. J., Grant, R. A., & Prescott, T. J. 2007. Feedback control in active sensing: Rat exploratory whisking is modulated by environmental contact. *Proceedings of the Royal Society B: Biological Sciences* 274: 1035–1041.

Miura, S. K., Scanziani, M. 2022. Distinguishing externally from saccade-induced motion in visual cortex. *Nature* 610: 135–142.

Mizrahi, A., Shalev, A., & Nelken, I. 2014. Single neuron and population coding of natural sounds in auditory cortex. *Current Opinion in Neurobiology* 24: 103–110.

Mobbs, D., Trimmer, P. C., Blumstein, D. T., & Dayan, P. 2018. Foraging for foundations in decision neuroscience: Insights from ethology. *Nature Reviews Neuroscience* 19: 419–427.

Mogensen, F., & Møhl, B. 1979. Sound radiation patterns in the frequency domain of cries from a vespertilionid bat. *Journal of Comparative Physiology* 134: 165–171.

Monaco, J. D., Rao, G., Roth, E. D., & Knierim, J. J. 2014. Attentive scanning behavior drives one-trial potentiation of hippocampal place fields. *Nature Neuroscience* 17: 725–731.

Mongillo, G., Barak, O., & Tsodyks, M. 2008. Synaptic theory of working memory. *Science* 319: 1543–1546.

Montello, D. R., Hagarty, M., Richardson, A. E., & Waller, D. 2004. Spatial memory of real environments, virtual environments, and maps. In *Human spatial memory: Remembering where*, ed. G. L. Allen, pp. 251–285. Laurence Erlbaum.

Moore, J. D., Deschênes, M., Furuta, T., et al. 2013. Hierarchy of orofacial rhythms revealed through whisking and breathing. *Nature* 497: 205–210.

Moran, J., & Desimone, R. 1985. Selective attention gates visual processing in the extrastriate cortex. *Science* 229: 782–784.

Morris, A. P., Krekelberg, B. 2019. A stable visual world in primate primary visual cortex. *Current Biology* 29: 1471–1480.

Morris, R. G. M. 2007. Stress and the hippocampus. In *The hippocampus book*, ed. P, Andersen, R. G. M. Morris, D. Amaral, T. V. Bliss, & J. O'Keefe, pp. 751–768. Oxford University Press.

Morris, R. G. M., Garrud, P., Rawlins, J. N., & O'Keefe, J. 1982. Place navigation impaired in rats with hippocampal lesions. *Nature* 297: 681–683.

Morton, E. S. 1998. Degradation and signal ranging in birds: Memory matters. *Behavioral Ecology and Sociobiology* 42: 135–137.

Morton, E. S., Howlett, J., Kopysh, N. C., & Chiver, I. 2006. Song ranging by incubating male blue-headed vireos: The importance of song representation in repertoires and implications for song delivery patterns and local/foreign dialect discrimination. *Journal of Field Ornithology* 77: 291–301.

Moser, E. I., Moser, M.-B., & McNaughton, B. L. 2017. Spatial representation in the hippocampal formation: A history. *Nature Neuroscience* 20: 1448–1464.

Moss, C. F., & Surlykke, A. 2010. Probing the natural scene by echolocation in bats. *Frontiers in Behavioral Neuroscience* 4: 33.

Mountcastle, V. B. 1957. Modality and topographic properties of single neurons of cat's somatic sensory cortex. *Journal of Neurophysiology* 20: 408–434.

Mouritsen, H. 2018. Long-distance navigation and magnetoreception in migratory animals. *Nature* 558: 50–59.

Movshon, J. A., & Lennie, P. 1979. Pattern-selective adaptation in visual cortical neurones. *Nature* 278: 850–852.

Movshon, J. A., & Simoncelli, E. P. 2014. Representation of naturalistic image structure in the primate visual cortex. *Cold Spring Harbor Symposia on Quantitative Biology* 79: 115–122.

Murray, T., & Zeil, J. 2017. Quantifying navigational information: The catchment volumes of panoramic snapshots in outdoor scenes. *PLoS One* 12: e0187226.

Murugan, M., Jang, H. J., Park, M., et al. 2017. Combined social and spatial coding in a descending projection from the prefrontal cortex. *Cell* 171: 1663–1677.

Musall, S., Kaufman, M. T., Juavinett, A. L., Gluf, S., & Churchland, A. K. 2019. Single-trial neural dynamics are dominated by richly varied movements. *Nature Neuroscience* 22: 1677–1686.

Musk, E., & Neuralink. 2019. An integrated brain-machine interface platform with thousands of channels. *Journal of Medical Internet Research* 21: e16194.

Muzzu, T., & Saleem, A. B. 2022. Feature selectivity can explain mismatch signals in mouse visual cortex. *Cell Reports* 38: 110413.

Nagel, G., Szellas, T., Huhn, W., et al. 2003. Channelrhodopsin-2, a directly light-gated cation-selective membrane channel. *Proceedings of the National Academy of Sciences of the USA* 100: 13940–13945.

Nakayama, K. 1994. James J. Gibson: An appreciation. *Psychological Review* 101: 329–335.

Nastase, S. A., Goldstein, A., & Hasson, U. 2020. Keep it real: Rethinking the primacy of experimental control in cognitive neuroscience. *NeuroImage* 222: 117254.

Nath, T., Mathis, A., Chen, A. C., Patel, A., Bethge, M., & Mathis, M. W. 2019. Using DeepLabCut for 3D markerless pose estimation across species and behaviors. *Nature Protocols* 14: 2152–2176.

Nauhaus, I., & Nielsen, K. J. 2014. Building maps from maps in primary visual cortex. *Current Opinion in Neurobiology* 24: 1–6.

Neisser, U. 1967. *Cognitive psychology.* Appleton-Century-Crofts.

Neisser, U. 1976. *Cognition and reality: Principles and implications of cognitive psychology.* Freeman.

Neisser, U. 1978. Memory: What are the important questions? In *Practical aspects of memory,* ed. M. M. Gruneberg, E. E. Morris, & R. N. Sykes, pp. 3–24. Academic Press.

Nelken, I. 2017. Neural population coding of natural sounds in non-flying mammals. In *Oxford research encyclopedia of neuroscience*. Oxford University Press.

Nelken, I., & Bar-Yosef, O. 2008. Neurons and objects: the case of auditory cortex. *Frontiers in Neuroscience* 2: 107–113.

Nelken, I., Bizley, J., Shamma, S. A., & Wang, X. 2014. Auditory cortical processing in real-world listening: The auditory system going real. *Journal of Neuroscience* 34: 15135–15138.

Nelken, I., Rotman, Y., & Bar Yosef, O. 1999. Responses of auditory-cortex neurons to structural features of natural sounds. *Nature* 397: 154–157.

Nelken, I., & Young, E. D. 1996. Why do cats need a dorsal cochlear nucleus? *Journal of Basic and Clinical Physiology and Pharmacology* 7: 199–220.

Nelson, M. E., & MacIver, M. A. 2006. Sensory acquisition in active sensing systems. *Journal of Comparative Physiology A* 192: 573–586.

Nesse, R. M. 2013. Tinbergen's four questions, organized: A response to Bateson and Laland. *Trends in Ecology & Evolution* 28: 681–682.

Nestler, E. J., & Hyman, S. E. 2010. Animal models of neuropsychiatric disorders. *Nature Neuroscience* 13: 1161–1169.

Niell, C. M., & Stryker, M. P. 2010. Modulation of visual responses by behavioral state in mouse visual cortex. *Neuron* 65: 472–479.

Nielson, D. M., Smith, T. A., Sreekumar, V., Dennis, S., & Sederberg, P. B. 2015. Human hippocampus represents space and time during retrieval of real-world memories. *Proceedings of the National Academy of Sciences of the USA* 112: 11078–11083.

Nitz, D. A. 2006. Tracking route progression in the posterior parietal cortex. *Neuron* 49: 747–756.

Nitz, D. A. 2012. Spaces within spaces: Rat parietal cortex neurons register position across three reference frames. *Nature Neuroscience* 15: 1365–1367.

Noble, D. 2008. *The music of life: Biology beyond genes.* Oxford University Press.

Noel, J. P., Balzani, E., Avila, E., et al. 2022. Coding of latent variables in sensory, parietal, and frontal cortices during closed-loop virtual navigation. *eLife* 11: e80280.

Norman, J. 2001. Ecological psychology and the two visual systems: Not to worry! *Ecological Psychology* 13: 135–145.

Norman, J. 2002. Two visual systems and two theories of perception: An attempt to reconcile the constructivist and ecological approaches. *Behavioral and Brain Sciences* 25: 73–96.

Nottebohm, F., & Liu, W. C. 2010. The origins of vocal learning: New sounds, new circuits, new cells. *Brain and Language* 115: 3–17.

Nyberg, N., Duvelle, É., Barry, C., & Spiers, H. J. 2022. Spatial goal coding in the hippocampal formation. *Neuron* 110: 394–422.

O'Keefe, J. 1976. Place units in the hippocampus of the freely moving rat. *Experimental Neurology* 51: 78–109.

O'Keefe, J., & Conway, D. H. 1978. Hippocampal place units in the freely moving rat: Why they fire where they fire. *Experimental Brain Research* 31: 573–590.

O'Keefe, J., & Dostrovsky, J. 1971. The hippocampus as a spatial map: Preliminary evidence from unit activity in the freely-moving rat. *Brain Research* 34: 171–175.

O'Keefe, J., & Nadel, L. 1978. *The hippocampus as a cognitive map.* Oxford University Press.

Ohki, K., Chung, S., Ch'ng, Y. H., Kara, P., & Reid, R. C. 2005. Functional imaging with cellular resolution reveals precise micro-architecture in visual cortex. *Nature* 433: 597–603.

Ohki, K., Chung, S., Kara, P., Hübener, M., Bonhoeffer, T., & Reid, R. C. 2006. Highly ordered arrangement of single neurons in orientation pinwheels. *Nature* 442: 925–928.

Ohlemiller, K. K., Kanwal, J. S., & Suga, N. 1996. Facilitative responses to species-specific calls in cortical FM-FM neurons of the mustached bat. *Neuroreport* 7: 1749–1755.

Okuyama, T., Kitamura, T., Roy, D. S., Itohara, S., & Tonegawa, S. 2016. Ventral CA1 neurons store social memory. *Science* 353: 1536–1541.

Oliva, A., Fernández-Ruiz, A., Leroy, F., & Siegelbaum, S. A. 2020. Hippocampal CA2 sharp-wave ripples reactivate and promote social memory. *Nature* 587: 264–269.

Olshausen, B. A., & Field, D. J. 2005. How close are we to understanding V1? *Neural Computation* 17: 1665–1699.

Omer, D. B., Fekete, T., Ulchin, Y., Hildesheim, R., & Grinvald, A. 2019. Dynamic patterns of spontaneous ongoing activity in the visual cortex of anesthetized and awake monkeys are different. *Cerebral Cortex* 29: 1291–1304.

Omer, D. B., Maimon, S. R., Las, L., & Ulanovsky, N. 2018. Social place-cells in the bat hippocampus. *Science* 359: 218–224.

O'Neill, W. E., & Suga, N. 1979. Target range-sensitive neurons in the auditory cortex of the mustache bat. *Science* 203: 69–73.

Op de Beeck, H. P., Dicarlo, J. J., Goense, J. B., et al. 2008. Fine-scale spatial organization of face and object selectivity in the temporal lobe: Do functional magnetic resonance imaging, optical imaging, and electrophysiology agree? *Journal of Neuroscience* 28: 11796–11801.

O'Regan, J. K., & Noë, A. 2001. A sensorimotor account of vision and visual consciousness. *Behavioral and Brain Sciences* 24: 939–973.

Ormond, J., & O'Keefe, J. 2022. Hippocampal place cells have goal-oriented vector fields during navigation. *Nature* 607: 741–746.

Ortega-Martinez, A., Rogers, D., Anderson, J., et al. 2023. How much do time-domain functional near-infrared spectroscopy (fNIRS) moments improve estimation of brain activity over traditional fNIRS? *Neurophotonics* 10: 013504.

Otchy, T. M., Wolff, S. B., Rhee, J. Y., et al. 2015. Acute off-target effects of neural circuit manipulations. *Nature* 528: 358–363.

Padamsey, Z., Katsanevaki, D., Dupuy, N., & Rochefort, N. L. 2022. Neocortex saves energy by reducing coding precision during food scarcity. *Neuron* 110: 280–296.

Papinutto, M., Lao, J., Lalanne, D., & Caldara, R. 2020. Watchers do not follow the eye movements of walkers. *Vision Research* 176: 130–140.

Parisi, G., Urbani, P., & Zamponi, F. 2020. *Theory of simple glasses.* Cambridge University Press.

Park, J. L., & Donaldson, D. I. 2019. Detecting the neural correlates of episodic memory with mobile EEG: Recollecting objects in the real world. *NeuroImage* 193: 1–9.

Park, J. L., Dudchenko, P. A., & Donaldson, D. I. 2018. Navigation in real-world environments: New opportunities afforded by advances in mobile brain imaging. *Frontiers in Human Neuroscience* 12: 361.

Parker, A. J. 2007. Binocular depth perception and the cerebral cortex. *Nature Reviews Neuroscience* 8: 379–391.

Patihis, L., Frenda, S. J., LePort, A. K. R., et al. 2013. False memories in highly superior autobiographical memory individuals. *Proceedings of the National Academy of Sciences of the USA* 110: 20947–20952.

Paudel, P., Vogt, C., Liu, C., Sheehan, M. J., Fernández-Ruiz, A., & Oliva, A. 2023. Social dynamics and memory consolidation: A study in semi-natural outdoor settings. *Society for Neuroscience Annual Meeting*: 373.308.

Payne, H. L., Lynch, G. F., & Aronov, D. 2021. Neural representations of space in the hippocampus of a food-caching bird. *Science* 373: 343–348.

Payne, R. S. 1971. Acoustic location of prey by barn owls (Tyto alba). *Journal of Experimental Biology* 54: 535–573.

Pearl, J. 2009. *Causality: Models, reasoning, and inference*. Cambridge University Press.

Pearson, J. M., Watson, K. K., & Platt, M. L. 2014. Decision making: The neuroethological turn. *Neuron* 82: 950–965.

Peebles, P. Z. 2008. *Radar principles*. Wiley.

Pereira, T. D., Aldarondo, D. E., Willmore, L., et al. 2019. Fast animal pose estimation using deep neural networks. *Nature Methods* 16: 117–125.

Pereira, T. D., Shaevitz, J. W., & Murthy, M. 2020. Quantifying behavior to understand the brain. *Nature Neuroscience* 23: 1537–1549.

Pereira, T. D., Tabris, N., Matsliah, A., et al. 2022. SLEAP: A deep learning system for multi-animal pose tracking. *Nature Methods* 19: 486–495.

Peters, A. J., Liu, H., & Komiyama, T. 2017. Learning in the rodent motor cortex. *Annual Review of Neuroscience* 40: 77–97.

Petersen, C. C. 2014. Cortical control of whisker movement. *Annual Review of Neuroscience* 37: 183–203.

Peterson, M. F., Lin, J., Zaun, I., & Kanwisher, N. 2016. Individual differences in face-looking behavior generalize from the lab to the world. *Journal of Vision* 16: 12.

Peterson, R. E., Choudhri, A., Mitelut, C. C., et al. 2023. Unsupervised discovery of family specific vocal usage in the Mongolian gerbil. *eLife*. https://doi.org/10.7554/eLife.89892.1.

Petreanu, L., Gutnisky, D. A., Huber, D., et al. 2012. Activity in motor-sensory projections reveals distributed coding in somatosensation. *Nature* 489: 299–303.

Pfeiffer, B. E., & Foster, D. J. 2013. Hippocampal place-cell sequences depict future paths to remembered goals. *Nature* 497: 74–79.

Pinti, P., Aichelburg, C., Gilbert, S., et al. 2018. A review on the use of wearable functional near-infrared spectroscopy in naturalistic environments. *Japanese Psychological Research* 60: 347–373.

Pinti, P., Tachtsidis, I., Hamilton, A., et al. 2020. The present and future use of functional near-infrared spectroscopy (fNIRS) for cognitive neuroscience. *Annals of the New York Academy of Sciences* 1464: 5–29.

Pitkow, X., & Angelaki, D. E. 2017. Inference in the brain: Statistics flowing in redundant population codes. *Neuron* 94: 943–953.

Poeppel, D. 2014. The neuroanatomic and neurophysiological infrastructure for speech and language. *Current Opinion in Neurobiology* 28: 142–149.

Poletti, M., Aytekin, M., & Rucci, M. 2015. Head-eye coordination at a microscopic scale. *Current Biology* 25: 3253–3259.

Pollack, G. S., Mason, A. C., Popper, A. N., & Fay, R. R. 2016. *Insect hearing*. Springer.

Polley, D. B., Kvasnák, E., & Frostig, R. D. 2004. Naturalistic experience transforms sensory maps in the adult cortex of caged animals. *Nature* 429: 67–71.

Popoli, M., Yan, Z., McEwen, B. S., & Sanacora, G. 2011. The stressed synapse: The impact of stress and glucocorticoids on glutamate transmission. *Nature Reviews Neuroscience* 13: 22–37.

Popper, K. R. 1966. Of clouds and clocks: An approach to the problem of rationality and the freedom of man. Lecture presented at Washington University, St. Louis.

Porter, J., Craven, B., Khan, R. M., et al. 2007. Mechanisms of scent-tracking in humans. *Nature Neuroscience* 10: 27–29.

Portfors, C. V., & Roberts, P. D. 2014. Mismatch of structural and functional tonotopy for natural sounds in the auditory midbrain. *Neuroscience* 258: 192–203.

Poulet, J. F., & Petersen, C. C. 2008. Internal brain state regulates membrane potential synchrony in barrel cortex of behaving mice. *Nature* 454: 881–885.

Prat, Y., & Yovel, Y. 2020. Decision making in foraging bats. *Current Opinion in Neurobiology* 60: 169–175.

Pravosudov, V. V. 1985. Food searching and storing by Parus-cinctus-lapponicus and Parus-montanus-borealis (Paridae). *Zoologicheskii Zhurnal* 64: 1036–1043.

Prescott, T. J., Diamond, M. E., & Wing, A. M. 2011. Active touch sensing. *Philosophical Transactions of the Royal Society B: Biological Sciences* 366: 2989–2995.

Prins, H. H. T. 1996. *Ecology and behaviour of the African buffalo: Social inequality and decision making.* Springer.

Qasim, S. E., Fried, I., & Jacobs, J. 2021. Phase precession in the human hippocampus and entorhinal cortex. *Cell* 184: 3242–3255.

Qazi, R., Gomez, A. M., Castro, D. C., et al. 2019. Wireless optofluidic brain probes for chronic neuropharmacology and photostimulation. *Nature Biomedical Engineering* 3: 655–669.

Quiroga, R. Q., Reddy, L., Kreiman, G., Koch, C., & Fried, I. 2005. Invariant visual representation by single neurons in the human brain. *Nature* 435: 1102–1107.

Quist, B. W., & Hartmann, M. J. 2012. Mechanical signals at the base of a rat vibrissa: The effect of intrinsic vibrissa curvature and implications for tactile exploration. *Journal of Neurophysiology* 107: 2298–2312.

Raducanu, B. C., Yazicioglu, R. F., Lopez, C. M., et al. 2017. Time multiplexed active neural probe with 1356 parallel recording sites. *Sensors (Basel)* 17: 2388.

Ranck, J. B., Jr. 1985. Head direction cells in the deep cell layer of dorsal presubiculum in freely moving rats. In *Electrical activity of the archicortex*, ed. G. Buzsáki & C. H. Vanderwolf, pp. 217–220. Akademiai Kiado.

Rao, R. P., & Ballard, D. H. 1999. Predictive coding in the visual cortex: A functional interpretation of some extra-classical receptive-field effects. *Nature Neuroscience* 2: 79–87.

Raposo, D., Kaufman, M. T., & Churchland, A. K. 2014. A category-free neural population supports evolving demands during decision-making. *Nature Neuroscience* 17: 1784–1792.

Ray, S., Yona, I., Elami, N., et al. 2024. Hippocampal coding of identity, sex, hierarchy and affiliation in a social group of wild fruit bats. Science (in press).

Ray, S., Yona, I., Las, L., & Ulanovsky, N. 2022. Hippocampal representations during natural social behaviours in a bat colony. *Society for Neuroscience Annual Meeting*, San Diego.

Reardon, S. 2016. A mouse's house may ruin experiments. *Nature* 530: 264.

Rees, G. 2004. On Francis Bacon's originality. *Intellectual News* 14: 69–73.

Reinhold, A. S., Sanguinetti-Scheck, J. I., Hartmann, K., & Brecht, M. 2019. Behavioral and neural correlates of hide-and-seek in rats. *Science* 365: 1180–1183.

Requarth, T., & Sawtell, N. B. 2011. Neural mechanisms for filtering self-generated sensory signals in cerebellum-like circuits. *Current Opinion in Neurobiology* 21: 602–608.

Reynolds, J. H., & Chelazzi, L. 2004. Attentional modulation of visual processing. *Annual Review of Neuroscience* 27: 611–647.

Riccio, M., Cole, S., & Balcetis, E. 2013. Seeing the expected, the desired, and the feared: Influences on perceptual interpretation and directed attention. *Social and Personality Psychology Compass* 7: 401–414.

Rich, P. D., Liaw, H. P., & Lee, A. K. 2014. Large environments reveal the statistical structure governing hippocampal representations. *Science* 345: 814–817.

Richards, B. A., Lillicrap, T. P., Beaudoin, P., et al. 2019. A deep learning framework for neuroscience. *Nature Neuroscience* 22: 1761–1770.

Richardson, A. E., Montello, D. R., & Hegarty, M. 1999. Spatial knowledge acquisition from maps and from navigation in real and virtual environments. *Memory & Cognition* 27: 741–750.

Rieke, F., Bodnar, D. A., & Bialek, W. 1995. Naturalistic stimuli increase the rate and efficiency of information transmission by primary auditory afferents. *Proceedings of the Royal Society B: Biological Sciences* 262: 259–265.

Rieke, F., Warland, D., de Ruyter van Steveninck, R., & Bialek, W. 1996. *Spikes: Exploring the neural code.* MIT Press.

Riggs, L. A., Armington, J. C., & Ratliff, F. 1954. Motions of the retinal image during fixation. *Journal of the Optical Society of America* 44: 315–321.

Rigotti, M., Barak, O., Warden, M. R., et al. 2013. The importance of mixed selectivity in complex cognitive tasks. *Nature* 497: 585–590.

Rinberg, D., & Davidowitz, H. 2000. Do cockroaches "know" about fluid dynamics? *Nature* 405: 756.

Rinberg, D., Koulakov, A., & Gelperin, A. 2006. Sparse odor coding in awake behaving mice. *Journal of Neuroscience* 26: 8857–8865.

Rito Lima, I., Haar, S., Di Grassi, L., & Faisal, A. A. 2020. Neurobehavioural signatures in race car driving: A case study. *Scientific Reports* 10: 11537.

Ritt, J. T., Andermann, M. L., & Moore, C. I. 2008. Embodied information processing: Vibrissa mechanics and texture features shape micromotions in actively sensing rats. *Neuron* 57: 599–613.

Rizzolatti, G., & Sinigaglia, C. 2010. The functional role of the parieto-frontal mirror circuit: Interpretations and misinterpretations. *Nature Reviews Neuroscience* 11: 264–274.

Robbers, Y., Koster, E. A., Krijbolder, D. I., Ruijs, A., van Berloo, S., & Meijer, J. H. 2015. Temporal behaviour profiles of Mus musculus in nature are affected by population activity. *Physiology & Behavior* 139: 351–360.

Roberts, P. D., & Portfors, C. V. 2015. Responses to social vocalizations in the dorsal cochlear nucleus of mice. *Frontiers in Systems Neuroscience* 9: 172.

Robie, A. A., Hirokawa, J., Edwards, A. W., et al. 2017. Mapping the neural substrates of behavior. *Cell* 170: 393–406.

Roediger, H. L., 3rd, & McDermott, K. B. 2013. Two types of event memory. *Proceedings of the National Academy of Sciences of the USA* 110: 20856–20857.

Rolls, E. T., Deco, G., Huang, & C. C., Feng, J. 2023. Multiple cortical visual streams in humans. *Cerebral Cortex* 33: 3319–3349.

Romanski, L. M., Tian, B., Fritz, J., Mishkin, M., Goldman-Rakic, P. S., & Rauschecker, J. P. 1999. Dual streams of auditory afferents target multiple domains in the primate prefrontal cortex. *Nature Neuroscience* 2: 1131–1136.

Romero-Ferrero, F., Bergomi, M. G., Hinz, R. C., Heras, F. J. H., & de Polavieja, G. G. 2019. idtracker.ai: Tracking all individuals in small or large collectives of unmarked animals. *Nature Methods* 16: 179–182.

Rose, M. C., Styr, B., Schmid, T. A., Elie, J. E., & Yartsev, M. M. 2021. Cortical representation of group social communication in bats. *Science* 374: eaba9584.

Rost, B. R., Schneider-Warme, F., Schmitz, D., & Hegemann, P. 2017. Optogenetic tools for subcellular applications in neuroscience. *Neuron* 96: 572–603.

Roth, M. M., Dahmen, J. C., Muir, D. R., Imhof, F., Martini, F. J., & Hofer, S. B. 2016. Thalamic nuclei convey diverse contextual information to layer 1 of visual cortex. *Nature Neuroscience* 19: 299–307.

Rubin, A., Sheintuch, L., Brande-Eilat, N., et al. 2019. Revealing neural correlates of behavior without behavioral measurements. *Nature Communications* 10: 4745.

Rubin, A., Yartsev, M. M., & Ulanovsky, N. 2014. Encoding of head direction by hippocampal place cells in bats. *Journal of Neuroscience* 34: 1067–1080.

Rubin, J., Ulanovsky, N., Nelken, I., & Tishby, N. 2016. The representation of prediction error in auditory cortex. *PLoS Computational Biology* 12: e1005058.

Rucci, M., Ahissar, E., & Burr, D. 2018. Temporal coding of visual space. *Trends in Cognitive Sciences* 22: 883–895.

Rucci, M., Iovin, R., Poletti, M., & Santini, F. 2007. Miniature eye movements enhance fine spatial detail. *Nature* 447: 851–854.

Rucci, M., & Victor, J. D. 2015. The unsteady eye: An information-processing stage, not a bug. *Trends in Neurosciences* 38: 195–206.

Russell, J. C., McMorland, A. J. C., & MacKay, J. W. B. 2010. Exploratory behaviour of colonizing rats in novel environments. *Animal Behaviour* 79: 159–164.

Rust, N. C., & Movshon, J. A. 2005. In praise of artifice. *Nature Neuroscience* 8: 1647–1650.

Sabbah, S., Gemmer, J. A., Bhatia-Lin, A., et al. 2017. A retinal code for motion along the gravitational and body axes. *Nature* 546: 492–497.

Safi, K., & Dechmann, D. K. 2005. Adaptation of brain regions to habitat complexity: A comparative analysis in bats (Chiroptera). *Proceedings of the Royal Society B: Biological Sciences* 272: 179–186.

Sahani, M., & Linden, J. F. 2002. How linear are auditory cortical responses? *Advances in Neural Information Processing Systems* 15: 125–132.

Saito, D. N., Tanabe, H. C., Izuma, K., et al. 2010. "Stay tuned": Inter-individual neural synchronization during mutual gaze and joint attention. *Frontiers in Integrative Neuroscience* 4: 127.

Saleem, A. B., Ayaz, A., Jeffery, K. J., Harris, K. D., & Carandini, M. 2013. Integration of visual motion and locomotion in mouse visual cortex. *Nature Neuroscience* 16: 1864–1869.

Saleem, A. B., Diamanti, E. M., Fournier, J., Harris, K. D., & Carandini, M. 2018. Coherent encoding of subjective spatial position in visual cortex and hippocampus. *Nature* 562: 124–127.

Salwiczek, L. H., Watanabe, A., & Clayton, N. S. 2010. Ten years of research into avian models of episodic-like memory and its implications for developmental and comparative cognition. *Behavioural Brain Research* 215: 221–234.

Salzman, C. D., Britten, K. H., & Newsome, W. T. 1990. Cortical microstimulation influences perceptual judgements of motion direction. *Nature* 346: 174–177.

Sandi, C., & Haller, J. 2015. Stress and the social brain: behavioural effects and neurobiological mechanisms. *Nature Reviews Neuroscience* 16: 290–304.

Sarel, A., Finkelstein, A., Las, L., & Ulanovsky, N. 2017. Vectorial representation of spatial goals in the hippocampus of bats. *Science* 355: 176–180.

Sarel, A., Palgi, S., Blum, D., Aljadeff, J., Las, L., & Ulanovsky, N. 2022. Natural switches in behaviour rapidly modulate hippocampal coding. *Nature* 609: 119–127.

Sargolini, F., Fyhn, M., Hafting, T., et al. 2006. Conjunctive representation of position, direction, and velocity in entorhinal cortex. *Science* 312: 758–762.

Savelli, F., Yoganarasimha, D., & Knierim, J. J. 2008. Influence of boundary removal on the spatial representations of the medial entorhinal cortex. *Hippocampus* 18: 1270–1282.

Saxena, S., & Cunningham, J. P. 2019. Towards the neural population doctrine. *Current Opinion in Neurobiology* 55: 103–111.

Schacter, D. L., & Loftus, E. F. 2013. Memory and law: What can cognitive neuroscience contribute? *Nature Neuroscience* 16: 119–123.

Scherrer, J. R., Lynch, G. F., Zhang, J. J., & Fee, M. S. 2023. An optical design enabling lightweight and large field-of-view head-mounted microscopes. *Nature Methods* 20: 546–549.

Schilbach, L., Timmermans, B., Reddy, V., et al. 2013. Toward a second-person neuroscience. *Behavioral and Brain Sciences* 36: 393–414.

Schmidt, K. E., & Wolf, F. 2021. Punctuated evolution of visual cortical circuits? Evidence from the large rodent Dasyprocta leporina, and the tiny primate Microcebus murinus. *Current Opinion* in *Neurobiology* 71: 110–118.

Schneider, D. M., & Mooney, R. 2018. How movement modulates hearing. *Annual Review of Neuroscience* 41: 553–572.

Schneider, D. M., Nelson, A., & Mooney, R. 2014. A synaptic and circuit basis for corollary discharge in the auditory cortex. *Nature* 513: 189–194.

Schneider, D. M., Sundararajan, J., & Mooney, R. 2018. A cortical filter that learns to suppress the acoustic consequences of movement. *Nature* 561: 391–395.

Schnitzler, H.-U., Moss, C. F., & Denzinger, A. 2003. From spatial orientation to food acquisition in echolocating bats. *Trends in Ecology & Evolution* 18: 386–394.

Schnupp, J. W., Mrsic-Flogel, T. D., & King, A. J. 2001. Linear processing of spatial cues in primary auditory cortex. *Nature* 414: 200–204.

Schnupp, J. W., Nelken, I., & King, A. J. 2012. *Auditory neuroscience: Making sense of sound.* MIT Press.

Schregardus, D. S., Pieneman, A. W., Ter Maat, A., Jansen, R. F., Brouwer, T. J., & Gahr, M. L. 2006. A lightweight telemetry system for recording neuronal activity in freely behaving small animals. *Journal of Neuroscience Methods* 155: 62–71.

Schwarz, D. A., Lebedev, M. A., Hanson, T. L., et al. 2014. Chronic, wireless recordings of large-scale brain activity in freely moving rhesus monkeys. *Nature Methods* 11: 670–676.

Schweinfurth, M. K., & Call, J. 2019. Revisiting the possibility of reciprocal help in non-human primates. *Neuroscience & Biobehavioral Reviews* 104: 73–86.

Schyns, P. G., Snoek, L., & Daube, C. 2022. Degrees of algorithmic equivalence between the brain and its DNN models. *Trends in Cognitive Sciences* 26: 1090–1102.

Scott, B. B., Constantinople, C. M., Erlich, J. C., Tank, D. W., & Brody, C. D. 2015. Sources of noise during accumulation of evidence in unrestrained and voluntarily head-restrained rats. *eLife* 4: e11308.

Scoville, W. B., & Milner, B. 1957. Loss of recent memory after bilateral hippocampal lesions. *Journal of Neurology, Neurosurgery, and Psychiatry* 20: 11–21.

Sedivy, J. 2019. *Language in mind: An introduction to psycholinguistics.* 2nd ed. Sinauer.

Seelig, J. D., & Jayaraman, V. 2015. Neural dynamics for landmark orientation and angular path integration. *Nature* 521: 186–191.

Seger, C. A., & Miller, E. K. 2010. Category learning in the brain. *Annual Review of Neuroscience* 33: 203–219.

Sejnowski, T. J., Koch, C., & Churchland, P. S. 1988. Computational neuroscience. *Science* 241: 1299–1306.

Seymour, R. A., Alexander, N., Mellor, S., et al. 2021. Using OPMs to measure neural activity in standing, mobile participants. *NeuroImage* 244: 118604.

Sharot, T., Martorella, E. A., Delgado, M. R., & Phelps, E. A. 2007. How personal experience modulates the neural circuitry of memories of September 11. *Proceedings of the National Academy of Sciences of the USA* 104: 389–394.

Sheintuch, L., Rubin, A., Brande-Eilat, N., et al. 2017. Tracking the same neurons across multiple days in Ca^{2+} imaging data. *Cell Reports* 21: 1102–1115.

Shemesh, Y., Sztainberg, Y., Forkosh, O., Shlapobersky, T., Chen, A., & Schneidman, E. 2013. High-order social interactions in groups of mice. *eLife* 2: e00759.

Sherborne, A. L., Thom, M. D., Paterson, S., et al. 2007. The genetic basis of inbreeding avoidance in house mice. *Current Biology* 17: 2061–2066.

Sherry, D. F. 1984. Food storage by black-capped chickadees: Memory for the location and contents. *Animal Behaviour* 32: 451–464.

Sherry, D. F., & Vaccarino, A. L. 1989. Hippocampus and memory for food caches in black-capped chickadees. *Behavioral Neuroscience* 103: 308–318.

Shimamura, A. P., & Squire, L. R. 1988. Long-term memory in amnesia: Cued recall, recognition memory, and confidence ratings. *Journal of Experimental Psychology: Learning, Memory, and Cognition* 14: 763–770.

Shimomura, O. 2006. Discovery of green fluorescent protein. *Methods of Biochemical Analysis* 47: 1–13.

Shmuelof, L., & Zohary, E. 2005. Dissociation between ventral and dorsal fMRI activation during object and action recognition. *Neuron* 47: 457–470.

Shuler, M. G., & Bear, M. F. 2006. Reward timing in the primary visual cortex. *Science* 311: 1606–1609.

Shuman, T., Aharoni, D., Cai, D. J., et al. 2020. Breakdown of spatial coding and interneuron synchronization in epileptic mice. *Nature Neuroscience* 23: 229–238.

Shusterman, R., Smear, M. C., Koulakov, A. A., & Rinberg, D. 2011. Precise olfactory responses tile the sniff cycle. *Nature Neuroscience* 14: 1039–1044.

Siemers, B. M., & Schnitzler, H.-U. 2004. Echolocation signals reflect niche differentiation in five sympatric congeneric bat species. *Nature* 429: 657–661.

Sieu, L. A., Bergel, A., Tiran, E., et al. 2015. EEG and functional ultrasound imaging in mobile rats. *Nature Methods* 12: 831–834.

Simoncelli, E. P., & Olshausen, B. A. 2001. Natural image statistics and neural representation. *Annual Review of Neuroscience* 24: 1193–1216.

Singla, S., Dempsey, C., Warren, R., Enikolopov, A. G., & Sawtell, N. B. 2017. A cerebellum-like circuit in the auditory system cancels responses to self-generated sounds. *Nature Neuroscience* 20: 943–950.

Sinha, S. R., & Moss, C. F. 2007. Vocal premotor activity in the superior colliculus. *Journal of Neuroscience* 27: 98–110.

Skaggs, W. E., McNaughton, B. L., Wilson, M. A., & Barnes, C. A. 1996. Theta phase precession in hippocampal neuronal populations and the compression of temporal sequences. *Hippocampus* 6: 149–172.

Skolnik, M. I. 2001. *Introduction to radar systems*. McGraw-Hill.

Skyberg, R. J., & Niell, C. M. 2024. Natural visual behavior and active sensing in the mouse. *Current Opinion in Neurobiology* 86: 102882.

Sliwa, J., & Freiwald, W. A. 2017. A dedicated network for social interaction processing in the primate brain. *Science* 356: 745–749.

Smear, M., Shusterman, R., O'Connor, R., Bozza, T., & Rinberg, D. 2011. Perception of sniff phase in mouse olfaction. *Nature* 479: 397–400.

Smith, E. C., & Lewicki, M. S. 2006. Efficient auditory coding. *Nature* 439: 978–982.

Snodderly, D. M. 2016. A physiological perspective on fixational eye movements. *Vision Research* 118: 31–47.

Snodderly, D. M., Kagan, I., & Gur, M. 2001. Selective activation of visual cortex neurons by fixational eye movements: Implications for neural coding. *Visual Neuroscience* 18: 259–277.

Solstad, T., Boccara, C. N., Kropff, E., Moser, M.-B., & Moser, E. I. 2008. Representation of geometric borders in the entorhinal cortex. *Science* 322: 1865–1868.

Sorrells, S. F., Paredes, M. F., Cebrian-Silla, A., et al. 2018. Human hippocampal neurogenesis drops sharply in children to undetectable levels in adults. *Nature* 555: 377–381.

Spalding, K. L., Bergmann, O., Alkass, K., et al. 2013. Dynamics of hippocampal neurogenesis in adult humans. *Cell* 153: 1219–1227.

Speakman, J. R., & Racey, P. A. 1991. No cost of echolocation for bats in flight. *Nature* 350: 421–423.

Spiers, H. J., & Maguire, E. A. 2006. Thoughts, behaviour, and brain dynamics during navigation in the real world. *NeuroImage* 31: 1826–1840.

Spiers, H. J., & Maguire, E. A. 2007. Decoding human brain activity during real-world experiences. *Trends in Cognitive Sciences* 11: 356–365.

Squire, L. R. 1992. Memory and the hippocampus: A synthesis from findings with rats, monkeys, and humans. *Psychological Review* 99: 195–231.

Squire, L. R., & Kandel, E. 2008. *Memory: From mind to molecules.* 2nd ed. Roberts and Company.

Sridhar, V. H., Li, L., Gorbonos, D., et al. 2021. The geometry of decision-making in individuals and collectives. *Proceedings of the National Academy of Sciences of the USA* 118: e2102157118.

Srinivasan, M. V. 2011. Honeybees as a model for the study of visually guided flight, navigation, and biologically inspired robotics. *Physiological Reviews* 91: 413–460.

Srinivasan, M. V., Lehrer, M., Kirchner, W. H., & Zhang, S. W. 1991. Range perception through apparent image speed in freely flying honeybees. *Visual Neuroscience* 6: 519–535.

Srinivasan, M. V., Zhang, S., Altwein, M., & Tautz, J. 2000a. Honeybee navigation: Nature and calibration of the "odometer." *Science* 287: 851–853.

Srinivasan, M. V., Zhang, S. W., Chahl, J. S., Barth, E., & Venkatesh, S. 2000b. How honeybees make grazing landings on flat surfaces. *Biological Cybernetics* 83: 171–183.

Stander, P. E. 1992. Cooperative hunting in lions: The role of the individual. *Behavioral Ecology and Sociobiology* 29: 445–454.

Stangl, M., Maoz, S. L., & Suthana, N. 2023. Mobile cognition: Imaging the human brain in the "real world." *Nature Reviews Neuroscience* 24: 347–362.

Stangl, M., Topalovic, U., Inman, C. S., et al. 2021. Boundary-anchored neural mechanisms of location-encoding for self and others. *Nature* 589: 420–425.

Stark, E., Drori, R., & Abeles, M. 2006. Partial cross-correlation analysis resolves ambiguity in the encoding of multiple movement features. *Journal of Neurophysiology* 95: 1966–1975.

Steen, J. B., Mohus, I., Kvesetberg, T., & Walløe, L. 1996. Olfaction in bird dogs during hunting. *Acta Physiologica Scandinavica* 157: 115–119.

Steinman, R. M., & Levinson, J. Z. 1990. The role of eye movement in the detection of contrast and spatial detail. *Reviews of Oculomotor Research* 4: 115–212.

Steinmetz, N. A., Aydin, C., Lebedeva, A., et al. 2021. Neuropixels 2.0: A miniaturized high-density probe for stable, long-term brain recordings. *Science* 372: eabf4588.

Stemmler, M., Mathis, A., & Herz, A. V. 2015. Connecting multiple spatial scales to decode the population activity of grid cells. *Science Advances* 1: e1500816.

Stephens, D. W., Brown, J. S., & Ydenberg, R. C. 2007. *Foraging: Behavior and ecology.* University of Chicago Press.

Stephens, D. W., & Krebs, J. R. 1987. *Foragining theory.* Princeton University Press.

Stephens, G. J., Johnson-Kerner, B., Bialek, W., & Ryu, W. S. 2008. Dimensionality and dynamics in the behavior of *C. elegans. PLoS Computational Biology* 4: e1000028.

Stephens, G. J., Johnson-Kerner, B., Bialek, W., & Ryu, W. S. 2010a. From modes to movement in the behavior of Caenorhabditis elegans. *PLoS One* 5: e13914.

Stephens, G. J., Osborne, L. C., & Bialek, W. 2011. Searching for simplicity in the analysis of neurons and behavior. *Proceedings of the National Academy of Sciences of the USA* 108 (Suppl 3): 15565–15571.

Stephens, G. J., Silbert, L. J., & Hasson, U. 2010b. Speaker-listener neural coupling underlies successful communication. *Proceedings of the National Academy of Sciences of the USA* 107: 14425–14430.

Sterbing-D'Angelo, S., Chadha, M., Chiu, C., et al. 2011. Bat wing sensors support flight control. *Proceedings of the National Academy of Sciences of the USA* 108: 11291–11296.

Stevens, J. C., Banks, G. T., Festing, M. F., & Fisher, E. M. 2007. Quiet mutations in inbred strains of mice. *Trends in Molecular Medicine* 13: 512–519.

Stevens, J. K., Emerson, R. C., Gerstein, G. L., et al. 1976. Paralysis of the awake human: Visual perceptions. *Vision Research* 16: 93–98,

Stevenson, I. H., & Körding, K. P. 2011. How advances in neural recording affect data analysis. *Nature Neuroscience* 14: 139–142.

Stewart, M., & Fox, S. E. 1991. Hippocampal theta activity in monkeys. *Brain Research* 538: 59–63.

Stowers, J. R., Hofbauer, M., Bastien, R., et al. 2017. Virtual reality for freely moving animals. *Nature Methods* 14: 995–1002.

Strandburg-Peshkin, A., Farine, D. R., Couzin, I. D., & Crofoot, M. C. 2015. Shared decision-making drives collective movement in wild baboons. *Science* 348: 1358–1361.

Strappini, F., Wilf, M., Karp, O., et al. 2019. Resting-state activity in high-order visual areas as a window into natural human brain activations. *Cerebral Cortex* 29: 3618–3635.

Stringer, C., Pachitariu, M., Steinmetz, N., Reddy, C. B., Carandini, M., & Harris, K. D. 2019. Spontaneous behaviors drive multidimensional, brainwide activity. *Science* 364: eaav7893.

Strogatz, S. H. 2015. *Nonlinear dynamics and chaos: With applications to physics, biology, chemistry, and engineering.* 2nd ed. CRC Press.

Stuart, T., Jeang, W. J., Slivicki, R. A., et al. 2023. Wireless, battery-free implants for electrochemical catecholamine sensing and optogenetic stimulation. *ACS Nano* 17: 561–574.

Stürzl, W., Grixa, I., Mair, E., Narendra, A., & Zeil, J. 2015. Three-dimensional models of natural environments and the mapping of navigational information. *Journal of Comparative Physiology A* 201: 563–584.

Stüttgen, M. C., Kullmann, S., & Schwarz, C. 2008. Responses of rat trigeminal ganglion neurons to longitudinal whisker stimulation. *Journal of Neurophysiology* 100: 1879–1884.

Suga, N. 1989. Principles of auditory information-processing derived from neuroethology. *Journal of Experimental Biology* 146: 277–286.

Suga, N. 1990. Biosonar and neural computation in bats. *Scientific American* 262: 60–68.

Suga, N., & O'Neill, W. E. 1979. Neural axis representing target range in the auditory cortex of the mustache bat. *Science* 206: 351–353.

Suga, N., O'Neill, W. E., & Manabe, T. 1978. Cortical neurons sensitive to combinations of information-bearing elements of biosonar signals in the mustache bat. *Science* 200: 778–781.

Szabo, V., Ventalon, C., De Sars, V., Bradley, J., & Emiliani, V. 2014. Spatially selective holographic photoactivation and functional fluorescence imaging in freely behaving mice with a fiberscope. *Neuron* 84: 1157–1169.

Szuts, T. A., Fadeyev, V., Kachiguine, S., et al. 2011. A wireless multi-channel neural amplifier for freely moving animals. *Nature Neuroscience* 14: 263–269.

Szwed, M., Bagdasarian, K., & Ahissar, E. 2003. Encoding of vibrissal active touch. *Neuron* 40: 621–630.

Takahashi, S., Hombe, T., Matsumoto, S., Ide, K., & Yoda, K. 2022. Head direction cells in a migratory bird prefer north. *Science Advances* 8: eabl6848.

Tanaka, K., & Saito, H. 1989. Analysis of motion of the visual field by direction, expansion/contraction, and rotation cells clustered in the dorsal part of the medial superior temporal area of the macaque monkey. *Journal of Neurophysiology* 62: 626–641.

Tarsitano, M. S., & Jackson, R. R. 1994. Jumping spiders make predatory detours requiring movement away from prey. *Behavior* 131: 65–73.

Taube, J. S. 2007. The head direction signal: Origins and sensory-motor integration. *Annual Review of Neuroscience* 30: 181–207.

Taube, J. S., Muller, R. U., & Ranck, J. B., Jr. 1990. Head-direction cells recorded from the postsubiculum in freely moving rats. I. Description and quantitative analysis. *Journal of Neuroscience* 10: 420–435.

Taube, J. S., & Shinder, M. E. 2020. On the absence or presence of 3D tuned head direction cells in rats: A review and rebuttal. *Journal of Neurophysiology* 123: 1808–1827.

Taube, J. S., Valerio, S., & Yoder, R. M. 2013. Is navigation in virtual reality with fMRI really navigation? *Journal of Cognitive Neuroscience* 25: 1008–1019.

Taylor, K. D. 1978. Range of movement and activity of common rats (Rattus norvegicus) on agricultural land. *Journal of Applied Ecology* 15: 663–677.

Ter Hofstede, H. M., & Ratcliffe, J. M. 2016. Evolutionary escalation: The bat-moth arms race. *Journal of Experimental Biology* 219: 1589–1602.

Terborgh, J., & Estes, J. A. 2010. *Trophic cascades: Predators, prey, and the changing dynamics of nature.* Island Press.

Testard, C., Tremblay, S., Parodi, F., et al. 2024. Neural signatures of natural behaviour in socializing macaques. *Nature* 628: 381–390.

Theunissen, F. E., & Elie, J. E. 2014. Neural processing of natural sounds. *Nature Reviews Neuroscience* 15: 355–366.

Thornton, A., & McAuliffe, K. 2006. Teaching in wild meerkats. *Science* 313: 227–229.

Thorpe, S., Fize, D., & Marlot, C. 1996. Speed of processing in the human visual system. *Nature* 381: 520–522.

Tian, B., Reser, D., Durham, A., Kustov, A., & Rauschecker, J. P. 2001. Functional specialization in rhesus monkey auditory cortex. *Science* 292: 290–293.

Tinbergen, N. 1951. *The study of instinct.* Oxford University Press.

Tinbergen, N. 1963. On aims and methods of ethology. *Zeitschrift für Tierpsychologie* 20: 410–433.

Tingley, D., McClain, K., Kaya, E., Carpenter, J., & Buzsáki, G. 2021. A metabolic function of the hippocampal sharp wave-ripple. *Nature* 597: 82–86.

Todd, J. G., Kain, J. S., & de Bivort, B. L. 2017. Systematic exploration of unsupervised methods for mapping behavior. *Physical Biology* 14: 015002.

Toledo, S., Shohami, D., Schiffner, I., et al. 2020. Cognitive map-based navigation in wild bats revealed by a new high-throughput tracking system. *Science* 369: 188–193.

Topalovic, U., Aghajan, Z. M., Villaroman, D., et al. 2020. Wireless programmable recording and stimulation of deep brain activity in freely moving humans. *Neuron* 108: 322–334.

Traxler, M. J. 2011. *Introduction to psycholinguistics: Understanding language science*. Wiley-Blackwell.

Triblehorn, J. D., Ghose, K., Bohn, K., Moss, C. F., & Yager, D. D. 2008. Free-flight encounters between praying mantids (Parasphendale agrionina) and bats (Eptesicus fuscus). *Journal of Experimental Biology* 211: 555–562.

Trotter, Y., & Celebrini, S. 1999. Gaze direction controls response gain in primary visual-cortex neurons. *Nature* 398: 239–242.

Truccolo, W., Eden, U. T., Fellows, M. R., Donoghue, J. P., & Brown, E. N. 2005. A point process framework for relating neural spiking activity to spiking history, neural ensemble, and extrinsic covariate effects. *Journal of Neurophysiology* 93: 1074–1089.

Tsao, D. Y., Freiwald, W. A., Tootell, R. B., & Livingstone, M. S. 2006. A cortical region consisting entirely of face-selective cells. *Science* 311: 670–674.

Tsoar, A., Nathan, R., Bartan, Y., Vyssotski, A., Dell'omo, G., & Ulanovsky, N. 2011. Large-scale navigational map in a mammal. *Proceedings of the National Academy of Sciences of the USA* 108: E718–724.

Tulving, E. 1983. *Elements of episodic memory*. Oxford University Press.

Tulving, E. 1991. Memory research is not a zero-sum game. *American Psychologist* 46: 41–42.

Tyree, T. J., Metke, M., & Miller, C. T. 2023. Cross-modal representation of identity in the primate hippocampus. *Science* 382: 417–423.

Uchida, N., Kepecs, A., & Mainen, Z. F. 2006. Seeing at a glance, smelling in a whiff: Rapid forms of perceptual decision making. *Nature Reviews Neuroscience* 7: 485–491.

Uchida, N., & Mainen, Z. F. 2003. Speed and accuracy of olfactory discrimination in the rat. *Nature Neuroscience* 6: 1224–1229.

Ulanovsky, N. 2004. *Neuronal adaptation in cat auditory cortex*. PhD thesis, Hebrew University of Jerusalem.

Ulanovsky, N., Las, L., Farkas, D., & Nelken, I. 2004. Multiple time scales of adaptation in auditory cortex neurons. *Journal of Neuroscience* 24: 10440–10453.

Ulanovsky, N., Las, L., & Nelken, I. 2003. Processing of low-probability sounds by cortical neurons. *Nature Neuroscience* 6: 391–398.

Ulanovsky, N., & Moss, C. F. 2007. Hippocampal cellular and network activity in freely moving echolocating bats. *Nature Neuroscience* 10: 224–233.

Ulanovsky, N., & Moss, C. F. 2008. What the bat's voice tells the bat's brain. *Proceedings of the National Academy of Sciences of the USA* 105: 8491–8498.

Ulanovsky, N., & Moss, C. F. 2011. Dynamics of hippocampal spatial representation in echolocating bats. *Hippocampus* 21: 150–161.

Ullman, S. 1980. Against direct perception. *Behavioral and Brain Sciences* 3: 373–415.

Ullman, S. 1996. *High-level vision: Object recognition and visual cognition*. MIT Press.

Ulusoy, A. O., Black, M. J., & Geiger, A. 2016. *Patches, planes and probabilities: A non-local prior for volumetric 3D reconstruction*. Presented at the IEEE Conference on Computer Vision and Pattern Recognition (CVPR), Las Vegas.

Ungerleider, L. G., & Mishkin, M. 1982. Two cortical visual systems. In *Analysis of visual behavior*, ed. M. A. Goodale, D. J. Ingle, & R. J. W. Mansfield, pp. 549–586. MIT Press.

Urai, A. E., Doiron, B., Leifer, A. M., & Churchland, A. K. 2022. Large-scale neural recordings call for new insights to link brain and behavior. *Nature Neuroscience* 25: 11–19.

van Gelder, T., & Port, R. 1995. It's about time: An overview of the dynamical approach to cognition. In *Mind as motion: explorations in the dynamics of cognition*, ed. T. van Gelder & R. Port, pp. 1–43. MIT Press.

Van Hooser, S. D., Heimel, J. A., Chung, S., Nelson, S. B., & Toth, L. J. 2005. Orientation selectivity without orientation maps in visual cortex of a highly visual mammal. *Journal of Neuroscience* 25: 19–28.

Varela, C., & Wilson, M. A. 2020. mPFC spindle cycles organize sparse thalamic activation and recently active CA1 cells during non-REM sleep. *eLife* 9: e48881.

Vélez-Fort, M., Bracey, E. F., Keshavarzi, S., et al. 2018. A circuit for integration of head- and visual-motion signals in layer 6 of mouse primary visual cortex. *Neuron* 98: 179–191.

Vetter, P., & Newen, A. 2014. Varieties of cognitive penetration in visual perception. *Consciousness and Cognition* 27: 62–75.

Vickerstaff, R. J., & Cheung, A. 2010. Which coordinate system for modelling path integration? *Journal of Theoretical Biology* 263: 242–261.

Vidal, J. R., Ossandón, T., Jerbi, K., et al. 2010. Category-specific visual responses: An intracranial study comparing gamma, beta, alpha, and ERP response selectivity. *Frontiers in Human Neuroscience* 4: 195.

Vinepinsky, E., Cohen, L., Perchik, S., Ben-Shahar, O., Donchin, O., & Segev, R. 2020. Representation of edges, head direction, and swimming kinematics in the brain of freely-navigating fish. *Scientific Reports* 10: 14762.

Vinje, W. E., & Gallant, J. L. 2000. Sparse coding and decorrelation in primary visual cortex during natural vision. *Science* 287: 1273–1276.

Voigts, J., Newman, J. P., Wilson, M. A., & Harnett, M. T. 2020. An easy-to-assemble, robust, and lightweight drive implant for chronic tetrode recordings in freely moving animals. *Journal of Neural Engineering* 17: 026044.

von Heimendahl, M., Rao, R. P., & Brecht, M. 2012. Weak and nondiscriminative responses to conspecifics in the rat hippocampus. *Journal of Neuroscience* 32: 2129–2141.

von Uexküll, J. J. 1920. *Theoretische biologie (Theoretical biology)*. Springer.

Voss, R. F., & Clarke, J. 1975. "1/f noise" in music and speech. *Nature* 258: 317–318.

Voss, R. F., & Clarke, J. 1978. "1/f noise" in music: Music from 1/f noise. *Journal of the Acoustical Society of America* 63: 258–263.

Vyas, S., Golub, M. D., Sussillo, D., & Shenoy, K. V. 2020. Computation through neural population dynamics. *Annual Review of Neuroscience* 43: 249–275.

Vyssotski, A. L., Serkov, A. N., Itskov, P. M., et al. 2006. Miniature neurologgers for flying pigeons: Multichannel EEG and action and field potentials in combination with GPS recording. *Journal of Neurophysiology* 95: 1263–1273.

Wachowiak, M. 2011. All in a sniff: Olfaction as a model for active sensing. *Neuron* 71: 962–973.

Wallace, D. J., Greenberg, D. S., Sawinski, J., Rulla, S., Notaro, G., & Kerr, J. N. 2013. Rats maintain an overhead binocular field at the expense of constant fusion. *Nature* 498: 65–69.

Wallace, D. J., & Kerr, J. N. D. 2019. Circuit interrogation in freely moving animals. *Nature Methods* 16: 9–11.

Wallach, A., Marom, S., & Ahissar, E. 2016. Closing Dewey's circuit. In *Closed loop neuroscience*, ed. A. El Hady, pp. 93–100. Academic Press.

Wallraff, H. G. 2005. *Avian navigation: Pigeon homing as a paradigm*. Springer.

Walls, G. L. 1942. *The vertebrate eye and its adaptive radiations*. Cranbrook Institute.

Walter, T., & Couzin, I. D. 2021. TRex, a fast multi-animal tracking system with markerless identification, and 2D estimation of posture and visual fields. *eLife* 10: e64000.

Walum, H., & Young, L. J. 2018. The neural mechanisms and circuitry of the pair bond. *Nature Reviews Neuroscience* 19: 643–654.

Wang, C., Chen, X., Lee, H., et al. 2018. Egocentric coding of external items in the lateral entorhinal cortex. *Science* 362: 945–949.

Wang, M. E., Wann, E. G., Yuan, R. K., Ramos Álvarez, M. M., Stead, S. M., & Muzzio, I. A. 2012. Long-term stabilization of place cell remapping produced by a fearful experience. *Journal of Neuroscience* 32: 15802–15814.

Wang, X. 2000. On cortical coding of vocal communication sounds in primates. *Proceedings of the National Academy of Sciences of the USA* 97: 11843–11849.

Wang, X., & Kadia, S. C. 2001. Differential representation of species-specific primate vocalizations in the auditory cortices of marmoset and cat. *Journal of Neurophysiology* 86: 2616–2620.

Wang, Y., & Frost, B. J. 1992. Time to collision is signalled by neurons in the nucleus rotundus of pigeons. *Nature* 356: 236–238.

Warwick, R. A., Riccitelli, S., Heukamp, A. S., et al. 2022. Top-down modulation of the retinal code via histaminergic neurons of the hypothalamus. *bioRxiv*. https://doi.org/10.1101/2022.04.26.489509.

Wathey, J. C., Pettigrew, J. D. 1989. Quantitative analysis of the retinal ganglion cell layer and optic nerve of the barn owl Tyto alba. *Brain, Behavior and Evolution* 33: 279–292.

Weber, J. N., Peterson, B. K., & Hoekstra, H. E. 2013. Discrete genetic modules are responsible for complex burrow evolution in Peromyscus mice. *Nature* 493: 402–405.

Wehner, R. 1987. "Matched filters"—neural models of the external world. *Journal of Comparative Physiology A* 161: 511–531.

Wehner R. 2020. *Desert navigator: The journey of an ant*. Belknap Press.

Wehr, M., & Zador, A. M. 2005. Synaptic mechanisms of forward suppression in rat auditory cortex. *Neuron* 47: 437–445.

Wei, Z., Lin, B. J., Chen, T. W., Daie, K., Svoboda, K., & Druckmann, S. 2020. A comparison of neuronal population dynamics measured with calcium imaging and electrophysiology. *PLoS Computational Biology* 16: e1008198.

Weinberger, N. M. 1995. Dynamic regulation of receptive fields and maps in the adult sensory cortex. *Annual Review of Neuroscience* 18: 129–158.

Weinreb, C., Pearl, J. E., Osman, M. A. M., et al. 2024. Keypoint-MoSeq: parsing behavior by linking point tracking to pose dynamics. *Nature Methods* 21: 1329–1339.

Weisberg, S. M., Schinazi, V. R., Newcombe, N. S., Shipley, T. F., & Epstein, R. A. 2014. Variations in cognitive maps: Understanding individual differences in navigation. *Journal of Experimental Psychology: Learning, Memory, and Cognition* 40: 669–682.

Weissbrod, A., Shapiro, A., Vasserman, G., et al. 2013. Automated long-term tracking and social behavioural phenotyping of animal colonies within a semi-natural environment. *Nature Communications* 4: 2018.

Welker, W. I. 1964. Analysis of sniffing of the albino rat. *Behaviour* 22: 223–244.

Westheimer, G. 1981. Visual hyperacuity. In *Progress in sensory physiology*, ed. H. Autrum, E. R. Perl, R. F. Schmidt, & D. Ottoson, pp. 1–30. Springer.

Westheimer, G., & McKee, S. P. 1977. Integration regions for visual hyperacuity. *Vision Research* 17: 89–93.

Whiten, A. 2021. The burgeoning reach of animal culture. *Science* 372: eabe6514.

Whitlock, J. R., Pfuhl, G., Dagslott, N., Moser, M.-B., & Moser, E. I. 2012. Functional split between parietal and entorhinal cortices in the rat. *Neuron* 73: 789–802.

Whitlock, J. R., Sutherland, R. J., Witter, M. P., Moser, M.-B., & Moser, E. I. 2008. Navigating from hippocampus to parietal cortex. *Proceedings of the National Academy of Sciences of the USA* 105: 14755–14762.

Widloski, J., & Foster, D. J. 2022. Flexible rerouting of hippocampal replay sequences around changing barriers in the absence of global place field remapping. *Neuron* 110: 1547–1558.

Wiesel, T. N., & Hubel, D. H. 1963. Single-cell responses in striate cortex of kittens deprived of vision in one eye. *Journal of Neurophysiology* 26: 1003–1017.

Wigderson, E., Nelken, I., & Yarom, Y. 2016. Early multisensory integration of self and source motion in the auditory system. *Proceedings of the National Academy of Sciences of the USA* 113: 8308–8313.

Wilf, M., Strappini, F., Golan, T., Hahamy, A., Harel, M., & Malach, R. 2017. Spontaneously emerging patterns in human visual cortex reflect responses to naturalistic sensory stimuli. *Cerebral Cortex* 27: 750–763.

Wilkinson, G. S. 1984. Reciprocal food sharing in the vampire bat. *Nature* 308: 181–184.

Wilkinson, G. S., Carter, G. G., Bohn, K. M., & Adams, D. M. 2016. Non-kin cooperation in bats. *Philosophical Transactions of the Royal Society B: Biological Sciences* 371: 20150095.

Williams, A. H., Kim, T. H., Wang, F., et al. 2018. Unsupervised discovery of demixed, low-dimensional neural dynamics across multiple timescales through tensor component analysis. *Neuron* 98: 1099–1115.

Williams, A. H., & Linderman, S. W. 2021. Statistical neuroscience in the single trial limit. *Current Opinion in Neurobiology* 70: 193–205.

Williamson, R. S., Ahrens, M. B., Linden, J. F., & Sahani, M. 2016. Input-specific gain modulation by local sensory context shapes cortical and thalamic responses to complex sounds. *Neuron* 91: 467–481.

Wills, T. J., Cacucci, F., Burgess, N., & O'Keefe, J. 2010. Development of the hippocampal cognitive map in preweanling rats. *Science* 328: 1573–1576.

Wilson, D. M. 1961. The central nervous control of flight in a locust. *Journal of Experimental Biology* 38: 471–490.

Wilson, M. A., & McNaughton, B. L. 1993. Dynamics of the hippocampal ensemble code for space. *Science* 261: 1055–1058.

Wiltschko, A. B., Johnson, M. J., Iurilli, G., et al. 2015. Mapping sub-second structure in mouse behavior. *Neuron* 88: 1121–1135.

Wiltschko, W., & Wiltschko, R. 2005. Magnetic orientation and magnetoreception in birds and other animals. *Journal of Comparative Physiology A* 191: 675–693.

Wirth, S., Soumier, A., Eliava, M., et al. 2021. Territorial blueprint in the hippocampal system. *Trends in Cognitive Sciences* 25: 831–842.

Wohlgemuth, M. J., Kothari, N. B., & Moss, C. F. 2018. Functional organization and dynamic activity in the superior colliculus of the echolocating bat, Eptesicus fuscus. *Journal of Neuroscience* 38: 245–256.

Wolbers, T., & Wiener, J. M. 2014. Challenges for identifying the neural mechanisms that support spatial navigation: The impact of spatial scale. *Frontiers in Human Neuroscience* 8: 571.

Wolfe, J., Hill, D. N., Pahlavan, S., Drew, P. J., Kleinfeld, D., & Feldman, D. E. 2008. Texture coding in the rat whisker system: Slip-stick versus differential resonance. *PLoS Biology* 6: e215.

Wollberg, Z., & Newman, J. D. 1972. Auditory cortex of squirrel monkey: Response patterns of single cells to species-specific vocalizations. *Science* 175: 212–214.

Wong, A. A., & Brown, R. E. 2006. Visual detection, pattern discrimination and visual acuity in 14 strains of mice. *Genes, Brain and Behavior* 5: 389–403.

Wood, E. R., Dudchenko, P. A., Robitsek, R. J., & Eichenbaum, H. 2000. Hippocampal neurons encode information about different types of memory episodes occurring in the same location. *Neuron* 27: 623–633.

Woolsey, T. A., & Van der Loos, H. 1970. The structural organization of layer IV in the somatosensory region (SI) of mouse cerebral cortex. The description of a cortical field composed of discrete cytoarchitectonic units. *Brain Research* 17: 205–242.

Wu, A., Roy, N. A., Keeley, S., & Pillow, J. W. 2017. Gaussian process based nonlinear latent structure discovery in multivariate spike train data. *Advances in Neural Information Processing Systems* 30: 3496–3505.

Wu, Y., Wu, M., Vázquez-Guardado, A., et al. 2022. Wireless multi-lateral optofluidic microsystems for real-time programmable optogenetics and photopharmacology. *Nature Communications* 13: 5571.

Wu, Y. E., Dang, J., Kingsbury, L., et al. 2021. Neural control of affiliative touch in prosocial interaction. *Nature* 599: 262–267.

Wurtz, R. H. 1969. Visual receptive fields of striate cortex neurons in awake monkeys. *Journal of Neurophysiology* 32: 727–742.

Xie, H., Karipidis, I. I., Howell, A., et al. 2020. Finding the neural correlates of collaboration using a three-person fMRI hyperscanning paradigm. *Proceedings of the National Academy of Sciences of the USA* 117: 23066–23072.

Xiong, Q., Znamenskiy, P., & Zador, A. M. 2015. Selective corticostriatal plasticity during acquisition of an auditory discrimination task. *Nature* 521: 348–351.

Yamins, D. L., & DiCarlo, J. J. 2016. Eight open questions in the computational modeling of higher sensory cortex. *Current Opinion in Neurobiology* 37: 114–120.

Yang, B., Karigo, T., & Anderson, D. J. 2022. Transformations of neural representations in a social behaviour network. *Nature* 608: 741–749.

Yang, Y., Wu, M., Vázquez-Guardado, A., et al. 2021. Wireless multilateral devices for optogenetic studies of individual and social behaviors. *Nature Neuroscience* 24: 1035–1045.

Yarbus, A. L. 1967. *Eye movements and vision*. Plenum.

Yaron, A., Hershenhoren, I., & Nelken, I. 2012. Sensitivity to complex statistical regularities in rat auditory cortex. *Neuron* 76: 603–615.

Yartsev, M. M. 2017. The emperor's new wardrobe: Rebalancing diversity of animal models in neuroscience research. *Science* 358: 466–469.

Yartsev, M. M., & Ulanovsky, N. 2013. Representation of three-dimensional space in the hippocampus of flying bats. *Science* 340: 367–372.

Yartsev, M. M., Witter, M. P., & Ulanovsky, N. 2011. Grid cells without theta oscillations in the entorhinal cortex of bats. *Nature* 479: 103–107.

Yeshurun, Y., Nguyen, M., & Hasson, U. 2021. The default mode network: Where the idiosyncratic self meets the shared social world. *Nature Reviews Neuroscience* 22: 181–192.

Yoo, S. B. M., Hayden, B. Y., & Pearson, J. M. 2021. Continuous decisions. *Philosophical Transactions of the Royal Society B: Biological Sciences* 376: 20190664.

Young, E. D., Nelken, I., & Conley, R. A. 1995. Somatosensory effects on neurons in dorsal cochlear nucleus. *Journal of Neurophysiology* 73: 743–765.

Yovel, Y., Falk, B., Moss, C. F., & Ulanovsky, N. 2010. Optimal localization by pointing off axis. *Science* 327: 701–704.

Yovel, Y., Geva-Sagiv, M., & Ulanovsky, N. 2011. Click-based echolocation in bats: Not so primitive after all. *Journal of Comparative Physiology A* 197: 515–530.

Yovel, Y., & Ulanovsky, N. 2017. Bat navigation. In *Learning and memory: A comprehensive reference* (2nd ed,). ed. J. H. Byrne, pp. 333–345. Academic Press.

Yu, Y., Romero, R., & Lee, T. S. 2005. Preference of sensory neural coding for 1/f signals. *Physical Review Letters* 94: 108103.

Zeigler, H. P., & Marler, P. 2012. *Neuroscience of birdsong.* Reprint ed. Cambridge University Press.

Zeil, J. 2023. Visual navigation: Properties, acquisition and use of views. *Journal of Comparative Physiology A* 209: 499–514.

Zhang, W., & Yartsev, M. M. 2019. Correlated neural activity across the brains of socially interacting bats. *Cell* 178: 413–428.

Zhou, T., Zhu, H., Fan, Z., et al. 2017. History of winning remodels thalamo-PFC circuit to reinforce social dominance. *Science* 357: 162–168.

Ziemba, C. M., Perez, R. K., Pai, J., et al. 2019. Laminar differences in responses to naturalistic texture in macaque V1 and V2. *Journal of Neuroscience* 39: 9748–9756.

Zilkha, N., Scott, N., & Kimchi, T. 2017. Sexual dimorphism of parental care: From genes to behavior. *Annual Review of Neuroscience* 40: 273–305.

Zilkha, N., Sofer, Y., Beny, Y., & Kimchi, T. 2016. From classic ethology to modern neuroethology: Overcoming the three biases in social behavior research. *Current Opinion in Neurobiology* 38: 96–108.

Zilkha, N., Sofer, Y., Kashash, Y., & Kimchi, T. 2021. The social network: Neural control of sex differences in reproductive behaviors, motivation, and response to social isolation. *Current Opinion in Neurobiology* 68: 137–151.

Ziv, Y., Burns, L. D., Cocker, E. D., et al. 2013. Long-term dynamics of CA1 hippocampal place codes. *Nature Neuroscience* 16: 264–266.

Zong, W., Wu, R., Chen, S., et al. 2021. Miniature two-photon microscopy for enlarged field-of-view, multiplane and long-term brain imaging. *Nature Methods* 18: 46–49.

Zong, W., Wu, R., Li, M., et al. 2017. Fast high-resolution miniature two-photon microscopy for brain imaging in freely behaving mice. *Nature Methods* 14: 713–719.

Zuber, B. L., Stark, L., & Cook, G. 1965. Microsaccades and the velocity-amplitude relationship for saccadic eye movements. *Science* 150: 1459–1460.

Zupanc, G. K. H. 2019. *Behavioral neurobiology: An integrative approach.* 3rd ed. Oxford University Press.

Zynyuk, L., Huxter, J., Muller, R. U., & Fox, S. E. 2012. The presence of a second rat has only subtle effects on the location-specific firing of hippocampal place cells. *Hippocampus* 22: 1405–1416.

Index

Note: Photos, Figures and Tables are indicated by italicized page numbers.